A short table of function–Laplace transform pairs

$f(t)$	$F(s)$	$f(t)$	
$u(t)$	$\dfrac{1}{s}$	$e^{-at}\cos \omega t$	$\dfrac{s+a}{(s+a)^2+\omega^2}$
$r(t)=t$	$\dfrac{1}{s^2}$	$t\sin \omega t$	$\dfrac{2\omega s}{(s^2+\omega^2)^2}$
C	$\dfrac{C}{s}$	$t\cos \omega t$	$\dfrac{s^2-\omega^2}{s^2+\omega^2}$
e^{-at}	$\dfrac{1}{s+a}$	te^{-at}	$\dfrac{1}{(s+a)^2}$
$\sin \omega t$	$\dfrac{\omega}{s^2+\omega^2}$	t^2	$\dfrac{2}{s^3}$
$\cos \omega t$	$\dfrac{s}{s^2+\omega^2}$	t^3	$\dfrac{3!}{s^4}$
$\delta(t)$	1	$t^n, n=1, 2, 3, \ldots$	$\dfrac{n!}{s^{n+1}}$
$e^{-at}\sin \omega t$	$\dfrac{\omega}{(s+a)^2+\omega^2}$	$t^n, n \neq -1, -2, -3, \ldots$	$\dfrac{\Gamma(n+1)}{s^{n+1}}$

Greek alphabet

A	α	Alpha	N	ν	Nu
B	β	Beta	Ξ	ξ	Xi
Γ	γ	Gamma	O	o	Omicron
Δ	δ	Delta	Π	π	Pi
E	ϵ	Epsilon	P	ρ	Rho
Z	ζ	Zeta	Σ	σ	Sigma
H	η	Eta	T	τ	Tau
Θ	θ	Theta	Υ	υ	Upsilon
I	ι	Iota	Φ	ϕ	Phi
K	κ	Kappa	X	χ	Chi
Λ	λ	Lambda	Ψ	ψ	Psi
M	μ	Mu	Ω	ω	Omega

CIRCUIT ANALYSIS

CIRCUIT ANALYSIS

Allan D. Kraus Naval Postgraduate School

WEST PUBLISHING COMPANY

St. Paul   New York   Los Angeles   San Francisco

Composition: *Syntax International*
Interior design: *Technical Texts, Inc.*
Copyediting: *Technical Texts, Inc.*
Illustrations: *Visual Graphic Systems, Ltd.*
Cover: *Visual Graphic Systems, Ltd.*

The representation of screens from PSpice and its options used by permission of MicroSim Corporation. Copyright 1990 MicroSim Corporation. This material is proprietary to MicroSim Corporation. Unauthorized use, copying, or duplication is strictly prohibited.

PSpice is a registered trademark of MicroSim Corporation.

Library of Congress Cataloging-in-Publication Data

Kraus, Allan D.
    Circuit analysis/Allan D. Kraus.
      p. cm.
    Includes bibliographical references and index.
    ISBN 0-314-79500-6 (hard)
     1. Electric circuit analysis.   I. Title.
TK454.K73 1991
621.319′2—dc20   90-24431
                                              CIP

To the threatened sea mammals:
the whales, dolphins, and sea otters.
Long may they flourish in their natural habitat.

CONTENTS

PREFACE

In preparing *Circuit Analysis*, I have been motivated by the desire to present some material that is currently not included in the other available texts. Moreover, I have tried to incorporate some of the ideas that I have used in the classroom.

I believe that this subject is at the core of the entire electrical engineering curriculum and that the book that supports it must lay the foundation for a great deal of what is to follow. It is here that the student receives the first exposure and possibly the only training in the Laplace transform and in Fourier analysis. The treatment of these important topics must be lucid and complete and steps must be taken to maintain the students' interest during the presentation.

PREREQUISITES

The only prerequisites for the use of this book is a knowledge of elementary (*Freshman*) physics and differential and integral calculus. A knowledge of differential equations, although helpful, is not required. The book has been designed for a two-semester or three-quarter course sequence that begins in the sophomore year.

ORGANIZATION

The book is organized into seven parts. A brief description of each part follows:

- Part I deals with direct current and resistive networks. Fundamentals are introduced in Chapter 1 and the Kirchhoff laws with their applications are considered in Chapter 2. Node, mesh, and loop analysis are treated in Chapter 3 and superposition, reciprocity, and maximum power transfer, as well as the network theorems of Thevenin, Norton, and Tellegen, are presented in Chapter 4. Part I concludes with a discussion, in Chapter 5, of signals, singularity functions, exponential and sinusoidal functions, and signal average and effective values.
- Part II is devoted to the differential equations of network theory; first order networks in Chapter 6 and second order networks in Chapter 7. A knowledge of differential equations by the reader is **not** assumed and Chapter 6 contains sufficient material to permit the reader to proceed properly. The operational amplifier is introduced as a circuit element in Chapter 8 and several examples of its use are provided in Chapter 8 and through the balance of the book.
- AC analysis is contained in Part III where the heritage from resistive networks is never overlooked and where the six chapters contained in this part are introduced with a discussion, in Chapter 9, of complex frequency and the phasor.

The traditional ac "network" topics of ladder networks, node, mesh and loop analysis, superposition, and the network theorems of Thevenin and Norton are contained in Chapter 10. Complex power and power factor correction are contained in Chapter 11 along with the three conditions for maximum power transfer. Chapter 12 is devoted to three-phase power systems and includes a discussion of system performance with unbalanced loads. Frequency selectivity and resonance are treated in Chapter 13 and mutual inductance and transformers are considered in Chapter 14.

- The theory of the Laplace transformation is presented in Chapter 15 of Part IV. This is followed immediately by first order second order, and advanced applications in Chapter 16. The important topic of convolution is treated in Chapter 17.
- Part V is devoted to two-port networks, transfer functions, frequency response (including Bode asymptotic gain and phase plots) and filters, both passive, and active using the operational amplifier.
- The two chapters in Part VI begin with a definition of the Fourier series (Chapter 21) and progress through the sine–cosine, amplitude-phase angle and exponential forms through amplitude-phase spectra and the Fourier integral and Fourier transform to a consideration of the Laplace transform from the Fourier transform at the end of Chapter 22.
- Part VII contains four appendixes. Appendix A considers complex numbers and the algebra of complex numbers. Its placement in Part VII was motivated by a desire to keep this necessary diversion from interrupting the flow of Chapters 9 and 10. Comprehensive exposition of determinants and matrices is contained in Appendix B. The user's manual for the SPICE program, along with several examples, is presented in Appendix C, and Appendix D contains the customary answers to odd-numbered problems.

PEDAGOGICAL INNOVATIONS

The pedagogy includes several innovations:

- Each chapter is introduced with its objectives and concludes with a summary of what is covered.
- The formulation of problems that involve node, mesh, or loop analysis contain a parallel matrix representation. The use of matrices throughout the text is supported by a copious appendix. In many cases, the same circuit analysis problem is solved in many different ways.
- Because there is an unmistakeable trend toward the use of the computer, computer aided design and analysis is featured via the SPICE program wherever germaine. An appendix with a detailed treatment of the use of the SPICE program is included.
- The treatment of the differential equations of network theory occurs early in the book (just after the discussion of dc networks and signals) and, in this treatment, it is presumed that the student has had no prior training in differential equations. This treatment considers many types of forcing functions and is not limited to the step function (dc battery) or the sinusoid. Zero input and zero state response are given equal emphasis with natural and forced response.
- The treatment of the Laplace transform, which occurs after sinusoidal steady state analysis, includes the initial and final value theorems, as well as the ability

to assign an initial condition to an integral in the form of charge or flux accumulation. The employment of several types of forcing functions is maintained.

- The coverage of Fourier methods includes the decomposition of waveforms into odd and even functions and shows how to exploit network symmetry in the writing of the Fourier series.

- Copious illustrative examples are included and, in most cases, these are followed immediately by drill exercises, with answers, to be attempted by the student.

- The operational amplifier is introduced in Chapter 8 and examples of the use of this important *circuit element* are frequent.

- In recognition of the ABET desire for more *design* in the curriculum, many problems have been added that contain a design twist. These are marked with a logo, shown at top left, and are in addition to the computer aided design problems that can be handled with the SPICE program and which are also marked with a logo, shown at bottom left.

- A flexible organization that allows for the treatment of the traditional topics with coverage of the *enhancements* such as state variable theory, network stability, and the matrices of network theory in an *Instructor's Resource Manual* which is distributed free of charge. The instructor may make copies of these rather comprehensive sections and use them in the classroom as desired. Also included in the resource manual are additional chapter by chapter topics, additional SPICE problems, and additional problems with solutions.

ACKNOWLEDGEMENTS

I have received a great deal of help from several professors who have reviewed the manuscript for this text. I have tried to follow their advice and have incorporated many of the suggestions that they have made. I feel that this book is much the richer for their efforts and I would like to list them here.

Eric Adler McGill University	Fawzi Emad University of Maryland
Jon Anderson University of Washington	Joseph Frank New Jersey Institute of Technology
William A. Blackwell Virginia Polytechnic Institute	N. Thomas Gaarder University of Hawaii at Manoa
Leonard Bobrow University of Massachusetts	Arjun Godhwani Southern Illinois University
Eugene Bradley University of Kentucky	Alan L. Haase University of Texas, Dallas
Pankaj Das Rensselaer Polytechnic Institute	Bernie Hutchins Cornell University
Howard Deck University of South Alabama	W. J. Jameson Montana State University
William Eccles University of South Carolina	Richard D. Klafter Temple University
Louis W. Eggers California State University, Los Angeles	Russell H. Krackhardt Worcester Polytechnic Institute

Richard Kwor
University of Colorado

Gary Lebby
North Carolina A and T University

J. Venn Leeds
Rice University

Stanislaw Legowski
University of Wyoming

Roy Mattson
National Technological University

E. L. McMahon
University of Michigan

Evan Moustakas
San Jose State University

Paul Neudorfer
Seattle University

J. Eldon Steelman
New Mexico State University

Massoud Tabib-Azar
Case Western Reserve University

L. H. Tabrizi
University of Detroit

Karan Watson
Texas A and M University

Mark A. Yoder
Rose-Hulman Institute

I owe a special debt to Professor Jon Anderson of the University of Washington who provided a line by line review of the entire manuscript and to two very worthy students at the Naval Postgraduate School, Janet Pande, who prepared the solutions manual and Lt. John Eremic, who prepared the SPICE problems. Jim Healzer and Brian Paulson of Intercept Software prepared the disk containing the Matrix and EEtools codings.

Pat and Jim Allen of Allen Computype were always ready to answer my questions on how to type the manuscript. Deliveries could never have been made on schedule without the willing and able assistance of Alicia Arnold. The artwork was rendered on the Macintosh by Christine Bentley and Edward Rose of Visual Graphic Systems; they converted many rough sketches and raw data into exceptional figures. Michael Slaughter of West Publishing Company has always been there to keep me going with his contagious faith in this project and I have been very lucky to have been able to work with Deanna Quinn of West Publishing Company during the production phase of this book. Finally, I acknowledge the assistance of my wife Ruth, fondly known as *the Buckeye*, for her continuous help during this four-year endeavor.

Allan Kraus

RESISTIVE NETWORKS AND DIRECT CURRENT

I

INTRODUCTION

OBJECTIVES

The objectives of this chapter are to:

- Employ the principles of elementary physics to develop the quantities of interest in electric network theory: charge, current, potential, voltage, work (energy), and power.

- Show the relationships between current and voltage for some of the elements that are interconnected to form electric networks.

- Introduce the concepts of linearity and time invariance and discuss what is meant by lumped, linear, and bilateral network elements.

- Consider power and energy flow to and from the network elements; resistors, inductors, and capacitors.

- Introduce the concept of the continuity of stored energy.

NETWORK THEORY   SECTION 1.1

Engineering, in general, is concerned with the design of systems to perform in accordance with certain prescribed specifications. Electrical engineering is the branch of engineering that utilizes, among other things, electromagnetic phenomena for the purposes of information processing and the transfer and control of energy.

A network or circuit is an interconnected collection of elements such as resistors, inductors, capacitors or devices such as electronic devices, and transformers. The network is the principal component of an electrical or electronic system, and network design is the art of combining the elements of the proper kind to meet a specification. Moreover, network analysis is often useful, in an analog sense, to other forms of engineering.

Network theory is derived from the more encompassing electromagnetic field theory, and the subject of network analysis is studied because electric network models can be used, in varying forms of complexity, to represent devices or entire systems. The subject is studied in its own right because attention is most often focused on currents passing through and voltages appearing across network elements; and other information of interest such as energy, power, and charge distribution can readily be determined from current and voltage.

However, the study of networks involves certain approximations that are not necessarily included in the more general study of electromagnetic field theory. These approximations involve, among others to be treated in this chapter, the concepts of lumped, linear, time-invariant, and bilateral network elements as well as the idealizations for current and voltage sources. The fact is that analyses of networks based on these approximations can be verified in the laboratory. Thus, the approximations do not negate the enormous contributions of the pioneering early experimenters such as Ampère, Oersted, Coulomb, Volta, Ohm, Faraday, Henry and, of course, Maxwell, who, many think, brought all of the contributions together.

The electric field is set up by electric charges that, in combination, exert forces on one another. The forces in the electric field are dependent on the location and magnitude of these electric charges. The magnetic field is also derived from electric charges, but in this case, the forces are derived from the movement (the velocity) of the charges.

Network theory, by definition, involves the systemizing and generalizing of the relations between currents and voltages associated with the elements of an electric network. Note that the emphasis is first on current and voltage and then on the network element that has a certain position in the network. So one can see that network theory, although following electromagnetic field theory, does indeed present a directed emphasis and is subject to certain restrictions involving the network elements themselves.

SECTION 1.2   UNITS

A unit may be defined as a basic quantity adopted as a standard of measurement. In the study of networks, the *SI* (*Système International*) system of units is employed. This is often referred to as the *MKS* system, which uses the meter as the unit of length, the kilogram as the unit of mass, and the second as the unit of time. The meter, kilogram, and second are fundamental units that are used in all disciplines of engineering. But to consider all phenomena, one needs three more fundamental units: the unit of current, the ampere; the unit of temperature, the kelvin; and the unit of luminous intensity, the candela. All six of these fundamental quantities, along with their traditional units and symbols, are displayed in Table 1.1.

Different systems of units exist. For example, consider the ft-lb$_m$-s system still used in the United States (1990). Units in this system must be converted to the MKS

TABLE 1.1   Fundamental quantities with their units

Quantity	Quantity Symbol	Unit Symbol	Unit
Mass	m	kg	kilogram
Length	l	m	meter
Time	s	s	second
Current	i	A	ampere
Temperature	T	K	kelvin
Luminous intensity	I	cd	candela

system for use in the electrical engineering arena, and the familiar conversions of 1 foot = 0.3048 meter, 1 inch = 25.4 millimeters, and 1 lb_m = 0.4536 kilogram may be noted. Table 1.1 provides the fundamental quantities with their units.

FUNDAMENTAL QUANTITIES   SECTION 1.3

1.3.1 Charge and Current

The concept of electric charge is based on considerations of atomic theory and conservation of matter. Matter is defined as anything that occupies space and possesses masss, and the law of conservation of matter states that matter cannot be created or destroyed (the fact that a study of atomic physics will show that mass may be transformed into energy is not germane to this discussion).

All matter is composed of molecules, which, in turn, are composed of atoms. The atom consists of a positively charged nucleus containing one or more positively charged protons and a number of neutrons that carry no charge. The nucleus is surrounded by a number of negatively charged particles, equal to the number of protons, called electrons. The electron is a form of matter and its charge was found by the American R.A. Milliken to be equal to 1.602×10^{-19} coulomb. Thus, the basic unit of charge is the coulomb, equal to the charge possessed by 6.24×10^{18} electrons. Because of the conservation of matter, charge is said to be conservative, and it is intimately connected with the electron, which cannot be created or destroyed.

Electric current is the time rate of change of charge passing through a specified cross-sectional area. Current, therefore, is charge flow; and because net charge may be composed of positive and negative charges,

$$q = n_p p + n_e e \tag{1.1}$$

where q is the net charge, n with subscript p is the number of positive charges, and n with subscript e is the number of negative charges (e for electron). The magnitude of the actual charges are equal, $e = -p = -1.602 \times 10^{-19}$ coulomb.

The current, in amperes, is by definition

$$i = \frac{dq}{dt} \tag{1.2}$$

and by convention, it is based on the net flow of positive charge. It is a scalar quantity; and it may be observed that a flow of positive charge in a specified direction, say to the left, may actually be caused by negative charges moving to the right. The cumulative effect of both actions is a positive charge flow to the left. This shows that eqs. (1.1) and (1.2) can be combined to yield

$$i = \frac{dq}{dt} = p \frac{dn_p}{dt} + e \frac{dn_e}{dt} \tag{1.3}$$

1.3.2 Force, Work, Energy, and Potential

Force is measured in newtons, and as discussed in the previous section, it is a derived quantity. It may be considered from the Newtonian $F = ma$ point of view; a force of 1 newton is required to cause a mass of 1 kilogram to change its velocity (to accelerate or decelerate) at a rate of 1 meter per second. The force in newtons may also be related to current flow; a force of 1 newton is obtained when exactly 1 ampere flows through each of two infinitely long conductors separated in vacuum by a distance of exactly 1 meter.

Coulomb observed experimentally that two charges of like sign, q_1 and q_2, repel each other with a force that is directly proportional to the product of the magnitude of the charges and inversely proportional to the square of the distance between them. These observations are described by *Coulomb's law*, which gives the magnitude[1]

$$F = \frac{q_1 q_2}{4\pi\epsilon r^2} \tag{1.4}$$

where both q_1 and q_2 are charges in coulombs, r, the distance between the charges, is in meters, and in the proportionality constant, $4\pi\epsilon$, ϵ is the permittivity of the medium in which the charges exist. Different media exhibit different permittivities, and it is customary to consider a relative permittivity

$$\epsilon_r = \frac{\epsilon}{\epsilon_0} \tag{1.5}$$

where ϵ_0 is the permittivity of free space:

$$\epsilon_0 = \frac{1}{36\pi \times 10^9} = 8.854 \times 10^{-12} \qquad \text{(farads per meter)}$$

Here, the farad is the unit of capacitance. The force in newtons is a vector quantity, and in eq. (1.4), the force is directed along the line separating the two charges.

Energy is defined as the capacity or capability to do work. Notice that energy and work are synonymous and that if a force is applied to a charge and the charge moves, work is done and energy is either expended or absorbed. Thus,

$$w = Fr \tag{1.6}$$

where w is the work or energy in joules (1 joule = 1 newton-meter). Moreover, energy may neither be created nor destroyed but may be converted or transformed from one form to another. This is a statement of the familiar law of conservation of energy.

In the case of two charges, Coulomb's law,

$$F = \frac{q_1 q_2}{4\pi\epsilon r^2} \tag{1.4}$$

shows that the two charges react with each other (they attract or they repel), and it must be assumed that the space around a single charge must be influenced, in some manner, by the charge itself. If this effect is due to the charge $q_1 = q$, the effect can be evaluated by probing with a very small test charge $q_2 = dq$ so that the effect that q has upon dq can be observed. The magnitude of the force on the test charge will be

$$dF = \frac{q\,dq}{4\pi\epsilon r^2}$$

---

[1] Magnitudes of vectors bear no special designation. Here, F is the magnitude of the force vector $\vec{F}$.

Electric field (arrows) of strength or intensity E set up by charge q

FIGURE 1.1

The charge q_a is located at the point a that is specified by a position vector r_a from the origin of the coordinate system to point a.

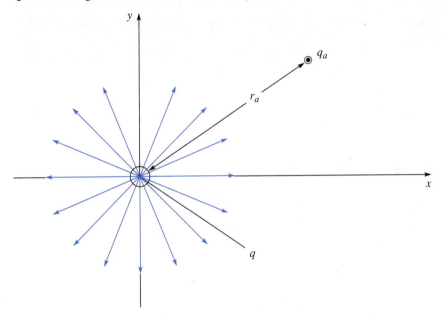

and one may define an *electric field intensity*, a vector quantity that has the magnitude[2]

$$E = \frac{dF}{dq} = \frac{q}{4\pi\epsilon r^2} \qquad (1.7)$$

so that Coulomb's law may be written for charge $q = q_2$:

$$F = Eq_2 \qquad (1.8)$$

Figure 1.1 shows a charge q_a located at a point identified by a position vector from the origin of the coordinate system. An electric field of intensity E is set up by the positive charge q. If it is desired to move q_a in the electric field along path dr in a direction toward q, the differential work done, dw, is the product of the applied force, $F_a = Eq_a$, and the distance to be traversed,

$$dw = -F_a\,dr = -q_a E\,dr$$

Here, the minus sign indicates that positive work is done (energy is expended) in opposition to the electric field. Because E varies with distance from the charge source that creates it, an integration is required to find the actual work done or energy expended in moving q_a between the points designated by the position vectors

---

[2] Here, E is the magnitude of the electric field intensity vector $\vec{E}$.

r_1 and r_2:

$$w = -q_a \int_{r_1}^{r_2} E \, dz$$

This energy is potential energy, which is energy by virtue of position, location, or configuration.

Potential or voltage is the work or energy per unit charge required to move the charge from infinity (some point that is very, very far away) to a point r in the electric field:

$$v = \frac{w}{q_a} = -\int_{\infty}^{r} E \, dz \qquad \text{(volts)} \tag{1.9}$$

and it is observed that a volt is a joule per coulomb.

Suppose a charge q_a is brought from infinity to point r_1 in an electric field of magnitude E set up by a charge q. Then, the voltage or electric potential at point r_1 will be

$$v_1 = \frac{w_1}{q_a} = -\int_{\infty}^{r_1} \frac{q}{4\pi\epsilon r^2} \, dz$$

or

$$v_1 = \frac{q}{4\pi\epsilon r_1}$$

If the charge q_a is brought to point r_2 $(r_2 > r_1)$, then it is observed that

$$v_2 = \frac{q}{4\pi\epsilon r_2}$$

and that a potential or voltage difference

$$\Delta v = v_2 - v_1 = \frac{q}{4\pi\epsilon} \left(\frac{1}{r_2} - \frac{1}{r_1} \right)$$

exists between the points located at r_1 and r_2. Notice that when a unit positive charge moves from a point of lower potential to a point of higher potential, it must gain energy; and when the charge moves from a point of higher potential to a point of lower potential, it gives up or delivers energy. These energies are, as stated previously, potential energies, and this shows that the use of the terms *potential* or *voltage* to indicate energy per unit charge (per unit coulomb),

$$v = \frac{w}{q}$$

is a very logical choice. Thus, if 1 joule is required to move 1 coulomb from b to a, then $v_{ab} = 1$ V; and it is to be noted that in this double-subscript notation, point a is at a higher potential than point b.

1.3.3 Power

Power is the rate of doing work or the rate of energy transfer,

$$p = \frac{dw}{dt} \quad \text{(watts)}$$

where 1 watt = 1 joule per second. By definition, $v = dw/dq$ and $i = dq/dt$, so that the instantaneous power is

$$p = \frac{dw}{dt} \cdot \frac{dq}{dq} = \frac{dw}{dq}\frac{dq}{dt}$$

and hence

$$p = vi \quad \text{(watts)} \qquad\qquad (1.10)$$

The instantaneous power absorbed or delivered by a network element depends on the direction of current flow through and the voltage across the element. Reference directions will be discussed in section 1.5.3 where it will be shown that these reference directions are established with the current flowing through the element in the direction of the voltage drop.

The four possible alternatives for the voltage across and the current through a network element are indicated in Fig. 1.2. In Figs. 1.2a and 1.2b, the element is

---

Network-element power transfer

FIGURE 1.2

In (a) and (b) the element is absorbing power from the network. In (c) and (d) the element is delivering power to the network.

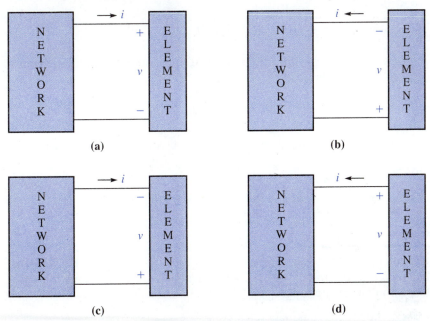

drawing or absorbing power from the network because the current is flowing through the network in the direction of the voltage drop. In Figs. 1.2c and 1.2d, the element is delivering power to the network because the element voltage drop is not in the direction of the current flow.

■ **EXAMPLE 1.1**

A charge of $q_a = 10$ nC exists at the origin of the xy plane in air, which has a relative permittivity of $\epsilon_r = 1$.

a. What is the magnitude of the electric field intensity at a point $r_2 = 540$ mm away from the origin?

Suppose a charge of $q_b = -40$ nC is brought from a point r_1 a considerable distance from the origin (a point at infinity) to the point r_2.

b. What is the voltage at point r_2?
c. How much work is expended to move charge q_b to point r_2?
d. What is the magnitude of the force exerted by q_a on q_b?

Solution

a. Because $\epsilon_r = 1$, $\epsilon = \epsilon_0 = 8.854 \times 10^{-12}$ F/m, and the magnitude of the electric field intensity is

$$E = \frac{q_a}{4\pi\epsilon_0 r_2^2} = \frac{10 \times 10^{-9}}{4\pi(8.854 \times 10^{-12})(0.54)^2} = 308.2 \text{ N/C or V/m}$$

b. The potential or voltage at point r_2 is

$$V = \frac{q_a}{4\pi\epsilon_0 r_2} = \frac{10 \times 10^{-9}}{4\pi(8.854 \times 10^{-12})(0.54)} = 166.4 \text{ V}$$

c. The work or energy expended in *retarding* the motion of q_b (unlike charges attract each other) is

$$w = vq_a = (166.4)(10 \times 10^{-9}) = 1.66 \text{ } \mu\text{J}$$

d. The magnitude of the force of attraction between the two charges is

$$F = \frac{q_a q_b}{4\pi\epsilon_0 r_2^2} = \frac{(10 \times 10^{-9})(40 \times 10^{-9})}{4\pi(8.854 \times 10^{-12})(0.54)^2} = 12.33 \text{ } \mu\text{N}$$ ■

EXERCISE 1.1

If a force of 1 newton is equivalent to a force of 0.2248 pound, find the magnitude of the force (in tons) between two charges of 1 coulomb each located in air 1 meter apart.

Answer 1.01×10^6 tons.

FIGURE 1.3

A point p located r meters from a differential length of conductor carrying a current i

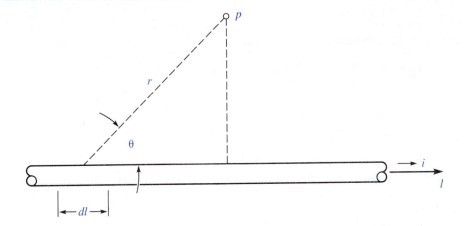

1.3.4 The Magnetic Field

Either a permanent magnet or a current can produce a static magnetic field. Almost everyone has, at one time or another, observed the experiment where iron filings have been used to show that "lines of force" exist between unlike poles of two magnets or the two poles of a horseshoe magnet. These "lines of force" represent the magnetic flux ϕ, in webers (V·s), a scalar quantity.

If a current i flows in a thin wire of differential length dl,[3] as shown in Fig. 1.3, a magnetic field of magnetic field strength[4] dH will be set up at a point p located r meters away from dl at an angle θ to dl. The *Biot-Savart law* mathematically states that the magnetic field strength, a vector, has a magnitude

$$dH = \frac{i \sin \theta \, dl}{4\pi r^2} \quad \text{(amperes per meter)} \tag{1.11}$$

Notice that the magnetic field strength depends only on moving charges (a current) and is independent of the medium surrounding the current-carrying wire.

The magnetic field strength is analogous to the electric field strength, and the existence of magnetic flux has been experimentally observed. Ampère was able to describe a magnetic flux density, a vector quantity having a magnitude[5]

$$dB = \frac{\mu i \sin \theta \, dl}{4\pi r^2} \quad \text{(teslas)} \tag{1.12}$$

where 1 tesla = 1 weber per square meter and where μ is the permeability of the medium in which the magnetic field exists. As in the case of ϵ, the permittivity, different media exhibit different permeabilities, and it is customary to define a relative

[3] Here, l is the magnitude of the length vector $\vec{l}$.
[4] Here, H is the magnitude of the vector $\vec{H}$.
[5] Here, B is the magnitude of the vector $\vec{B}$.

permeability

$$\mu_r = \frac{\mu}{\mu_0} \tag{1.13}$$

where μ_0 is the permeability of free space,

$$\mu_0 = 4\pi \times 10^{-7} \quad \text{(henrys per meter)}$$

Equations (1.11) and (1.12) clearly show that the magnitudes of the magnetic field strength and the magnetic flux density are related by

$$B = \mu H$$

and because magnetic flux density is magnetic flux per unit area, the magnetic flux ϕ, a scalar, will be given by

$$\phi = \int B \, dA$$

where dA is the differential area normal to the flux density vector.

If a coil of wire contains n turns and either produces a magnetic field or exists in a magnetic field where the flux contained within the cross section of the turns is ϕ, the number of *flux linkages* for the coil is

$$\lambda = n\phi \quad \text{(weber-turns)} \tag{1.14}$$

This will be considered once again in Section 1.8, where Faraday's law, which relates the rate of change of flux linkage and induced voltage, is discussed.

A summary of the derived quantities with their units is presented in Table 1.2. Because the numerical values associated with both the basic (Table 1.1) and the

TABLE 1.2   Summary of derived units

Quantity	Symbol	Unit	Unit Abbreviation
Charge	q	coulomb	C
Force	F	newton	N
Work or energy	w	joule	J
Power	p	watt	W
Voltage	v	volt	V
Electric field strength	E	newton per coulomb	N/C
Magnetic flux	ϕ	weber	Wb
Flux linkages	λ	weber-turns	Wb·t
Magnetic field strength	H	ampere-turns per meter	A·t/m
Magnetic flux density	B	tesla	T

Notes:

1. Care should be exercised in the use of C to designate the charge in coulombs, because C is also used to designate capacitance (in farads).
2. Current is missing from this table because it is listed as a fundamental quantity.
3. Electric field strength or intensity can also be expressed in volts per meter (V/m).

			TABLE 1.3

Standard decimal prefixes

Prefix	Abbreviation	Multiplier
Larger Quantities		
deka	da	10^1
hecto	h	10^2
kilo	k	10^3
mega	M	10^6
giga	G	10^9
tera	T	10^{12}
Smaller Quantities		
deci	d	10^{-1}
centi	c	10^{-2}
milli	m	10^{-3}
micro	μ	10^{-6}
nano	n	10^{-9}
pico[6]	p	10^{-12}
femto	f	10^{-15}
atto	a	10^{-18}

derived (Table 1.2) units can vary over a wide range, a special notation based on the decimal system has been adopted. This notation involves the use of descriptive prefixes, examples of which are shown in Table 1.3.

THE NETWORK OR CIRCUIT

A complete electric network or circuit consists of at least one closed path composed of network elements in which current flow is confined. Figure 1.4 shows a very simple network, and the current is assumed to be flowing in the clockwise direction. This indicates a net positive clockwise charge flow (or a net negative counterclockwise charge flow). Notice that the network elements are tied or connected together at points or *nodes* a and b.

If the voltage at point a is higher than the voltage at point b, there is a voltage drop in element 2 from top to bottom at the right side of the figure. This means that the positive charges flowing from a to b through element 2 are giving up or losing energy in element 2. However, in flowing through element 1, the positive flow of charge (the current) experiences a voltage rise. This means that element 1 is considered as an *active device* or *element*, because it is some device that is supplying energy to the network. Thus, an active element is one that can deliver energy to a network, and because of this, in this case element 1 is considered as a source. Of course, the energy added to the network by element 1 must equal the energy received by element 2, because conservation of energy must prevail. Here, element 2 is referred to as a *passive element*, because it absorbs power.

---

[6] Pico is sometimes abbreviated as $\mu\mu$ for micro-micro.

FIGURE 1.4   A simple network showing two network elements and a circulating current

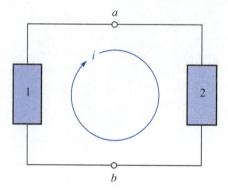

SECTION 1.5   **SEVEN BASIC TWO-TERMINAL NETWORK ELEMENTS**

Although it is not implied that there are only seven network elements that require consideration, seven network elements are discussed in this section. These network elements are those used in the construction of electric networks to be considered in this book. The seven network elements are the following:

- The resistor
- The inductor
- The capacitor
- The ideal voltage source
- The ideal current source
- The open circuit or zero-valued current source
- The short circuit or zero-valued voltage source

For the purposes of this study, the first three of these will be referred to as primary elements, and they are linear, time-invariant, lumped, and bilateral elements, which are characteristics that will be discussed in detail shortly. The two ideal sources may be referred to merely as independent sources (they will be considered active elements). All of the network elements listed may be described on the basis of a voltage-current characteristic.

Each network element possesses two terminals, and it is assumed that the current that enters one of the terminals leaves the other undiminished (with equal magnitude). Moreover, the element is small enough to render the time required for passage of current (the flow of charge) through it negligible. An element that exhibits this property is said to be a *lumped element*,[7] and this study is concerned with *lumped, two-terminal* elements.

Network elements, with the possible exception of controlled sources (to be discussed in Section 1.9), considered in this book will be assumed to be *bilateral*; their network behavior is not affected by interchanges in their terminal connections. The

---

[7] The lumped element is small in size with respect to the wavelength corresponding to the frequency of operation. For example, a resistor 2 cm long is a lumped element if it carries 60 Hz current having a wavelength of 5×10^6 meters.

primary elements (resistor, inductor, capacitor) will also be *time-invariant*; their values will not vary with time.

1.5.1 Linearity

Consider a system (which may consist of a single network element) represented by a block, as shown in Fig. 1.5, and observe that the system has an input designated by e (for excitation) and an output designated by r (for response). The system is considered to be *linear* if it satisfies the *homogeneity* and *superposition* conditions.

The Homogeneity Condition: If an arbitrary input to the system e causes a response r, then if ce is the input, the output is cr, where c is some arbitrary constant.

The Superposition Condition: If the input to the system e_1 causes a response r_1, and if an input to the system e_2 causes a response r_2, then a response $r_1 + r_2$ will occur when the input is $e_1 + e_2$.

If either the homogeneity condition or the superposition condition is not satisfied, the system is said to be *nonlinear*.

■ **EXAMPLE 1.2**

Consider a system such as the one shown in Fig. 1.5 whose output is related to its input by

$$r = \frac{a}{a + b} e$$

Determine whether the system is linear if (a) $b = a$ and (b) if $b = ae$.

Solution

a. If $a = b$,

$$r = \frac{1}{2} e$$

and for two excitations e_1 and e_2,

$$r_1 = \frac{1}{2} e_1$$

and

$$r_2 = \frac{1}{2} e_2$$

A simple system

FIGURE 1.5

$e\,(t) \longrightarrow$ SYSTEM $\longrightarrow r\,(t)$

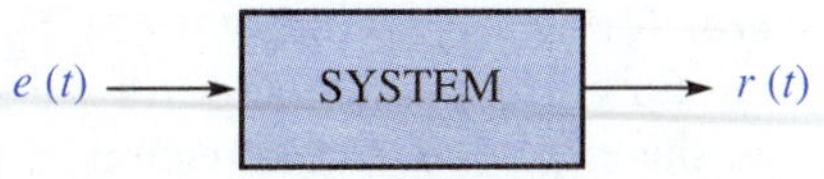

If the excitation is $e_3 = c_1 e_1 + c_2 e_2$, then

$$r_3 = \frac{1}{2}(c_1 e_1 + c_2 e_2) = \frac{1}{2} c_1 e_1 + \frac{1}{2} c_2 e_2 = r_1 + r_2$$

for any values of c_1, c_2, e_1, and e_2. The system is linear.

b. However, if $b = ae$, then

$$r = \frac{a}{a + ae} e = \frac{e}{1 + e}$$

Inputs e_1 and e_2 produce

$$r_1 = \frac{e_1}{1 + e_1}$$

and

$$r_2 = \frac{e_2}{1 + e_2}$$

If $e_3 = c_1 e_1 + c_2 e_2$,

$$r_3 = \frac{c_1 e_1 + c_2 e_2}{c_1 e_1 + c_2 e_2 + 1}$$

and the system is nonlinear, because

$$r_3 = \frac{c_1 e_1}{1 + c_1 e_1} + \frac{c_2 e_2}{1 + c_2 e_2} \neq c_1 r_1 + c_2 r_2$$ ■

1.5.2 Time Invariance

Consider a system in which an input e_1 produces an output r_1. Suppose a second input, e_2, which is a delayed version but otherwise is an exact replica of e_1, produces an output r_2. If the output r_2 caused by e_2 is a delayed version but an exact replica of r_1, the system is said to be *time-invariant*. Time invariance essentially means that the system is not changing with time.

1.5.3 Reference Directions for Voltage and Current

Figure 1.6 shows a network element with current passing through it from top to bottom. Notice that a current passes through the element and a voltage drop appears across the element; point a is at a higher potential than point b. Point a has been labeled with a plus ($+$) sign to indicate that it has a higher potential than point b, which has been labeled with a minus ($-$) sign.

An elementary discussion of network topology is contained in Chapter 3. There it is shown that a network branch is a network segment between two points in the network. The voltage v in Fig. 1.6 can be called a branch voltage, although some authors prefer the terminology *branch voltage drop* (or merely a voltage drop). Here, it is important to note that the reference directions for current and voltage (or voltage drop) are based on the current flowing through the element in the direction of the

Network element

FIGURE 1.6

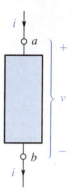

voltage drop. Indeed, it has been shown in section 1.3.3 that an element with the voltage polarity and the current direction shown in Fig. 1.6 is absorbing power and may be referred to as a passive element.

1.5.4 Through and Across Variables and Elemental Equations

Notice in Fig. 1.6 that the current i flows through and that the voltage v appears across the network element. In network analysis, current is designated as the through variable and voltage is the across variable. It is often necessary to consider integrated through and across variables. The integrated through variable is the charge,

$$q = \int_0^t i \, dz + q(0) \qquad (1.15)$$

and the integrated across variable is the flux linkage,

$$\lambda = \int_0^t v \, dz + \lambda(0) \qquad (1.16)$$

The relationship between the through and across variables depends on the element under consideration, and for this reason, this relationship is called an elemental or constitutive equation.

1.5.5 Three Primary Elements

The three ideal circuit elements of the resistor, the inductor and the capacitor can be defined if the functional relationship between the current and the voltage can be obtained. The defining relationships are called *elemental equations*, and these can be determined from considerations of electrostatic or electromagnetic fields or merely from experimental observations. The elemental equation represents a mathematical model for the ideal element.

A network element may be depicted as a box with two leads, as shown in Fig. 1.7. The two leads are called the terminals, and the box is often referred to as a *single terminal pair* or *one-port*. Observe that a voltage v appears across the terminals and a current i flows through the box. The linear characteristic between i and v will determine the elemental equations and define three network elements, the *resistor*, the *inductor*, and the *capacitor*.

FIGURE 1.7   A single terminal pair element

This type of element is also referred to as a one-port.

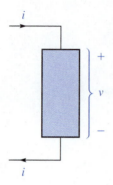

The Resistor   Suppose that in the configuration of Fig. 1.7, measurements are taken of the current i and the voltage v. If a linear characteristic is observed, then v is directly proportional to i ($v \sim i$). The proportionality constant is R:

$$v = Ri \tag{1.17}$$

which serves to define the resistive element

$$R = \frac{v}{i} \quad \text{(ohms)} \tag{1.18}$$

where 1 ohm = 1 volt per ampere. The linear characteristic is plotted in Fig. 1.8a, where it is seen that R is the slope of the v–i characteristic. Note that R is the constant in the homogeneity condition for linearity.

FIGURE 1.8   Linear characteristics of three network elements: (a) resistor, (b) inductor, and
                              (c) capacitor

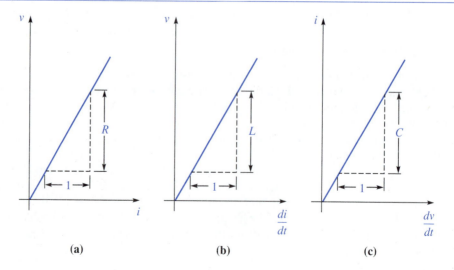

Equation (1.17) is one of the elemental equations for the resistive element or resistor. The other is

$$i = \frac{v}{R} \qquad (1.19)$$

Both eqs. (1.17) and (1.19) are expressions of *Ohm's law*.

The reciprocal of resistance is called the *conductance*,[8]

$$G = \frac{1}{R} \quad \text{(mhos)} \qquad (1.20)$$

and use of the conductance leads to two alternative forms of Ohm's law:

$$i = Gv \qquad (1.21a)$$

and

$$v = \frac{i}{G} \qquad (1.21b)$$

All materials possess a physical property called the *conductivity*, σ, in mhos per meter ($\mho$/m). The conductance is directly proportional to the conductivity and the material's cross-sectional area and inversely proportional to the length of the conductor:

$$G = \frac{\sigma A}{L} \qquad (1.22)$$

The *resistivity* is the reciprocal of the conductivity:

$$\rho = \frac{1}{\sigma} \quad \text{(ohm-meters)}$$

and the resistance can be described by the simple relationship

$$R = \frac{\rho L}{A} \qquad (1.23)$$

The Inductor　In Fig. 1.7, let the rate of change of current di/dt and the voltage v be measured. Suppose that v is observed to be directly proportional to di/dt, $v \sim di/dt$. If the proportionality constant is L,

$$v = L\frac{di}{dt} \qquad (1.24)$$

---

[8] Conductance is measured in *siemens*, where the siemen is the reciprocal of the ohm ($S = 1/\Omega$). However, the term *mho* ($\mho$), which is equivalent to the siemen, dominates the literature in the United States and will be used throughout this book.

which defines the inductive element or inductor as

$$L = \frac{v}{di/dt} \qquad \text{(henrys)} \tag{1.25}$$

where 1 henry = 1 volt-second per ampere. The linear characteristic for the inductor is indicated in Fig. 1.8b, and the elemental equations for the inductor are eqs. (1.24) and

$$i = \frac{1}{L} \int_0^t v \, dz + i(0) \tag{1.26}$$

The Capacitor   Finally, if a changing voltage dv/dt and a linearly varying current i are observed, then $i \sim dv/dt$ and the constant of proportionality is C:

$$i = C \frac{dv}{dt} \tag{1.27}$$

This defines the capacitive element or the capacitor as

$$C = \frac{i}{dv/dt} \qquad \text{(farads)} \tag{1.28}$$

where 1 farad = 1 ampere-second per volt = 1 coulomb per volt. The linear characteristic of the capacitor is shown in Fig. 1.8c, and the elemental equations for the capacitor are eqs. (1.27) and

$$v = \frac{1}{C} \int_0^t i \, dz + v(0) \tag{1.29}$$

The elemental equations for the resistor, inductor, and capacitor, along with the symbols used to represent them, are displayed in Table 1.4. These equations lie at the core of this study and should be memorized.

TABLE 1.4   Summary of elemental equations for three network elements

Element	Symbol	$v = f(i)$	$i = f(v)$
Resistor, R	—⅏—	$v = Ri$	$i = \dfrac{v}{R}$
Inductor, L	—⌇⌇⌇—	$v = L\dfrac{di}{dt}$	$i = \dfrac{1}{L}\displaystyle\int_0^t v \, dz + i(0)$
Capacitor, C	—⊣⊢—	$v = \dfrac{1}{C}\displaystyle\int_0^t i \, dz + v(0)$	$i = C\dfrac{dv}{dt}$

Examples of the characteristics of some nonlinear resistive elements: (a) *pn* junction diode, (b) tunnel diode, and (c) gas diode

FIGURE 1.9

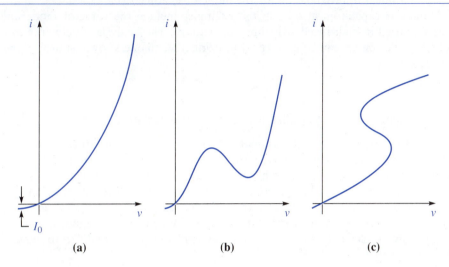

(a)   (b)   (c)

1.5.6 Nonlinear Elements

Nonlinear elements often appear in electric networks. Examples of nonlinear resistive elements include the tunnel diode, the gas diode, and the *pn* junction diode. Typical current-voltage characteristics for these devices are displayed in Fig. 1.9. All three of these devices are not *bilateral*: Not only are their characteristic curves not straight lines passing through the origin, but their curves are not symmetrical with respect to the origin. This means that the points (v, i) and $(-v, -i)$ do not both lie on the characteristic curve. None of these devices may be placed into a network in an arbitrary manner; their terminal designations must be known and given an account.

A plot of the flux-current characteristic of a typical real inductor[9] is shown in Fig. 1.10. Notice that for large currents in either direction, saturation is evident; that

Flux-current characteristic of a typical inductor

FIGURE 1.10

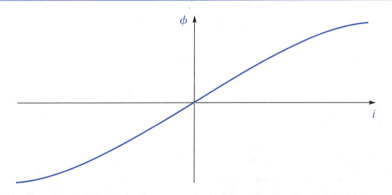

[9] As opposed to the ideal inductor discussed in Section 1.5.5.

is, the flux increases very slowly as the current increases. However, a considerable portion of the characteristic curve closely resembles a straight line, and this permits the use of a linear approximation for the inductor characteristic.

Nonlinear capacitors also exist and examples include the varactor diode and transistors used in high-speed-switching applications. The analysis of networks containing these elements can be effected by utilizing a technique known as *small-signal analysis.*

1.5.7 Linear Time-Varying Elements and a Nomenclature Convention

This study, up to this point, has made use of a definite nomenclature convention. Lowercase letters such as v or i have been employed to indicate quantities that vary with time. For brevity, no attempt has been made to represent v as $v(t)$ or i as $i(t)$.

Uppercase letters have been used to indicate quantities that represent the magnitude of vectors and network elements. In particular, the letters R, L, and C designate time-invariant resistors, inductors, and capacitors. When one is dealing with time-varying network elements, they can be represented by $R(t)$, $L(t)$, and $C(t)$ to indicate the time dependency.

The elemental equations for the time-varying resistor satisfy the linearity property but show the time variation. One writes either

$$v = R(t)i \qquad i = G(t)v$$

or

$$v(t) = R(t)i(t) \qquad i(t) = G(t)v(t)$$

Many examples of linear time-varying resistance are common. One of them stems from the fact that electric resistance is a function of temperature, in accordance with

$$R_T = R_o[1 + \alpha(T - 20)]$$

where R_T is the resistance at temperature T, R_o is the referenced resistance at 20°C, and α is a correction factor known as the *temperature coefficient of resistivity* (in $\Omega \cdot °C/m$). For example, a hard copper wire that possesses a resistance of 0.0208 Ω/m at 20°C and a value of $\alpha = 0.00393$ $\Omega \cdot °C/m$ will have a value of

$$R_T = 0.0208[1 + 0.00393(50 - 20)] = 0.0208(1.1179) = 0.0233 \ \Omega$$

at 50°C.

The foregoing calculation demonstrates that a resistor may well exhibit a time-varying characteristic if it is placed in an environment whose temperature is varying. Indeed, current flowing through a resistor generates heat, and this, in turn, causes a change in the value of the resistance with a subsequent change in current.

Examples of a linear time-varying capacitor and inductor are the familiar tuning wafers in a radio and the *variac* used in many laboratories. Although these may exhibit pronounced time-invariant characteristics when "set," they are considered as time varying in their adjustment phase. In this adjustment phase, the user provides the function of time in $C(t)$ and $L(t)$, so that for the capacitor

$$i = \frac{d}{dt}(Cv) = C\frac{dv}{dt} + \frac{dC}{dt}v$$

and for the inductor

$$v = \frac{d}{dt}(Li) = L\frac{di}{dt} + \frac{dL}{dt}i$$

It is important to note that for any of these linear time-varying elements, the slope of the characteristic curve is always a straight line passing through the origin at a particular value of time, but the slope itself will vary as a function of time.

IDEAL SOURCES, OPEN AND SHORT CIRCUITS

SECTION 1.6

1.6.1 Ideal Sources

Sources are active elements because they provide energy to an electric network. They are considered to be network elements and are often called generators. Although they provide a voltage difference and a current, a distinction is made between ideal voltage and ideal current sources.

An ideal voltage source is a voltage source whose terminal voltage $v_s(t)$ is a specified function of time regardless of the current that flows through it. Such a voltage source is difficult to obtain in real-world network applications, although a new battery provides a close approximation over a limited range. Its voltage-current characteristic is displayed in Fig. 1.11a. Indeed, regulated power supplies can be constructed to be even more effective as an ideal voltage source than the battery.

Characteristic curves for (a) a new storage battery and (b) the collector current in a bipolar junction transistor

FIGURE 1.11

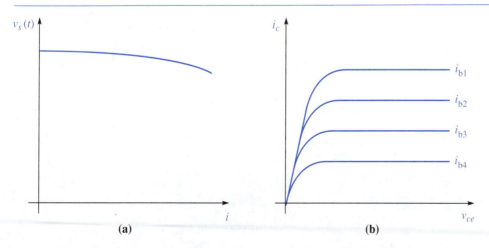

(a)   (b)

FIGURE 1.12   Ideal-source symbols: (a) the ideal voltage source, (b) the ideal current source, and
                                (c) the special case of an ideal dc voltage source

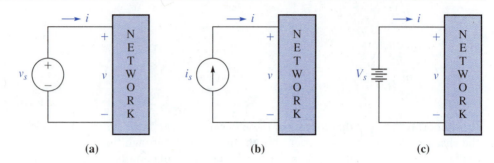

An ideal current source is a current source where the current flowing at its terminals is a specified function of time regardless of the voltage across its terminals. Here,
too, it is to be noted that such a source is difficult to construct, although the characteristics of a typical bipolar junction transistor (Fig. 1.11b) exhibit, for certain base
currents, i_B, a constant–collector current characteristic. If $v_s(t) = V$ or $i_s(t) = I$, the
sources are called *dc* (or *direct current*) sources.

The symbols for both the ideal voltage and ideal current sources are shown in
Fig. 1.12. Observe that for a direct-current (dc) battery or generator, a special symbol
may be employed, with the wide bar representing the positive terminal. Also observe
the use of lowercase letters to indicate instantaneous values and, in the case of the
dc battery, the use of a capital letter to indicate that the source voltage is constant.
But in Fig. 1.12c, the terminal voltage is also indicated by the lowercase letter v. In all
cases in Fig. 1.12, the subscript s is used to indicate that consideration is being given
to a source, and in all cases, the terminal value is equal to the source value.

1.6.2 Open and Short Circuits

Open and short circuits are also considered to be network elements. The open
circuit, shown in Fig. 1.13a, is equivalent to a zero-valued current source and has a
voltage across its terminals in the absence of current flow. Thus, it has a definite voltage across the terminals and a zero-current characteristic. The short circuit, shown
in Fig. 1.13b, is equivalent to a zero-valued voltage source and passes current without
a voltage across its terminals; it has a definite current and a zero-voltage characteristic.

FIGURE 1.13   Networks connected to (a) an open circuit and (b) a short circuit

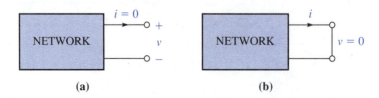

POWER AND ENERGY IN THE NETWORK ELEMENTS SECTION 1.7

In Section 1.2, it was shown that instantaneous power is the rate of doing work; and because work and energy are synonymous, the rate of energy transformation is

$$p = \frac{dw}{dt} \tag{1.30}$$

It was also shown that instantaneous power is always equal to the product of the current through the element and the voltage across the element,

$$p = vi \tag{1.10}$$

and from eq. (1.30), it is noted that the energy delivered, absorbed, stored, or dissipated (depending on the actual element) can be represented by the simple indefinite integral

$$w = \int p \, dz = \int vi \, dz \tag{1.31}$$

It is the intent of this section to examine the foregoing relationships in more detail.

1.7.1 The Resistor

It is a fact that a resistor cannot store energy and it converts electric energy, irreversibly, into dissipated heat. Because $v = Ri$, the instantaneous power drawn by a resistor is

$$p = vi = Ri^2 = \frac{v^2}{R} > 0 \tag{1.32}$$

and the power dissipated from time $t = t_0$ to time $t = t_1$ will be

$$w = \int_{t_0}^{t_1} vi \, dz = R \int_{t_0}^{t_1} i^2 \, dz = \frac{1}{R} \int_{t_0}^{t_1} v^2 \, dz \tag{1.33}$$

■ **EXAMPLE 1.3**

Determine the instantaneous power drawn by a 6-Ω resistor at $t = 4$ s and the energy dissipated by the resistor from $t = 0$ to $t = 4$ s if the resistor is connected to (a) a 12-V dc source and (b) a voltage source that provides $v_s(t) = 12t$ volts.

Solution

a. If $V_s = v = 12$ V, then $i = 12/6 = 2$ A, and

$$p = vi = 12(2) = 24 \text{ W}$$

Alternatively,

$$p = i^2 R = (2)^2 6 = 4(6) = 24 \text{ W}$$

and

$$p = \frac{v^2}{R} = \frac{(12)^2}{6} = \frac{144}{6} = 24 \text{ W}$$

The energy dissipated will be

$$w = \int_0^4 vi \, dt = \int_0^4 24 \, dt = 24t \Big|_0^4 = 96 \text{ J}$$

b. If $v_s(t) = 12t$ volts, then $i = 12t/6 = 2t$ amperes, and

$$p = vi = (12t)(2t) = 24t^2 \text{ W}$$

and at $t = 4$ s

$$p = 24(4)^2 = 384 \text{ W}$$

In this case, the energy dissipated will be

$$w = \int_0^4 vi \, dt = \int_0^4 24t^2 \, dt = \frac{24}{3} t^3 \Big|_0^4 = 512 \text{ J} \qquad \blacksquare$$

1.7.2 The Inductor

For the inductor, where $v = L \, di/dt$, the power is

$$p = vi = L\left(\frac{di}{dt}\right)i = Li\frac{di}{dt} \tag{1.34}$$

and the energy stored over a time interval from t_0 to t_1 will be

$$w_L = \int_{t_0}^{t_1} p \, dz = \int_{t_0}^{t_1} Li\frac{di}{dz} \, dz = \int_{i_0}^{i_1} Lz \, dz$$

where $i(t = t_0) = i_0$ and $i(t = t_1) = i_1$. Then

$$w_L = \int_{i_0}^{i_1} Lz \, dz$$

or

$$w_L = \frac{1}{2} L(i_1^2 - i_0^2) \tag{1.35}$$

For the specific case of the current $i_0 = 0$ at $t = t_0$, and $i_1 = i$ at $t = t$, eq. (1.35) reduces to

$$w_L = \frac{1}{2} Li^2$$

A series combination of a 10-Ω resistor and 2-H inductor connected to a current
source supplying $i_s = 3t^2$ amperes

FIGURE 1.14

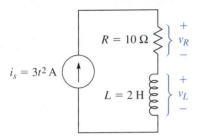

$i_s = 3t^2$ A

$R = 10\ \Omega$ $+$ v_R $-$

$L = 2$ H $+$ v_L $-$

■ **EXAMPLE 1.4**

It will be shown in Chapter 2 that a *series connection* of network elements is
a connection in which all of the elements are connected end to end so as to
provide a single path for the flow of current. In Fig. 1.14, the current source
$i_s = 3t^2$ amperes is applied to the series combination of a 10-Ω resistor and
a 2-H inductor. Find the instantaneous power drawn by the resistor and the
inductor at $t = 2$ s and the energy provided by the source to the resistor and
the inductor during the interval from $t = 0$ to 2 s.

Solution   It will be shown early in Chapter 2 that the same current flows
through each of the two elements in "series."

The voltages across the resistor and inductor are

$$v_R = Ri = 10(3t^2) = 30t^2$$

and with $di/dt = 6t$,

$$v_L = L\frac{di}{dt} = 2(6t) = 12t$$

The instantaneous powers at $t = 2$ s will be

$$p_R(t = 2) = v_R i = (30t^2)(3t^2) = 90t^4 = 90(16) = 1440 \text{ W}$$

and

$$p_L(t = 2) = v_L i = (12t)(3t^2)\Big|_{t=2} = 36t^3\Big|_{t=2} = 36(8) = 288 \text{ W}$$

The source, because of conservation of energy, must be delivering

$$p_s(t = 2) = 1440 + 288 = 1728 \text{ W}$$

The energy delivered between $t = 0$ and $t = 2$ s will be *dissipated* by the
resistor,

$$w_R = \int_0^2 v_R i\, dt = \int_0^2 90t^4\, dt = \frac{90}{5}t^5\Big|_0^2 = 18(32) = 576 \text{ J}$$

and stored in the inductor,

$$w_L = \int_0^2 v_L i\, dt = \int_0^2 36t^3\, dt = \left. \frac{36}{4} t^4 \right|_0^2 = 9(16) = 144 \text{ J}$$

Alternatively, for the inductor at $t = 2$ s when $i = 3(2)^2 = 12$ A,

$$w_L = \frac{1}{2} Li^2 = \frac{1}{2}(2)(12)^2 = 144 \text{ J}$$

and because of conservation of energy, the energy delivered by the source over the time period $t = 0$ to 2 s will be

$$w_s = w_R + w_L = 576 + 144 = 720 \text{ J}$$ ∎

1.7.3 The Capacitor

The reasoning employed in the consideration of the inductor can be applied to the capacitor. But the variable will be the capacitor voltage, which is the voltage across its terminals. Here, $i = C\, dv/dt$, so that the power drawn is

$$p = vi = v\left(C \frac{dv}{dt} \right) = Cv \frac{dv}{dt} \tag{1.36}$$

and the energy stored over the time interval from t_0 to t_1 will be

$$w_C = \int_{t_0}^{t_1} p\, dz = \int_{t_0}^{t_1} Cv \frac{dv}{dz}\, dz = \int_{v_0}^{v_1} Cz\, dz$$

where $v(t = t_0) = v_0$ and $v(t = t_1) = v_1$. Then

$$w_C = \int_{v_0}^{v_1} Cz\, dz$$

or

$$w_C = \frac{1}{2} C(v_1^2 - v_0^2) \tag{1.37}$$

and with $v_0 = 0$ and $v(t = t) = v$,

$$w_C = \frac{1}{2} Cv^2$$

EXERCISE 1.2

It will be shown in Chapter 2 that a parallel connection of network elements is one in which all elements are connected such that an identical voltage appears across all of the elements. In Fig. 1.15, the voltage source $v_s = 4t$ volts is applied to the parallel combination of a 4-Ω resistor and a 0.125-F capacitor. Determine the instantaneous power drawn by the resistor and the capacitor at $t = 4$ s and the energy handled by the resistor and the capacitor from $t = 0$ to $t = 3$ s.

A parallel combination of a 4-Ω resistor and a 0.125-F capacitor connected to
a voltage source supplying $v_s = 4t$ volts (Exercise 1.2)

FIGURE 1.15

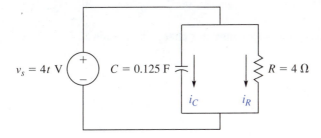

Answer $p_R = 64$ W, $p_C = 8$ W, $w_R = 36$ J, $w_C = 9$ J.

CONTINUITY OF STORED ENERGY

SECTION 1.8

Faraday's law relates the generation of a voltage v with the rate of change of flux
linkage,

$$v = \frac{d\lambda}{dt} \tag{1.38}$$

where, as indicated by eq. (1.14),

$$\lambda = n\phi \tag{1.14}$$

Here, λ represents the flux linkage produced by n turns of a coil with a flux of ϕ per
turn, and because $\int v \, dt = Li$,

$$\lambda = \int v \, dt = n\phi = Li \tag{1.39}$$

and because $v = L \, di/dt$, it is observed that an abrupt change in flux linkages can only
derive from an abrupt change in current. But eq. (1.38) shows that an abrupt change
in flux linkage would produce an infinite voltage which leads to a dilemma because
an infinite voltage cannot be generated. One resolution of the dilemma[10] is to pro-
hibit instantaneous changes of current in inductors and the law of continuity of stored
energy pertaining to the inductor, which is presented here in boldface for emphasis:

The current flow through an inductor may not change instantaneously.

Continuity of stored energy also applies to the capacitor. Here, the charge contained on the capacitor,

$$q = \int i \, dt = Cv$$

---

[10] Another resolution of the dilemma is through the use of the impulse function, which is introduced in
Chapter 5.

can only change instantaneously if the voltage changes instantaneously. But because $i = dq/dt$, an abrupt change of charge over a zero time interval would produce an infinite current. This too is impossible, and thus, continuity of stored energy also applies to the capacitor[11]:

> **The voltage across a capacitor may not change instantaneously.**

SECTION 1.9   CONTROLLED SOURCES

Ideal and independent voltage and current sources were introduced in Section 1.7. There it was shown that ideal voltage and current sources deliver source voltages or currents that are specified functions of time at their terminals. A network composed entirely of passive components (R's, L's, or C's) requires at least one independent source for its excitation; if no such source is present, the network remains "at rest," is passive, and exhibits no response in the form of currents through and voltages across the passive elements.

An ideal dependent or controlled source is a source whose output depends on some voltage or current elsewhere within the network. It too will deliver its source voltage or current at its terminals. However, it must be emphasized that because any voltage or current in a network derives from an independent source, no voltage or current can be produced by a dependent or controlled source acting alone.

There are four types of ideal dependent or controlled sources:

- The ideal voltage-controlled voltage source (VCVS)
- The ideal current-controlled voltage source (ICVS)
- The ideal voltage-controlled current source (VCIS)
- The ideal current-controlled current source (ICIS)

These sources are usually represented by a diamond, and their symbols are shown in Fig. 1.16. The multiplier K in each case is frequently called the *gain* and may bear units.

■ **EXAMPLE 1.5**

Six sources are indicated in Fig. 1.17. Classify each as independent or controlled. For those that are controlled, specify the type, the control, and the gain.

Solution   Sources 1 and 5 are ideal independent sources and are recognized by their representation as circles. Source 1 contains polarity markings and is an ideal independent voltage source. Source 5 contains a direction arrow and is an ideal independent current source.

All other sources are controlled sources because they are represented by a diamond. Sources 3 and 4 are recognized as controlled voltage sources, because the diamonds contain polarity markings. Source 3 is controlled with a gain of 80 Ω by i_c, the current through the 36-Ω resistor. Source 4 is controlled with a gain of 36 by v_b, the voltage across the 60-Ω resistor.

---

[11] As does an analysis using the impulse function.

A network with two independent sources and four dependent or controlled sources **FIGURE 1.16**

All sources are active sources and drive the network if the Ks are nonzero. Somewhere within the network, a current i_1 and a voltage v_1 exist. It is implied that the gain K_3 bears the units ohms and the gain K_4 bears the units mhos.

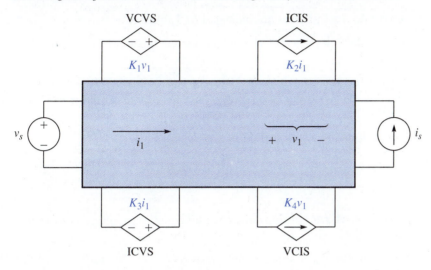

Sources 2 and 6 are controlled current sources, because the diamond contains a direction arrow. Source 2 is controlled with a gain of 16 by i_a, the current through the 80-Ω resistor. Source 6 is controlled with a gain of 30 Ω by v_d, the voltage across the 12-Ω resistor. ∎

A network with two independent sources and four dependent or controlled sources **FIGURE 1.17**

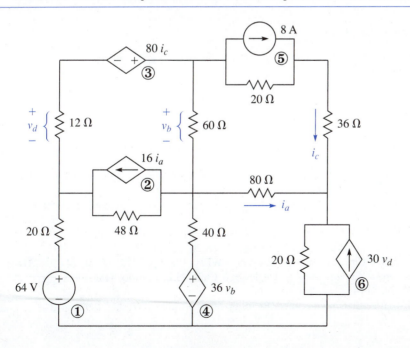

CHAPTER 1

SUMMARY

- Fundamental quantities and their units are displayed in Table 1.1.

- Derived quantities and their units are shown in Table 1.2.

- Standard decimal prefixes are indicated in Table 1.3.

- Current is the flow of electric charge; 1 ampere is the flow of 1 coulomb of charge in 1 second.

- Voltage is the potential difference between two points; 1 volt represents the potential difference between two points where 1 coulomb of charge will do 1 joule of work in moving from one point to the other.

- Power and energy are expressed by

$$p = \frac{dw}{dt} = vi \qquad w = \int p\,dt = \int vi\,dt$$

- The constitutive relationships or elemental equations (voltage-current relationships) for R, L, and C are summarized in Table 1.4.

- Energy is dissipated irreversibly in a resistor.

- The energy stored in an element must be a continuous function of time. The energy stored in an inductor and a capacitor is given, respectively, by

$$w_L = \frac{1}{2}Li^2 \qquad w_C = \frac{1}{2}Cv^2$$

- The current flowing through an inductor and the voltage across a capacitor cannot change instantaneously.

Additional Readings

Blackwell, W.A., and L.L. Grigsby. *Introductory Network Theory*. Boston: PWS Engineering, 1985, pp. 12–20, 25–40, 55–72, 106.

Bobrow, L.S. *Elementary Linear Circuit Analysis*. 2d ed. New York: Holt, Rinehart and Winston, 1987, pp. 1–11, 27–40, 160–172, 201–207.

Del Toro, V. *Engineering Circuits*. Englewood Cliffs, N.J.: Prentice-Hall, 1987, pp. 1–14, 17–19, 23–47, 157–159.

Dorf, R.C. *Introduction to Electric Circuits*. New York: Wiley, 1989, pp. 9–19, 32–38, 47–51, 166–174, 178–187.

Hayt, W.H., Jr., and J.E. Kemmerly. *Engineering Circuit Analysis*. 4th ed. New York: McGraw-Hill, 1986, pp. 3–21, 24–27, 116–130.

Irwin, J.D. *Basic Engineering Circuit Analysis*. 3d ed. New York: Macmillan, 1989, pp. 1–15, 20–24, 231–248.

Johnson, D.E., J.L. Hilburn, and J.R. Johnson. *Basic Electric Circuit Analysis*. 4th ed. Englewood Cliffs, N.J.: Prentice-Hall, 1989, pp. 2–16, 20–25, 60–63, 170–176, 180–185, 195–199.

Karni, S. *Applied Circuit Analysis*. New York: Wiley, 1988, pp. 1, 2, 11–21, 24–30, 153–165.

Madhu, S. *Linear Circuit Analysis*. Englewood Cliffs, N.J.: Prentice-Hall, 1988, pp. 2–23, 26, 27, 30–38, 41–43, 264–269, 472–484.

Nilsson, J.W. *Electric Circuits*. 3d ed. Reading, Mass.: Addison-Wesley, 1990, pp. 3–11, 16–21, 29–31, 176–192.

Paul, C.R. *Analysis of Linear Circuits*. New York: McGraw-Hill, 1989, pp. 3–11, 18–20, 32–38, 52–57, 196–210.
Sears, F.W., and M.W. Zemansky. *College Physics*. 2d ed. Reading, Mass.: Addison-Wesley, 1952, pp. 413–458.

CHAPTER 1   **PROBLEMS**

Section 1.3

1.1 How many electrons pass a given point in a conductor in 20 s if the conductor is carrying 10 A?

1.2 If 240 C pass through a wire in 12 s, what is the current?

1.3 What current flows through a conductor if 3.2×10^{21} electrons flow through the conductor in 16 s?

1.4 How much charge flows through a conductor from time $t = 0$ to time $t = 5$ s if the current varies in accordance with the relationship $i = 4t + 24(1 - e^{-4t})$ amperes?

1.5 How much work is done in moving 100 nC of charge a distance of 68 cm in the direction of a uniform electric field having a field strength of $E = 80$ kV/m?

1.6 A charge of 0.5 C is brought from infinity to a point. Assume that infinity is at 0 V, and determine the voltage at the terminal point if 14.5 J is required to move the charge.

1.7 If the potential difference between two points is 125 V, how much work is required to move a 3.2-C charge from one point to the other?

1.8 How many coulombs can be moved from point A to point B if $\Delta V_{AB} = 440$ V and if a maximum of 842 J can be expended?

1.9 In Problem 1.8, points A and B are separated by a distance of 62.5 cm. What is the magnitude of the electric field intensity if it has the same magnitude at all points between A and B?

1.10 A certain resistor dissipates heat at the rate of 8.12 kJ/min. If charge is passing through the resistor at a rate of 312.5 C/min, what is the potential difference across its terminals?

1.11 If 1 horsepower (hp) is equal to 0.746 kW, how much energy does a 20-hp motor deliver in 20 min?

1.12 If a 150-W incandescent bulb operates at 120 V, how many coulombs and electrons flow through the bulb in 1 h?

Section 1.5

1.13 A system output is related to the system input by the relationship $r = ae$. Determine whether the system is linear, time-invariant, both, or neither.

1.14 A system output is related to the system input by the relationship $r = ae + A$. Determine whether the system is linear, time-invariant, both, or neither.

1.15 A system output is related to the system input by the functional relationship

$$f(t) = \begin{cases} -K, & t < -a \\ \dfrac{V}{K}t, & -a < t < a \\ K, & t > a \end{cases}$$

Determine whether the system is linear, time-invariant, both, or neither.

1.16 A copper transmission line consists of 24 strands of copper wire, each 1.62 mm in diameter. If the resistivity of copper is 1.72×10^{-8} $\Omega \cdot$m, find the resistance of a 10-km length of wire.

1.17 What is the resistance between the circular faces of a right circular cylinder that is 1.8 m high and 1.88 cm in diameter if the cylinder is made from a metal having a conductivity of 8×10^6 ℧/m?

1.18 A sheet of foil that is 10.16 cm wide and 0.0108 cm thick must carry 4 amperes and is permitted to dissipate a maximum of 5.104 milliwatts. If its conductivity is 5.805×10^7 ℧/m, what is the maximum length of foil that can be used?

1.19 What is the length of a rectangular, aluminum bus bar having dimensions 16 by 1.4 cm, a conductivity of 3.61×10^4 ℧/m, and a resistance of 1.527 Ω?

1.20 A metal wire having a resistance of 0.1567 Ω at a temperature of 20°C is to be used in an application where its resistance can lie between 0.1314 and 0.1872 Ω. If its temperature coefficient of resistivity is 0.00314 $\Omega \cdot$°C/m, find the permitted temperature extremes.

1.21 If a 40-μF capacitor is subjected to a voltage of 120 cos 400t volts, what is the current through the capacitor?

1.22 What charge is flowing in the capacitor of Problem 1.21?

1.23 For a 40-μF capacitor subjected to a voltage of 100 cos 500t volts, determine relationships for the charge flow, the current, the power handled, and the energy stored. Then, evaluate the energy stored between $t = 0$ and $t = \pi/250$ s.

Section 1.6

1.24 If the current flowing into a 10-μF capacitor is described by the relationship

$$i(t) = \begin{cases} 2t \text{ A}, & 0 < t < 2 \text{ s} \\ 4 \text{ A}, & 2 < t < 6 \text{ s} \\ -2 \text{ A}, & 6 < t < 8 \text{ s} \\ 0 \text{ A}, & t > 8 \text{ s} \end{cases}$$

sketch the voltage across the capacitor over the interval $0 < t < 12$ s.

1.25 If the current flowing through a 4-H inductor is given by the waveform shown in Fig. P1.1, sketch the voltage across the inductor and evaluate the energy stored at 9 ms.

Figure P1.1

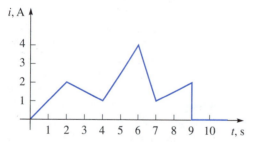

Section 1.7

1.26 The simple network shown in Fig. P1.2 is connected to a current source at $t = 0$. The current provided by the source, 4t amperes, flows in turn through the *series* combination of the resistor and inductor. Find the instantaneous power for the resistor, the inductor, and the source at $t = 5$ s. Then, determine the energy dissipated by the resistor, stored by the inductor, and delivered by the current source over the period $0 < t < 4$ s.

Figure P1.2

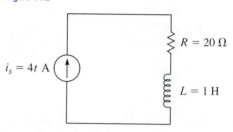

1.27 The simple network shown in Fig. P1.3 is connected to a voltage source at $t = 0$. The voltage provided by the source, 2$t + 8$ volts, appears across the terminals of each of the elements (the resistor and the capacitor) in this *parallel* combination. Find the instantaneous power for the resistor, the capacitor, and the source at $t = 4$ s. Then, determine the energy dissipated by the resistor, stored by the capacitor, and delivered by the voltage source over the period $0 < t < 4$ s.

Figure P1.3

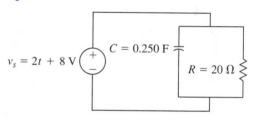

1.28 A 12-V automobile battery delivers 60 J of energy for 10 s. How much charge is moved, and how much current flows?

1.29 An electric heater takes 1.8 kW at 120 V for 20 min to boil a quantity of water. Find the current through the heater and its resistance.

1.30 If a light bulb takes 1.2 A at 120 V and operates for 8 h per day, what is the cost of its operation for 30 days if power costs 11.2¢/kWh.

1.31 If 1 calorie (1 cal) is equal to 4.184 J and it takes 1000 cal to raise 1 kg of water 1°C, how much current is carried by a 120-V heater if it is used to heat 4.82 kg of water from 25° to 45°C in 4 min?

1.32 What is the resistance of the heating element in Problem 1.31?

1.33 How much energy is stored in an initially un-charged $125\text{-}\mu\text{F}$ capacitor between 0 and 8 s when a voltage of $18e^{-t/10}$ volts is placed across its terminals?

1.34 The current through a 240-mH inductor is given by $i = 4\sin 400t$ amperes. Write expressions for the instantaneous power and the energy stored. Then, determine the instantaneous power at $t = \pi/200$ s and the energy stored during the period $0 < t < \pi/200$ s.

1.35 If the voltage across a $40\text{-}\mu\text{F}$ capacitor is given by $v = 4e^{-t} - 2e^{-2t}$ volts, how much energy will be stored or removed during the period $0 < t < 2$ s?

1.36 If the current through a 135-mH inductor is given by $i = 8t^2 + 4t + 2$ amperes, how much energy is stored or removed during the period $0 < t < 1.2$ s?

1.37 In Fig. P1.4, the current source is supplying 288 W at $t = 2$ s. What is the value of R?

Figure P1.4

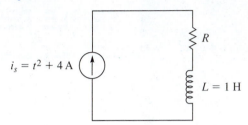

$i_s = t^2 + 4\,\text{A}$ R $L = 1\,\text{H}$

2

SIMPLE RESISTIVE NETWORKS

OBJECTIVES

The objectives of this chapter are to:

- Formulate the Kirchhoff current and voltage laws, using the principles of conservation of mass and energy. (KCL and KVL)

- Use the two Kirchhoff laws to develop relationships for the equivalence of elements in series and in parallel and for voltage and current division.

- Consider the departure from ideal to nonideal sources, and establish relationships for the transformation between nonideal voltage and current sources.

- Begin the analysis of resistive networks.

SECTION 2.1   **INTRODUCTION**

In the discussion of charge and current in Chapter 1, it was recalled from a study of physics that matter is any substance that occupies space and possesses mass. Charge was defined on the basis of the charge on the electron ($e = 1.602 \times 10^{-19}$ C), and since an electron is a particle having mass, charge is linked to mass. Because of this, the conservation of mass principle,

<blockquote>Mass may neither be created nor destroyed.</blockquote>

is linked to the conservation of charge principle:

<blockquote>Charge may neither be created nor destroyed.</blockquote>

The conservation of energy principle may also be recalled from a study of physics. This principle, also known as the first law of thermodynamics, is as follows:

<blockquote>Energy may neither be created nor destroyed but may be converted from one form to another.</blockquote>

Two very important laws that lie at the foundation of network theory are the Kirchhoff current and voltage laws. These will now be derived from considerations of conservation of mass and conservation of energy.

THE KIRCHHOFF LAWS   SECTION 2.2

Consider Fig. 2.1a, which is referred to as a *control volume*. Mass may flow into and out of this control volume; and if the flow is not uniform or steady, mass may accumulate within the control volume. Because mass generation is precluded by the conservation of mass (and there are no Einsteinian conversions of mass and energy), a mass balance may be written for the control volume:

    mass in = mass out + mass accumulation

If the flow is steady or uniform (conditions do not vary with time), there will be no mass accumulation:

    mass in = mass out

or

    mass in − mass out = 0

Because electrons are particles of mass that carry charge,

    electrons in − electrons out = 0

or

    charge in − charge out = 0

(a) Control volume used to derive the Kirchhoff current law and (b) a point in a network where six currents flow   **FIGURE 2.1**

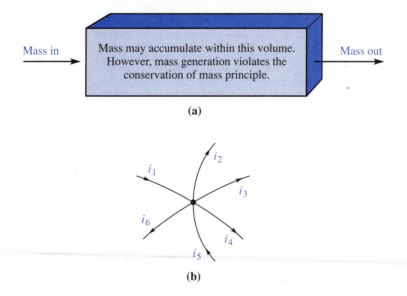

(a)

(b)

Over a period of time, charge will flow, so that

$$\text{charge flow in} - \text{charge flow out} = 0$$

and because charge flow is current,

$$\text{current in} - \text{current out} = 0$$

A slight revision gives

$$\text{current out} - \text{current in} = 0$$

The foregoing logical development involving mass, electrons, charge, and charge flow (current) leads to the statement of the *Kirchhoff current law*, hereinafter referred to merely as KCL.

 Kirchhoff Current Law: The algebraic sum of the currents flowing out of (away from) a point in a network at any instant is zero.

In Fig. 2.1b, KCL dictates that

$$-i_1 + i_2 + i_3 + i_4 - i_5 + i_6 = 0$$

or

$$i_1 - i_2 - i_3 - i_4 + i_5 - i_6 = 0$$

This shows that a "current in" is a negative "current out." In both cases, a slight algebraic adjustment yields

$$i_1 + i_5 = i_2 + i_3 + i_4 + i_6$$

This representation is a *continuity equation*, which is a relationship among through variables. The continuity equation in electric network analysis is the Kirchhoff current law.

Figure 2.2a shows four points, *A*, *B*, *C*, and *D*, in an electric field. Consider a charge at point *A*, and suppose that this charge is to be moved to point *D*. Two paths may be taken, one through point *B* and one through point *C*.

Energy is required to move charge in an electric field, and an energy expenditure may be assumed in moving the charge from *A* to *B*. With the use of a double-subscript notation, the first part of the subscript indicating *from* and the second part indicating *to*, four energy quantities can be described, w_{AB}, w_{BD}, w_{AC}, and w_{CD}. These are shown in Fig. 2.2b.

The energy expended in moving the charge from *A* to *B* does not depend on the path chosen, because a path dependency would infer a more favorable route, which is clearly in violation of the law of conservation of energy. Hence,

$$w_{AD} = w_{AB} + w_{BD}$$

$$w_{AD} = w_{AC} + w_{CD}$$

and

$$w_{AD} = w_{AB} + w_{BD} = w_{AC} + w_{CD}$$

(a) Four points in an electric field, (b) energy expenditures in moving a parcel of charge from point to point, and (c) voltage rises and drops between points that form a closed path or loop

FIGURE 2.2

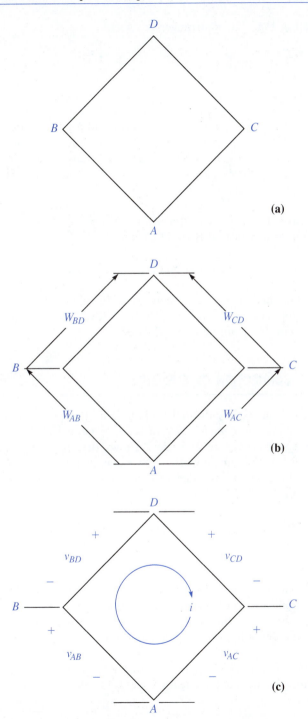

(a)

(b)

(c)

On a per unit charge basis, the potential or voltage increase is used (recall that $v = dw/dq$), so that

$$v_{AB} + v_{BD} = v_{AC} + v_{CD}$$

These voltages are displayed in Fig. 2.2c. Rearrangement gives

$$v_{AB} + v_{BD} - v_{AC} - v_{CD} = 0$$

or

$$v_{AC} + v_{CD} - v_{AB} - v_{BD} = 0$$

The selection of the clockwise loop in Fig. 2.2c causes v_{AB} and v_{BD} to be voltage rises and v_{AC} and v_{CD} to be voltage drops. With a voltage drop considered as a negative voltage rise, the preceding equations show that the algebraic sum of the voltage drops or the voltage rises around a closed path or loop is equal to zero. These facts lead to a general statement of the Kirchhoff voltage law:

> **Kirchhoff Voltage Law:** The algebraic sum of the voltages around a closed path or loop at any instant is equal to zero.

This also implies that the voltage between two points is independent of the path between the points.

The Kirchhoff voltage law, hereinafter referred to as KVL, is a *compatability equation*, which is a relationship among across variables. Both KVL and KCL lie at the foundation of network theory and can never be violated.

SECTION 2.3   **SERIES AND PARALLEL CONNECTION OF ELEMENTS**

A *series connection* of three elements is indicated in Fig. 2.3a. The points where the elements are tied together are called *nodes*. Use of KCL at node 1 shows that $i_1 - i_2 = 0$, or $i_1 = i_2$. The same procedure at node 2 shows that $i_2 - i_3 = 0$, or $i_2 = i_3$. Then, because quantities that are equal to the same quantity are equal to each other, $i_1 = i_2 = i_3$. This may be extended to n elements in series, and this extension provides the definition of a series connection of elements.

> A series connection of network elements is a connection in which all elements are connected end to end to provide a single path for the flow of current.

Figure 2.3b represents a *parallel connection* of three elements, where two closed paths or loops are evident. Application of KVL to each path shows that $v_1 - v_2 = 0$ and $v_2 - v_3 = 0$. Thus $v_1 = v_2 = v_3$, and this may be extended to n elements in parallel. This extension provides the definition of a parallel connection of elements.

> A parallel connection of network elements is a connection in which all elements are connected such that an identical voltage exists across all elements.[1]

---

[1] Another definition: a parallel connection of elements is a connection where the ends of the elements are tied together.

Elementary connection of network elements: (a) series and (b) parallel   **FIGURE 2.3**

The arrows in (b) indicate closed paths.

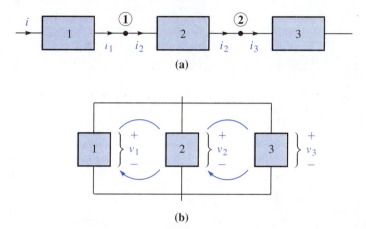

(a)

(b)

ONE-PORT NETWORKS AND THEIR EQUIVALENCE   **SECTION 2.4**

A *one-port network* is a network in which there are exactly two external terminals. Such a network is also referred to as a *one terminal pair* and is shown in Fig. 2.4. Two one-port networks[2] are said to be equivalent if they exhibit identical $v-i$ or $i-v$ relationships at the *driving point* (the terminals).

Figure 2.5 shows three one-ports and their equivalents. The specific cases of three R's, three L's and three C's are shown in Figs. 2.5a, 2.5b, and 2.5c, respectively. Observe that in each case, the voltage across the one-port and the current into the one-port are indicated for both the three-element case and the equivalent. In Fig. 2.5, the

A one-port network   **FIGURE 2.4**

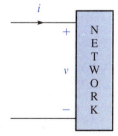

[2] The word *network* or *networks* is often omitted; *two one-ports* is the commonly used terminology.

FIGURE 2.5   Three elements connected in series and their equivalents

All six arrangements are one-ports. Equivalent component values are derived in Section 2.5.1.

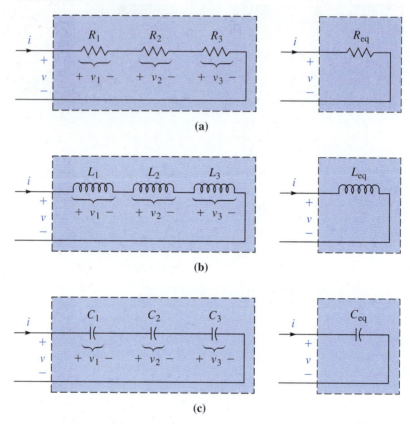

(a)

(b)

(c)

three elements are connected in series. A similar picture for three elements connected in parallel is shown in Fig. 2.6. The use of the concept of the equivalence of one-ports leads to the equivalence of elements in series and parallel.

SECTION 2.5   TWO APPLICATIONS OF KVL

2.5.1 Equivalences of Elements in Series

Figure 2.5a shows a one-port with three resistors in series and a one-port with a single-resistor equivalent. For the three resistors in series, KVL gives

$$-v + v_1 + v_2 + v_3 = 0$$

or

$$v = v_1 + v_2 + v_3 \tag{2.1}$$

Because the current is the same for each element in a series arrangement, each voltage is related to the current that flows through each resistor by Ohm's law, so that

Three elements connected in parallel and their equivalents

FIGURE 2.6

All six arrangements are one-ports. Equivalent component values are derived in Section 2.6.1.

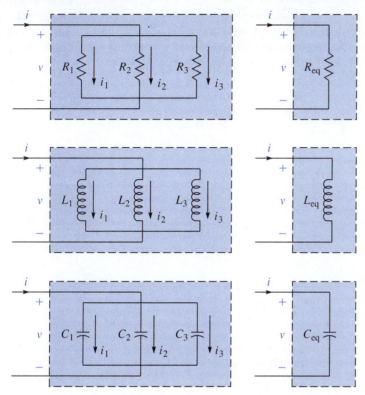

eq. (2.1) can be written as

$$v = R_1 i + R_2 i + R_3 i$$

or

$$v = (R_1 + R_2 + R_3)i$$

For the equivalent, which, by definition, must produce identical values of v and i,

$$v = R_{eq} i$$

This shows that for three resistances in series,

$$R_{eq} = R_1 + R_2 + R_3$$

and this may be extended to n resistances in series:

$$R_{eq} = \sum_{k=1}^{n} R_k \qquad (k = 1, 2, 3, \ldots, n) \qquad (2.2)$$

> The equivalent resistance of n resistors in series is equal to the sum of the resistances of the n resistors.

Although this chapter is entitled "Simple Resistive Networks," it is not out of place to consider the series arrangement of inductors and capacitors at this point. Just as all currents are equal in the three resistors in Fig. 2.5a, all rates of change of current must be equal in the three inductors connected in series in Fig. 2.5b. In this case, KVL demands that eq. (2.1) also hold; and with each voltage given by $v = L \, di/dt$, eq. (2.1) can be adjusted to

$$v = L_1 \frac{di}{dt} + L_2 \frac{di}{dt} + L_3 \frac{di}{dt}$$

or

$$v = (L_1 + L_2 + L_3) \frac{di}{dt}$$

For the equivalent,

$$v = L_{eq} \frac{di}{dt}$$

which shows that for three inductors in series,

$$L_{eq} = L_1 + L_2 + L_3$$

and for an extension to n inductors in series,

$$L_{eq} = \sum_{k=1}^{n} L_k \qquad (k = 1, 2, 3, \dots, n) \tag{2.3}$$

> The equivalent inductance of n inductors in series is equal to the sum of the inductances of the n inductors.

Three capacitors in series and the equivalent are shown in Fig. 2.5c. The voltage across each capacitor will be related to the charge contained in each capacitor. Thus, eq. (2.1), which governs, can be written as

$$v = \frac{1}{C_1} \int_{-\infty}^{t} i \, dz + \frac{1}{C_2} \int_{-\infty}^{t} i \, dz + \frac{1}{C_3} \int_{-\infty}^{t} i \, dz$$

and because the current flow is the same for all capacitors,

$$v = \left(\frac{1}{C_1} + \frac{1}{C_2} + \frac{1}{C_3} \right) \int_{-\infty}^{t} i \, dz$$

The equivalent is

$$v = \frac{1}{C_{eq}} \int_{-\infty}^{t} i \, dz$$

and this shows that

$$\frac{1}{C_{eq}} = \frac{1}{C_1} + \frac{1}{C_2} + \frac{1}{C_3}$$

or

$$C_{eq} = \frac{1}{1/C_1 + 1/C_2 + 1/C_3}$$

which may be extended to n capacitors in series:

$$C_{eq} = \frac{1}{\sum\limits_{k=1}^{n} 1/C_k} \qquad (k = 1, 2, 3, \ldots, n) \qquad (2.4)$$

The equivalent capacitance for n capacitors in series is equal to the reciprocal of the sum of the reciprocals of the capacitances of the n capacitors.

2.5.2 Voltage Division

Consider three resistors connected in series, as shown in Fig. 2.5a, and observe from eq. (2.1) that a certain fraction of the applied voltage will appear across each resistor. If the portion v_2 of the total voltage, v, is of interest, it is observed that

$$i = \frac{v_2}{R_2}$$

and

$$i = \frac{v}{R_1 + R_2 + R_3}$$

so that

$$\frac{v_2}{R_2} = \frac{v}{R_1 + R_2 + R_3}$$

and

$$v_2 = \frac{R_2}{R_1 + R_2 + R_3} v = \frac{R_2}{\Sigma R} v$$

where, of course, in this particular case, $\Sigma R = R_{eq}$. This may be extended to the case of n resistors in series to determine the voltage drop across the kth resistor:

$$v_k = \frac{R_k}{\Sigma R} v \qquad (2.5)$$

Equation (2.5) is the very useful *voltage division* relationship.

SECTION 2.6   **TWO APPLICATIONS OF KCL**

2.6.1 Equivalence of Elements in Parallel

For the three resistors connected in parallel displayed in Fig. 2.6a, KCL gives

$$i = i_1 + i_2 + i_3$$

Because the voltages across each resistor in the parallel combination are identical, Ohm's law can be used to yield

$$i = \frac{v}{R_1} + \frac{v}{R_2} + \frac{v}{R_3}$$

or

$$i = \left(\frac{1}{R_1} + \frac{1}{R_2} + \frac{1}{R_3} \right) v$$

For the equivalent, which must produce identical values of i and v,

$$i = \frac{1}{R_{eq}} v$$

This shows that

$$\frac{1}{R_{eq}} = \frac{1}{R_1} + \frac{1}{R_2} + \frac{1}{R_3}$$

or

$$R_{eq} = \frac{1}{1/R_1 + 1/R_2 + 1/R_3}$$

The extension to n resistors in parallel is obvious:

$$R_{eq} = \frac{1}{\sum\limits_{k=1}^{n} 1/R_k} \qquad (k = 1, 2, 3, \ldots, n) \tag{2.6}$$

> The equivalent resistance for n resistors in parallel is equal to the reciprocal of the sum of the reciprocals of the resistances of the n resistors.

It is often useful to think of a parallel combination of resistors as a combination of conductors. Observe from eq. (2.6) that because $G = 1/R$,

$$G_{eq} = G_1 + G_2 + G_3$$

thus for n resistors in parallel,

$$G_{eq} = \sum\limits_{k=1}^{n} G_k \qquad (k = 1, 2, 3, \ldots, n) \tag{2.7}$$

Combination of three resistors

FIGURE 2.7

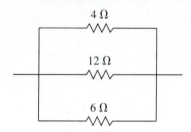

The combination of several resistors in parallel is facilitated by working with two of them at a time. If only two resistors are present, eq. (2.6) indicates that

$$R_{eq} = \frac{1}{1/R_1 + 1/R_2}$$

and a little algebraic manipulation yields

$$R_{eq} = \frac{R_1 R_2}{R_1 + R_2} \tag{2.8}$$

Note that the units here are $\Omega^2/\Omega = \Omega$, which is required.

Thus, the simplest possible parallel combination, that of two resistors, has an equivalent resistance determined by dividing the product of the two by the sum of the two. Notice that if the resistance values are equal, the equivalent resistance is just half the resistance of either resistor. Also notice that eqs. (2.6) and (2.8) clearly show that for resistors connected in parallel, *the equivalent resistance must always be less than the resistance of any resistor in the combination.*

■ **EXAMPLE 2.1**

Determine the equivalent resistance of the parallel combination of the three resistors shown in Fig. 2.7.

Solution   First, obtain the equivalent resistance of the 4-Ω and 12-Ω resistors and call this equivalent R_a.

$$R_a = \frac{4(12)}{4 + 12} = \frac{48}{16} = 3 \ \Omega$$

Then combine R_a with the 6-Ω resistor to obtain the equivalent:

$$R_{eq} = \frac{6R_a}{6 + R_a} = \frac{6(3)}{6 + 3} = \frac{18}{9} = 2 \ \Omega$$

■

■ **EXAMPLE 2.2**

Find the equivalent resistance of the complex resistive network shown in Fig. 2.8.

Solution   The network can be reduced to a single equivalent resistance by repeated applications of eqs. (2.2), (2.6), (2.7), and (2.8). First, consider the

FIGURE 2.8   An example of the reduction of a fairly complex resistive network through the
rules for addition of resistances in series and the combination of resistances in
parallel

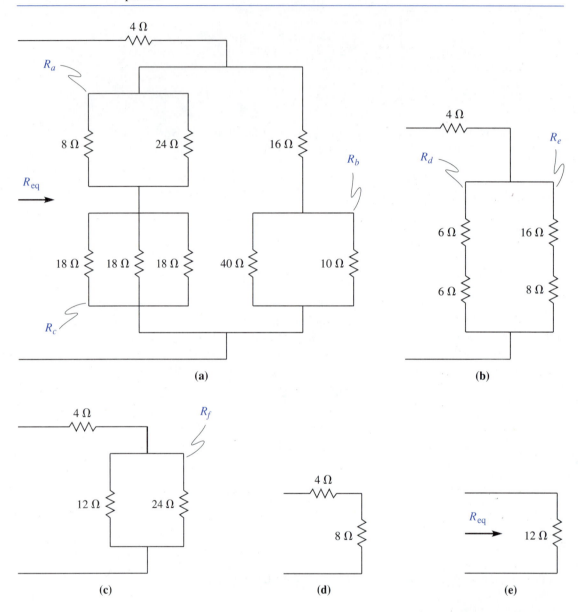

resistance combinations designated by R_a, R_b, and R_c in Fig. 2.8a, which, in
each case, represent a combination of resistances in parallel. By eq. (2.8),

$$R_a = \frac{8(24)}{8 + 24} = \frac{192}{32} = 6\ \Omega$$

$$R_b = \frac{40(10)}{40 + 10} = \frac{400}{50} = 8\ \Omega$$

and because R_c is composed of three 18-Ω resistors, $R_c = \frac{1}{3}(18) = 6\,\Omega$, or by eq. (2.6),

$$R_c = \frac{1}{1/18 + 1/18 + 1/18} = \frac{1}{3/18} = \frac{18}{3} = 6\,\Omega$$

Figure 2.8b incorporates these results and, in addition, shows two more combinations, R_d and R_e. Here, by eq. (2.2)

$$R_d = 6 + 6 = 12\,\Omega$$

and

$$R_f = 16 + 8 = 24\,\Omega$$

and these are incorporated in Fig. 2.8c.

In Fig. 2.8c, the combination indicated by R_f can be reduced to

$$R_f = \frac{12(24)}{12 + 24} = \frac{288}{36} = 8\,\Omega$$

and with this in Fig. 2.8d, it is observed that the entire network has been reduced to a pair of two resistors in series.

The equivalent is

$$R_{eq} = 4 + 8 = 12\,\Omega$$

as shown in Fig. 2.8e. ⬛

EXERCISE 2.1

Find the equivalent resistance looking into terminals a–b in Fig. 2.9.

Answer 110 Ω.

A series-parallel combination of resistors (Exercise 2.1)

FIGURE 2.9

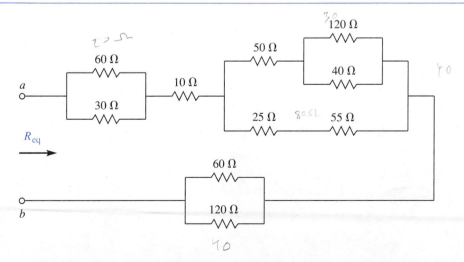

Three inductors in parallel and a single equivalent are indicated by the pair of one-ports in Fig. 2.6b. Again, KCL governs and because an identical v appears across each of the three inductors,

$$i = \frac{1}{L_1} \int_{-\infty}^{t} v \, dz + \frac{1}{L_2} \int_{-\infty}^{t} v \, dz + \frac{1}{L_3} \int_{-\infty}^{t} v \, dz$$

or

$$i = \left(\frac{1}{L_1} + \frac{1}{L_2} + \frac{1}{L_3} \right) \int_{-\infty}^{t} v \, dz$$

For the equivalent,

$$i = \frac{1}{L_{eq}} \int_{-\infty}^{t} v \, dz$$

which indicates that

$$\frac{1}{L_{eq}} = \frac{1}{L_1} + \frac{1}{L_2} + \frac{1}{L_3}$$

or

$$L_{eq.} = \frac{1}{1/L_1 + 1/L_2 + 1/L_3}$$

The extension to n inductors in parallel is

$$L_{eq} = \frac{1}{\sum\limits_{k=1}^{n} 1/L_k} \qquad (k = 1, 2, 3, \ldots, n) \tag{2.9}$$

The equivalent inductance for n inductors in parallel is equal the reciprocal of the sum of the reciprocals of the inductances of the n inductors.

For the case of three capacitors in parallel, shown as a one-port with an equivalent one-port in Fig. 2.6c, the voltage is the same across each of the three. An application of KCL gives

$$i = C_1 \frac{dv}{dt} + C_2 \frac{dv}{dt} + C_3 \frac{dv}{dt}$$

or

$$i = (C_1 + C_2 + C_3) \frac{dv}{dt}$$

For the equivalent,

$$i = C_{eq} \frac{dv}{dt}$$

so that for *n* capacitors in parallel,

$$C_{eq} = \sum_{k=1}^{n} C_k \qquad (k = 1, 2, 3, \ldots, n) \qquad\qquad (2.10)$$

The equivalent capacitance for *n* capacitors in parallel is the sum of the individual capacitance values.

2.6.2 Current Division

Current that flows in a parallel combination of network elements must divide among them in accordance with KCL. Figure 2.10 shows the case for two resistors, and here,

$$i = i_1 + i_2$$

Because $v = R_{eq}i$,

$$v = \frac{R_1 R_2}{R_1 + R_2} i$$

and

$$i_1 = \frac{v}{R_1} = \frac{R_2}{R_1 + R_2} i \qquad\qquad (2.11a)$$

A similar development shows that

$$i_2 = \frac{R_1}{R_1 + R_2} i \qquad\qquad (2.11b)$$

Equations (2.11) describe the current division principle.

Observe that in performing a current division for a parallel combination of two resistors, one first selects the current through one of the resistors. The fraction of current that flows through the selected resistor is the ratio of the resistance of the "other" resistor to the sum of the two resistances in the parallel combination.

Current division between two resistors in a parallel combination

FIGURE 2.10

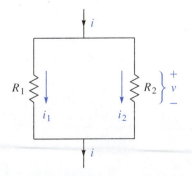

Some find it easier to perform current division by using conductance values. With $G = 1/R$ in eq. (2.11a),

$$i_1 = \frac{1/G_2}{1/G_1 + 1/G_2}\, i = \frac{1}{G_2}\left(\frac{G_1 G_2}{G_2 + G_1}\right)i$$

or

$$i_1 = \frac{G_1}{G_1 + G_2}\, i \tag{2.12a}$$

A similar procedure applied to eq. (2.11b) will provide i_2 in terms of i and the conductance values:

$$i_2 = \frac{G_2}{G_1 + G_2}\, i \tag{2.12b}$$

Equations (2.12) show that the fraction of current flowing through one of two conductances in parallel is the ratio of the conductance of the element whose current is to be determined to the total conductance in the parallel combination.

For current division among three or more resistors in parallel, it is best to work with the conductance rather than the resistance.

■ EXAMPLE 2.3

Figure 2.11 shows four resistors with subscripts corresponding to their resistance value. There is a series leg composed of R_2 and R_4, and this leg will be designated as R_s (s for series), where $R_s = 2 + 4 = 6\ \Omega$. The current that enters the parallel combination of R_3 and R_s is designated as i_p. The problem is to find the currents through and the voltages across the four resistors when a current of $i = 6$ A enters the combination.

FIGURE 2.11   A combination of four resistors

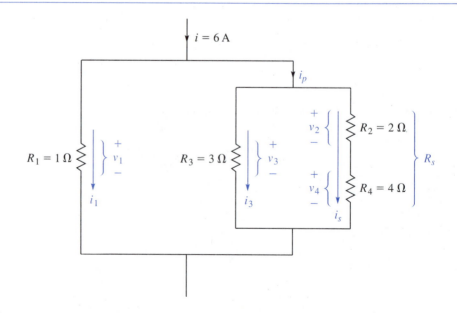

Solution   Designate R_a as the equivalent resistance of R_2, R_3, and R_4, with $R_s = 6\ \Omega$.

$$R_a = \frac{R_3 R_s}{R_3 + R_s} = \frac{3(6)}{3 + 6} = 2\ \Omega$$

Then by current division, as indicated in eqs. (2.11),

$$i_1 = \frac{R_a}{R_1 + R_a}\, i = \left(\frac{2}{1 + 2}\right)(6) = \frac{2}{3}(6) = 4\ \text{A}$$

and

$$i_p = \frac{R_1}{R_1 + R_a}\, i = \left(\frac{1}{1 + 2}\right)(6) = \frac{1}{3}(6) = 2\ \text{A}$$

Note that $v_1 = R_1 i_1 = 1(4) = 4$ V and that KCL could have been used to find i_p:

$$i_p = i - i_1 = 6 - 4 = 2\ \text{A}$$

Current division may once again be employed to find i_3 and i_s:

$$i_3 = \frac{R_s}{R_3 + R_s}\, i_p = \left(\frac{6}{3 + 6}\right)(2) = \frac{2}{3}(2) = \frac{4}{3}\ \text{A}$$

and

$$i_s = \frac{R_3}{R_3 + R_s}\, i_p = \left(\frac{3}{3 + 6}\right)(2) = \frac{1}{3}(2) = \frac{2}{3}\ \text{A}$$

Now, note that $v_3 = R_3 i_3 = 3(4/3) = 4$ V and $v_s = R_s i_s = 6(2/3) = 4$ V. This confirms that the voltages across all of the elements in parallel must be equal. Additional confirmation of a correct procedure can be obtained from KCL applied where i_p divides into i_3 and i_s:

$$i_p = i_3 + i_s = \frac{4}{3} + \frac{2}{3} = \frac{6}{3} = 2\ \text{A}$$

The voltages across R_2 and R_4 are seen to be $V_2 = R_2 i_s = 2(2/3) = 4/3$ V and $V_4 = R_4 i_s = 4(2/3) = 8/3$ V. These can also be obtained through the use of voltage division:

$$v_2 = \frac{R_2}{R_2 + R_1}\, v_s = \left(\frac{2}{2 + 4}\right)(4) = \frac{1}{3}(4) = \frac{4}{3}\ \text{V}$$

and

$$v_4 = \frac{R_4}{R_2 + R_4}\, v_s = \left(\frac{4}{2 + 4}\right)(4) = \frac{2}{3}(4) = \frac{8}{3}\ \text{V}$$

These voltages can be confirmed by KVL:

$$v_s = v_3 = v_2 + v_4 = \frac{4}{3} + \frac{8}{3} = \frac{12}{3} = 4 \text{ V}$$

■

EXERCISE 2.2

Find the currents through and the voltages across all of the resistors shown in Fig. 2.12.

Answer $v_4 = 12$ V, $i_4 = 3$ A; $v_5 = 20$ V, $i_5 = 4$ A; $v_6 = 48$ V, $i_6 = 8$ A; $v_8 = 32$ V, $i_8 = 4$ A; $v_{12} = 12$ V, $i_{12} = 1$ A.

SECTION 2.7   NONIDEAL SOURCES AND THEIR TRANSFORMATION

Practical voltage and current sources are not ideal and, as Fig. 2.13 shows, can be modeled by an ideal voltage source in series with a resistor or by an ideal current source in parallel with a resistor. In the voltage source, the source resistor accounts for a linear reduction in the terminal voltage as the current drawn from the source is increased. In the current source, the resistor provides a current leak and reduces the current delivered at the terminals as the voltage across the terminals is increased. The symbols for the voltage and current sources such as those shown in Fig. 2.13 always mean that the sources internal to the network are ideal. It is the resistors, R_s and R_p, that cause the departure of the overall network from the ideal because in the limit as $R_s \to 0$ and $R_p \to \infty$, the sources become ideal.

FIGURE 2.12   A combination of five resistors (Exercise 2.2)

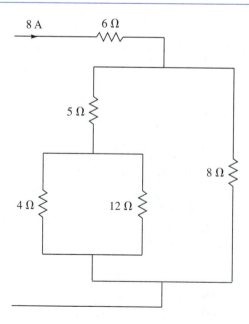

FIGURE 2.13

Two one-ports (a) a nonideal voltage source and (b) a nonideal current source

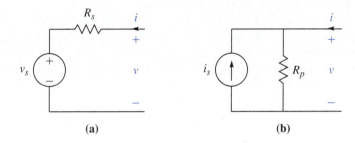

(a)   (b)

Conversion between the two models is possible. In Fig. 2.13a, an application of KVL yields

$$v = R_s i + v_s \tag{2.13}$$

and in Fig. 2.13b, KCL requires that

$$i = \frac{v}{R_p} - i_s$$

or

$$v = R_p i + R_p i_s \tag{2.14}$$

If there is to be an equivalence between the two nonideal sources shown in Fig. 2.13, then eqs. (2.13) and (2.14) must apply for either type of source. This will only occur if

$$R_p = R_s = R \tag{2.15a}$$
$$v_s = R i_s \tag{2.15b}$$

or

$$i_s = \frac{v_s}{R} \tag{2.15c}$$

Equations (2.15) are used for making the conversions.

Observe in Fig. 2.13 that an external current enters the terminals of the two one-ports used to describe the two nonideal sources. Note that if source excitations are nulled (made zero) for any reason, eqs. (2.13) and (2.14) show that

$$v = Ri$$

and

$$i = \frac{v}{R}$$

The v's and i's in the foregoing are terminal values of voltage and current. This leads to the conclusion that if an ideal voltage source ($R_s = 0$) is to be equal to zero, it is replaced by a short circuit. If an ideal current source ($R_p = \infty$) is to be equal to zero, it is replaced by an open circuit. Moreover, KVL prohibits the placement of a short circuit in parallel with an ideal voltage source, and KCL requires that an open circuit may not be placed in series with an ideal current source.

Combinations of ideal current sources of unequal value in series and ideal voltage sources of unequal value in parallel are prohibited by KCL and KVL. However, non-ideal sources can be combined to yield an equivalent and, as the preceding development shows, it is possible to convert an ideal voltage source in series with a resistor into an ideal current source in parallel with a resistor and vice versa.

Figure 2.14 shows a pair of one-ports. The first (Fig. 2.14a) contains two nonideal voltage sources in series. The second (Fig. 2.14b) contains two nonideal current sources in parallel.

KVL can be used to describe the voltage source pair in Fig. 2.14a:

$$v = v_{s1} + R_1 i + v_{s2} + R_2 i$$

or

$$v = v_{s1} + v_{s2} + (R_1 + R_2)i$$

which is in the form of eq. (2.13). In an extension to n nonideal voltage sources in series,

$$v = v_{s,\text{eq}} + R_{\text{eq}} i$$

FIGURE 2.14   Two one-ports containing (a) a pair of nonideal voltage sources arranged in series and (b) a pair of nonideal current sources arranged in parallel

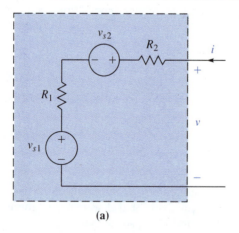

(a)

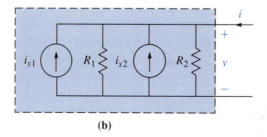

(b)

where

$$v_{s,eq} = \sum_{k=1}^{n} v_{s,k} \qquad (k = 1, 2, 3, \ldots, n) \qquad (2.16)$$

and where the resistances associated with each source are added in accordance with the rule applying to the addition of resistances in series.

KCL can be used to provide a similar development for the case of the two non-ideal current sources in parallel shown in Fig. 2.14b. Here,

$$i = \frac{v}{R_1} - i_{s1} + \frac{v}{R_2} - i_{s2}$$

$$i = v\left(\frac{1}{R_1} + \frac{1}{R_2}\right) - (i_{s1} + i_{s2})$$

$$i = \frac{v}{R_{eq}} - i_{s,eq}$$

or

$$v = R_{eq}i + R_{eq}i_{s,eq}$$

which is in the form of eq. (2.14). This may be extended to n nonideal current sources in parallel to give an equivalent nonideal current source with

$$i_{s,eq} = \sum_{k=1}^{n} i_{s,k} \qquad (k = 1, 2, 3, \ldots, n) \qquad (2.17)$$

and in this extension, R_{eq} is the parallel combination of the n resistances.

The clue to the combination of several nonideal sources in arrangements of varying degrees of complexity is to use the simple transformations provided by eqs. (2.15) to arrange voltage sources in series and current sources in parallel. Combinations in accordance with eqs. (2.16) and (2.17) can then be effected in a continuing procedure that terminates when the entire conglomerate is represented by a single voltage or a single current source.

■ **EXAMPLE 2.4**

Adjust the network composed exclusively of resistors and ideal voltage and ideal current sources in Fig. 2.15a to an equivalent of a single ideal voltage source and a single resistor.

Solution   In Fig. 2.15a, notice the current sources designated by a and b. These current sources can be transformed to voltage sources:

$$v_{sa} = R_a i_{sa} = 2(2) = 4 \text{ V}$$

and

$$v_{sb} = R_b i_{sb} = 4(2) = 8 \text{ V}$$

The result of these transformations is shown in Fig. 2.15b.

FIGURE 2.15

(a) A network containing four sources and (b)–(f) steps in the reduction of the network in (a) to the voltage source equivalent in (g) and the current source equivalent in (h)

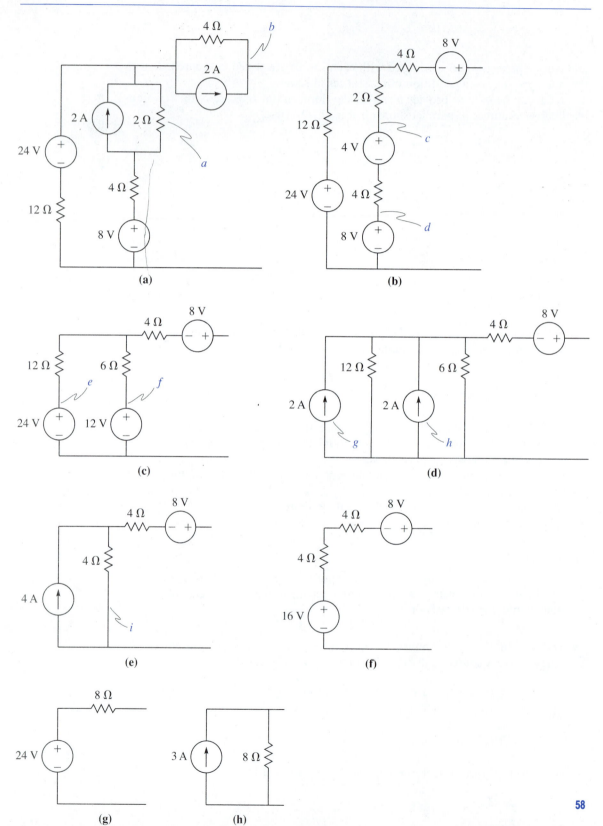

In Fig. 2.15b, sources c and d can be combined by using eq. (2.16):

$$v_{s,cd} = 4 + 8 = 12 \text{ V}$$

and

$$R_{cd} = 2 + 4 = 6 \text{ }\Omega$$

and this is shown in Fig. 2.15c.

In Fig. 2.15c, sources e and f can be combined if they are transformed to current sources:

$$i_{se} = \frac{v_e}{R_e} = \frac{24}{12} = 2 \text{ A}$$

and

$$i_{sf} = \frac{v_f}{R_f} = \frac{12}{6} = 2 \text{ A}$$

These are shown in Fig. 2.15d as current sources g and h. By eq. (2.17),

$$i_{s,gh} = 2 + 2 = 4 \text{ A}$$

and a parallel combination of R_g and R_h gives

$$R_{gh} = \frac{6(12)}{6 + 12} = \frac{72}{18} = 4 \text{ }\Omega$$

This is shown in Fig. 2.15e as current source i.

Current source i in Fig. 2.15e can be transformed to a voltage source:

$$v_{si} = R_i i_{si} = 4(4) = 16 \text{ V}$$

and the result in Fig. 2.15f leads to the single–voltage source equivalent:

$$v_{s,eq} = 16 + 8 = 24 \text{ V}$$

and

$$R_{eq} = 4 + 4 = 8 \text{ }\Omega$$

as indicated in Fig. 2.15g.

If a current source equivalent is desired, a simple voltage–to–current source transformation provides the result shown in Fig. 2.15h. ∎

EXERCISE 2.3

Convert the source configuration shown in Fig. 2.16 to a single current source.

Answer $i_{s,eq} = 4.5 \text{ A}$, $R_{eq} = 16 \text{ }\Omega$.

FIGURE 2.16   A combination of several voltage and current sources (Exercise 2.3)

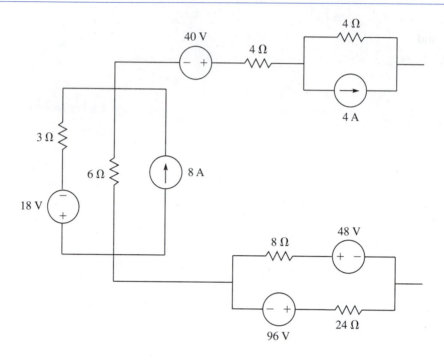

SECTION 2.8   **THE NETWORK MODEL**

This book is concerned with network analysis and, while there is a considerable amount of mathematical elegance that lies at the core of network analysis, the objective of this book is not mathematical rigor but the development and understanding of useful techniques for finding, as a minimum, voltages across and currents through network elements. There is a virtue in being able to "think electrically" and, because visualization problems have been known to be an obstacle to the learning process, it is worthwhile to consider how the network model is formulated.

As an example, consider the problem of the placement of 12 floodlights across the rear of a house to provide light for a social function in the yard. Some readers may know by intuition that the facts concerning the elements are:

- The light bulbs may be considered as ideal resistors and their brightness is proportional to the current passing through them.
- The power source, whether a wall receptacle or a portable generator, may be considered as an ideal voltage source.
- The "on-off" switch is either a short circuit (when in the "on" condition) or an open circuit (when in the "off" condition).
- The lengths of connecting wire are presumed short so that the resistance of the wire is not a consideration.

Several arrangements of the network elements may be considered. For example, all of the elements may be arranged in series as in Fig. 2.17a. In this arrangement,

Twelve light bulbs with switch and voltage source arranged (a) in series,
(b) in series-parallel 2 × 6, (c) in series-parallel 12 × 1, and (d) in a ladder network
configuration.

FIGURE 2.17

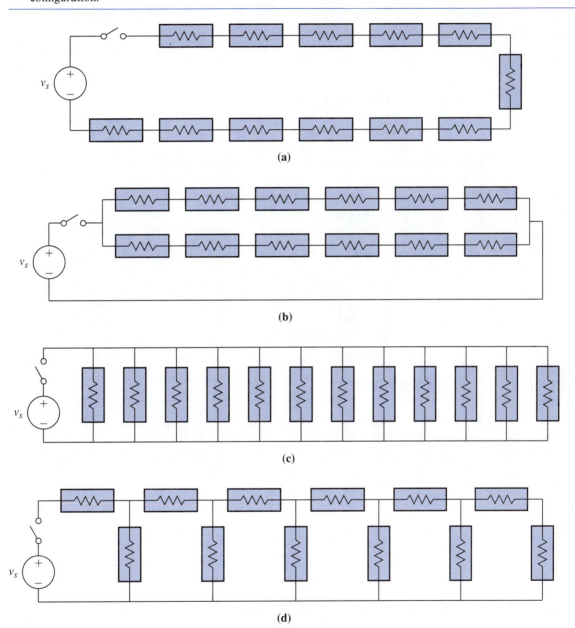

the total resistance is $12R$ and the current in each bulb will be

$$i = \frac{v_s}{12R}$$

Because the series arrangement leads to the maximum possible resistance, the current will be at a minimum and this "least current" configuration leads to minimum illumination. Moreover, if one bulb fails (burns out) creating an open circuit, the current to all bulbs is interrupted and the yard is plunged into darkness.

An improvement is the series-parallel arrangement of Fig. 2.17b. Here there are 6 bulbs in series in each of the two parallel legs. The resistance of each leg is $6R$ and each leg receives a current of

$$i = \frac{v_s}{6R}$$

This is twice the current in the series arrangement of Fig. 2.17a and leads to greater illumination. And if one bulb fails, six are still in operation.

Edison was not an engineer but a meticulous experimenter. Many feel that he intuitively realized that the best lighting system would derive from many bulbs arranged in parallel and this seems to be confirmed by the arrangement of the 12 bulbs in parallel shown in Fig. 2.17c. Here the current in each bulb is

$$i = \frac{v_s}{R}$$

There is no way to get more illumination with the voltage source at hand and, if one bulb burns out, the loss of illumination will hardly be noticed.

The question now becomes: suppose the bulbs are arranged as shown in Fig. 2.17d? The determination of the current in each bulb becomes more difficult. This is where additional techniques in network analysis enter the arena and this study now turns to a consideration of ladder networks.

SECTION 2.9   **LADDER NETWORKS**

Ladder networks, such as the network containing a single voltage source in Fig. 2.18, are close to being the simplest type of network that can be subjected to analysis. There are six *branches* in this network; five containing single resistors and the sixth

FIGURE 2.18   A ladder network

Notice that the voltage source is enclosed in a box. This nonideal source is composed of the 63-V ideal source in series with the 5-Ω resistor.

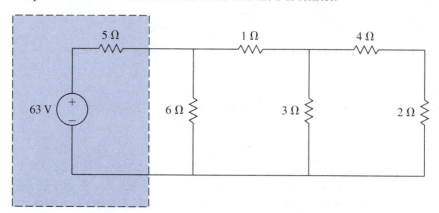

containing the nonideal voltage source composed of a 63-V ideal source and a 5-Ω resistor. The common point where the 2-, 3-, and 6-Ω resistors are connected is called the *ground* or *datum node*.

Observe in Fig. 2.18 that every network element is no more than one node removed from the datum or ground node. The word *element* is used here because in this book, the series combination of the ideal voltage source and the 5-Ω resistor will constitute a branch.

Ladder networks may be analyzed by a systematic application of any or all of the following previously discussed simple, fundamental principles:

- The addition of resistors in series
- The combination of resistors in parallel
- The Kirchhoff voltage law
- The Kirchhoff current law
- The voltage division principle
- The current division principle
- The voltage–to–current source transformation
- The current–to–voltage source transformation

These will be referred to as the fundamental network principles (but only for the balance of this chapter), and to be sure, more advanced methods—such as the methods of loop, mesh, or node analysis (to be introduced in Chapter 3)—can be employed. But at this stage, all of the fundamental principles have been discussed in detail, and it is well to proceed to an actual example.

The eight fundamental principles need not be applied in any fixed sequence, and most often, application of all of them is not necessary. However, the usual solution strategy first calls for a computation of the total resistance seen by the source. This resistance is determined from the principles of addition of resistances in series and combination of resistances in parallel.

The analysis of the ladder network in Fig. 2.18 will now be conducted by using the first four of the fundamental network principles. The network, with the additional designation of three closed paths, or *meshes*,[3] is repeated as Fig. 2.19, and the first part of the strategy is the determination of the resistance presented to the 63-V source. Its effect will be included in the establishment of the total resistance (equivalent resistance) of the network.

The addition of the resistors in series and the combination of the resistors in parallel are shown directly on the network in Fig. 2.19; and it can be noted that an equivalent resistance of 7 Ω is presented to the 63-V source, and the terminology that the 63-V ideal source is *looking into* a network with a driving-point resistance of 7 Ω is sometimes used.

With subscripts used to identify the currents through and the voltages across the resistors having the same resistance value as the subscript, it is noted that the current through the 5-Ω resistor is

$$i_5 = \frac{63}{7} = 9 \text{ A}$$

and the voltage across the 5-Ω resistor is

$$v_5 = 5i_5 = 5(9) = 45 \text{ V}$$

[3] Meshes are considered in detail in Chapter 3.

FIGURE 2.19

The ladder network of Fig. 2.18 showing the computation of resistance equivalents to arrive at a total equivalent of 7 Ω seen by the ideal 63-V source

The circles with roman numerals indicate closed paths, or meshes, used for applications of KVL in the ladder network analysis, and the numerals within circles represent nodes.

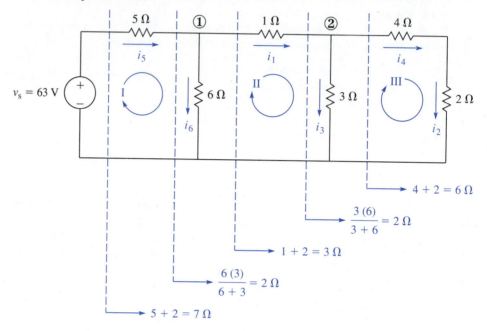

KVL may now be applied around mesh I in Fig. 2.19:

$$v_5 + v_6 - 63 = 0$$

or

$$v_6 = 63 - v_5 = 63 - 45 = 18 \text{ V}$$

and hence,

$$i_6 = \frac{v_6}{6} = \frac{18}{6} = 3 \text{ A}$$

Next, apply KCL at node 1 in Fig. 2.19:

$$i_1 + i_6 - i_5 = 0$$

or

$$i_1 = i_5 - i_6 = 9 - 3 = 6 \text{ A}$$

and

$$v_1 = 1i_1 = 1(6) = 6 \text{ V}$$

Once again, use KVL, but this time around mesh II in Fig. 2.19:

$$v_1 + v_3 - v_6 = 0$$

$$v_3 = v_6 - v_1 = 18 - 6 = 12 \text{ V}$$

and then

$$i_3 = \frac{v_3}{3} = \frac{12}{3} = 4 \text{ A}$$

Finally, by KCL at node 2 in Fig. 2.19,

$$i_4 + i_3 - i_1 = 0$$

$$i_4 = i_2 = i_1 - i_3 = 6 - 4 = 2 \text{ A}$$

and

$$v_4 = 4i_4 = 4(2) = 8 \text{ V}$$

$$v_2 = 2i_2 = 2(2) = 4 \text{ V}$$

Notice that KVL can be used in mesh III in Fig. 2.19 to check the result:

$$v_4 + v_2 - v_3 = 0$$

or

$$8 + 4 - 12 = 0$$

as it should.

An identical, but not necessarily more expeditious, result can be obtained by using current and voltage division instead of KVL and KCL. Once $i_5 = 9$ A and $v_5 = 45$ V are determined, a current division at node 1 in Fig. 2.19 yields

$$i_6 = \left(\frac{3}{6+3}\right)i_5 = \frac{1}{3}(9) = 3 \text{ A}$$

and

$$i_3 = \left(\frac{6}{6+3}\right)i_5 = \frac{2}{3}(9) = 6 \text{ A}$$

so that

$$v_6 = 6i_6 = 6(3) = 18 \text{ V}$$

$$v_1 = 1i_1 = 1(6) = 6 \text{ V}$$

Another current division at node 2 shows that

$$i_3 = \left(\frac{6}{6+3}\right)i_1 = \frac{2}{3}(6) = 4 \text{ A}$$

and

$$i_4 = i_2 = \left(\frac{3}{6+3}\right)i_1 = \frac{1}{3}(6) = 2 \text{ A}$$

which gives

$$v_3 = 3i_3 = 3(4) = 12 \text{ V}$$

Then, because $v_3 = 12$ V appears across the series combination of the 4- and 2-Ω resistors, voltage division can be used to show that

$$v_4 = \left(\frac{4}{4+2}\right)v_3 = \frac{2}{3}(12) = 8 \text{ V}$$

and

$$v_2 = \left(\frac{2}{4+2}\right)v_3 = \frac{1}{3}(12) = 4 \text{ V}$$

An alternative method that assumes a value for the current in the most remote branch, with subsequent application of Ohm's law and the two Kirchhoff laws, has many adherents. This method, sometimes referred to as the *ladder method*, will be illustrated in the example that follows.

■ **EXAMPLE 2.5**

Consider the ladder network shown in Fig. 2.19 and assume that the current through the 2-Ω resistor is 1 A. Determine the currents through and the voltages across each resistor.

Solution   With resistor values used as subscripts and $i_2 = 1$ A,

$$v_2 = 2(1) = 2 \text{ V}$$
$$i_4 = i_2 = 1 \text{ A}$$
$$v_4 = 4(1) = 4 \text{ V}$$

and by KVL,

$$v_3 = v_2 + v_4 = 2 + 4 = 6 \text{ V}$$

Then,

$$i_3 = \frac{v_3}{3} = \frac{6}{3} = 2 \text{ A}$$

and by KCL,

$$i_1 = i_3 + i_4 = 2 + 1 = 3 \text{ A}$$

Next,

$$v_1 = 1(3) = 3 \text{ V}$$

so that

$$v_6 = v_1 + v_3 = 3 + 6 = 9 \text{ V}$$

$$i_6 = \frac{v_6}{6} = \frac{9}{6} = \frac{3}{2} \text{ A}$$

and by KCL,

$$i_5 = i_6 + i_1 = \frac{3}{2} + 3 = \frac{9}{2} \text{ A}$$

with

$$v_5 = 5\left(\frac{9}{2}\right) = \frac{45}{2} \text{ V}$$

If $i_2 = 1$ A, the voltage v_s would be

$$v_s = v_5 + v_6 = \frac{45}{2} + 9 = \frac{63}{2} \text{ V}$$

which is just half the value of the ideal-source voltage (63 V). Because the network is linear, all of the foregoing currents and voltages may be adjusted upward by a factor of 2 to arrive at the correct result. ■

Ladder network analysis is quite straightforward, and there is a certain flexibility that the analyst may use in the eventual approach. However, it has been observed that deviations from this typical case can cause problems. The most common of these occurs when the ladder is driven by a single current source. And believe it or not, when the current source is at the right end of the ladder, further problems—perhaps due to visualization—occur. It is well, therefore, to look at an example that considers these two facets and shows that the location of the source on the right does not make the analysis more difficult.

There is no need to transform the current source in Fig. 2.20a to a voltage source. Observe how the resistances in series are added and how those in parallel are combined to arrive at an equivalent of a 6-Ω resistance in parallel with the 12-Ω resistance, which may be considered as part of the current source. Then at node 1, current division is used to obtain

$$i_4 = \left(\frac{12}{6+12}\right)(6) = \frac{2}{3}(6) = 4 \text{ A}$$

and

$$i_{12} = \left(\frac{6}{6+12}\right)(6) = \frac{1}{3}(6) = 2 \text{ A}$$

The voltages are

$$v_4 = 4i_4 = 4(4) = 16 \text{ V}$$

$$v_{12} = 12i_{12} = 12(2) = 24 \text{ V}$$

and the current source has a voltage of 24 V across its terminals.

FIGURE 2.20   (a) A ladder network driven by a current source and (b) the same ladder network
driven by a voltage source

The voltage source in (b) is equivalent to the current source in (a).

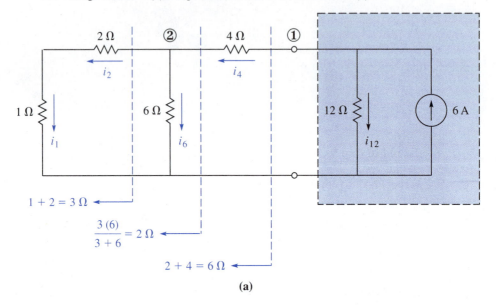

(a)

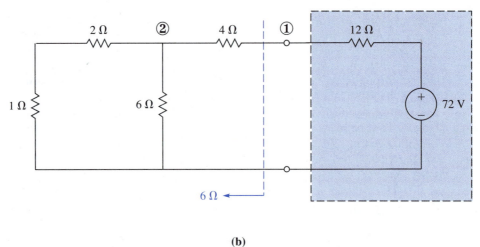

(b)

Another current division is used at node 2:

$$i_6 = \left(\frac{3}{3+6}\right) i_4 = \frac{1}{3}(4) = \frac{4}{3} \text{ A}$$

and

$$i_1 = \left(\frac{6}{3+6}\right) i_4 = \frac{2}{3}(4) = \frac{8}{3} \text{ A}$$

The voltages are

$$v_6 = 6i_6 = 6\left(\frac{4}{3}\right) = 8 \text{ V}$$

$$v_1 = 1i_1 = 1\left(\frac{8}{3}\right) = \frac{8}{3} \text{ V}$$

and because $i_1 = i_2$,

$$v_2 = 2i_2 = 2\left(\frac{8}{3}\right) = \frac{16}{3} \text{ V}$$

If it is desired to transform the current source to a voltage source, eq. (2.15) may be used, and the result is shown in Fig. 2.20b. In this case, an equivalent 18 Ω is presented to the 72-V ideal source, so that

$$i_{12} = i_4 = \frac{72}{18} = 4 \text{ A}$$

and the voltage source is delivering 4 A to the network (as did the current source). Here,

$$v_{12} = 12i_{12} = 12(4) = 48 \text{ V}$$

and it is seen that the terminal voltage of the nonideal voltage source is $72 - 48 = 24$ V.

Currents through and voltages across the other resistors in the network can be calculated in the same manner as for the case of the constant-current source.

There is some significance to a consideration of the current through and the voltage across the 12-Ω resistor when this resistor is considered as part of the ideal source. This is summarized in Fig. 2.21. Observe that in both cases, the network draws 4 A at a voltage of 24 V.

The sources in Fig. 2.20: (a) the current source and (b) the voltage source

FIGURE 2.21

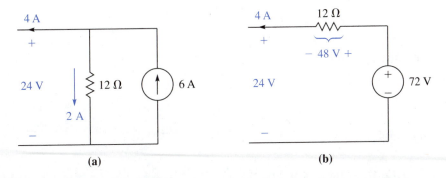

(a)   (b)

FIGURE 2.22   A ladder network (Exercise 2.4)

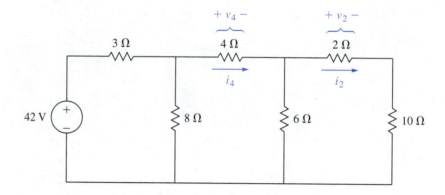

EXERCISE 2.4

Determine the current through and the voltage across the 2- and 4-Ω resistors in Fig. 2.22.

Answer $i_2 = 1$ A, $i_4 = 3$ A, $v_2 = 2$ V, and $v_4 = 12$ V.

SECTION 2.10   **A STEP FORWARD**

The network in Fig. 2.23 is not a ladder network, but its analysis is still quite straightforward as it is in a series-parallel structure. Note that the subscripts on the currents correspond to the resistance values, and the analysis objective is to determine the currents through and the voltages across every resistive element.

Here, too, the first step should be the determination of the total resistance "seen" by the source. Figure 2.23a indicates the first steps, and it is then observed that the value of R_p in Fig. 2.23b is

$$R_p = \frac{40(8 + 2)}{40 + (8 + 2)} = \frac{40(10)}{50} = 8\ \Omega$$

which makes the value of R_{eq} in Fig. 2.22c

$$R_{eq} = 1 + 3 + 8 = 12\ \Omega$$

The current in the 1-Ω resistor is

$$i_1 = \frac{48}{12} = 4\text{ A}$$

and the voltage is merely

$$v_1 = 1i_1 = 1(4) = 4\text{ V}$$

(a) A simple network that is not a ladder network, (b) the same network with
some simplifications, and (c) the network with just a source and an equivalent
resistance

FIGURE 2.23

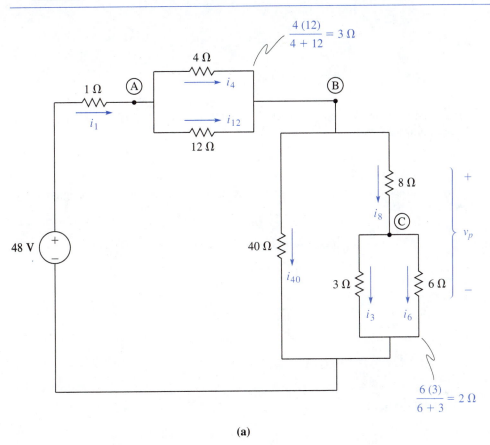

(a)

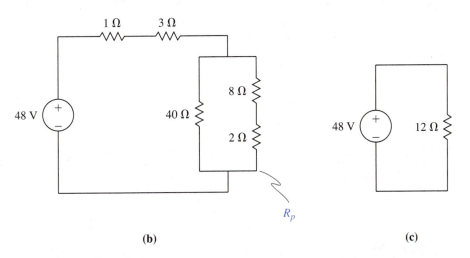

(b)

(c)

The parallel combination of the 4- and 12-Ω resistors handles a total of 4 A. By a current division,

$$i_4 = \left(\frac{12}{12+4}\right)(4) = \frac{3}{4}(4) = 3 \text{ A}$$

and

$$i_{12} = \left(\frac{4}{12+4}\right)(4) = \frac{1}{4}(4) = 1 \text{ A}$$

This can be confirmed by an application of KCL at node A. The voltage across this parallel combination is

$$v_4 = v_{12} = 3(4) = 1(12) = 12 \text{ V}$$

By a voltage division, the voltage across the parallel combination designated by R_p is seen to be

$$v_p = \left(\frac{8}{8+1+3}\right)(48) = \frac{2}{3}(48) = 32 \text{ V}$$

This means that

$$v_{40} = v_p = 32 \text{ V}$$

and

$$i_{40} = \frac{v_{40}}{40} = \frac{32}{40} = 0.80 \text{ A}$$

By KCL at node B,

$$i_8 = i_1 - i_{40} = 4.00 - 0.80 = 3.20 \text{ A}$$

so that

$$v_8 = 8i_8 = 8(3.20) = 25.60 \text{ V}$$

Then, another application of the voltage division principle gives

$$v_3 = v_6 = \left(\frac{2}{2+8}\right)v_p = 0.2(32) = 6.40 \text{ V}$$

from which

$$i_3 = \frac{v_3}{3} = \frac{6.40}{3} = 2.13 \text{ A}$$

and

$$i_6 = \frac{v_6}{6} = \frac{6.40}{6} = 1.07 \text{ A}$$

which can be confirmed by an application of KCL at node C.

A simple network that is not a ladder network (Exercise 2.5)

FIGURE 2.24

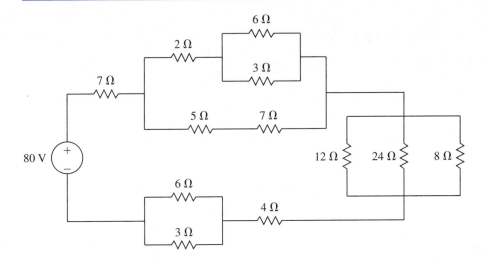

EXERCISE 2.5

Determine the currents through the 4- and 8-Ω resistors and the voltage drops across the 2- and 5-Ω resistors in Fig. 2.24.

Answer $i_4 = 4$ A, $i_8 = 2$ A, $v_2 = 6$ V, $v_5 = 5$ V.

THE TEE-PI (WYE-DELTA) TRANSFORMATION

The network shown in Fig. 2.25a is called a *tee network* (a *wye network* is shown in Fig. 2.25b). It may be transformed to a *pi* (or *delta*) *network* as indicated by Figs. 2.25c and 2.25d. All of these networks may be considered as multiports because they have three terminals.

The equivalence between the tee and pi networks can be established by considering the tee and pi resistances superimposed in the three-terminal network shown in Fig. 2.26. For terminals 1 and 2,

$$R_1 + R_2 = \frac{R_{12}(R_{23} + R_{31})}{R_{12} + R_{23} + R_{31}}$$

for terminals 2 and 3,

$$R_2 + R_3 = \frac{R_{23}(R_{31} + R_{12})}{R_{12} + R_{23} + R_{31}}$$

FIGURE 2.25   (a) Tee network, (b) wye network, (c) pi network, and (d) delta network

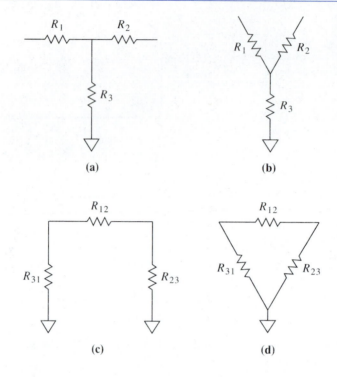

and for terminals 3 and 1,

$$R_3 + R_1 = \frac{R_{31}(R_{12} + R_{23})}{R_{12} + R_{23} + R_{31}}$$

FIGURE 2.26   Three-terminal network showing nomenclature for the development of the pi-tee and tee-pi transformation

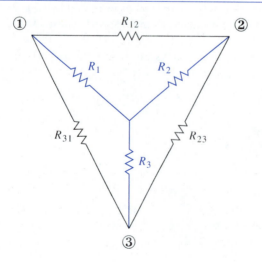

These three equations may be used in a laborious algebraic procedure to express the pi elements in terms of the tee elements,

$$R_{12} = \frac{R_1 R_2 + R_1 R_3 + R_2 R_3}{R_3} \tag{2.18a}$$

$$R_{23} = \frac{R_1 R_2 + R_1 R_3 + R_2 R_3}{R_1} \tag{2.18b}$$

and

$$R_{31} = \frac{R_1 R_2 + R_1 R_3 + R_2 R_3}{R_2} \tag{2.18c}$$

and the tee elements in terms of the pi elements,

$$R_1 = \frac{R_{12} R_{31}}{R_{12} + R_{23} + R_{31}} \tag{2.19a}$$

$$R_2 = \frac{R_{23} R_{12}}{R_{12} + R_{23} + R_{31}} \tag{2.19b}$$

and

$$R_3 = \frac{R_{31} R_{23}}{R_{12} + R_{23} + R_{31}} \tag{2.19c}$$

Equations (2.18) and (2.19), as well as similar equations involving conductances, are summarized in Fig. 2.27. The reader is cautioned that these equations pertain only to the resistance designations in the tee and pi networks shown in the figure. These designations may vary among the many available texts, and so may the appearance of the transformation equations.

There are two important observations to be made. The first is that it is somewhat easier to go from the tee to the pi using conductance values for the elements and from the pi to the tee using resistance values.

The second observation concerns the balanced case where all three branches in either the tee or the pi possess identical resistance values. Equations (2.18) show that if all branches in the tee have equal resistance values, the branches in the pi will also have equal resistance values, which are three times the tee values. If the branches of the pi have equal values of resistance, the branches of the tee, as shown by eqs. (2.19), will all have values equal to one-third of the pi values.

■ **EXAMPLE 2.6**

The network in Fig. 2.28a is not a ladder. The only current of interest is the current i flowing through the 8-Ω resistor.

Solution   Good technique demands that the first step be the search for balanced-tee or balanced-pi networks. In Fig. 2.28a, two balanced tee's are spotted immediately, and these may be converted to balanced pi's by multiplying the resistance values by 3. The result of this step is shown in Fig. 2.28b,

FIGURE 2.27   Summary of relationships for pi-to-tee and tee-to-pi transformations

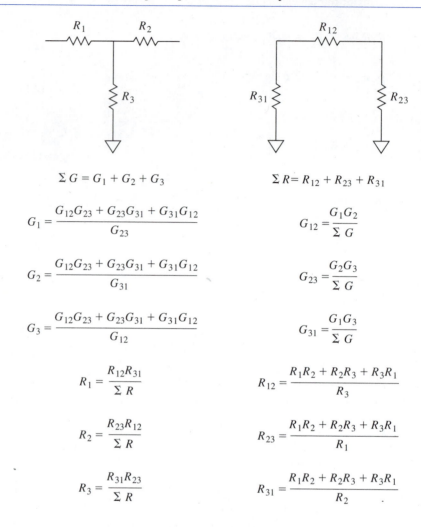

$$\Sigma\, G = G_1 + G_2 + G_3 \qquad\qquad \Sigma\, R = R_{12} + R_{23} + R_{31}$$

$$G_1 = \frac{G_{12}G_{23} + G_{23}G_{31} + G_{31}G_{12}}{G_{23}} \qquad\qquad G_{12} = \frac{G_1 G_2}{\Sigma\, G}$$

$$G_2 = \frac{G_{12}G_{23} + G_{23}G_{31} + G_{31}G_{12}}{G_{31}} \qquad\qquad G_{23} = \frac{G_2 G_3}{\Sigma\, G}$$

$$G_3 = \frac{G_{12}G_{23} + G_{23}G_{31} + G_{31}G_{12}}{G_{12}} \qquad\qquad G_{31} = \frac{G_1 G_3}{\Sigma\, G}$$

$$R_1 = \frac{R_{12}R_{31}}{\Sigma\, R} \qquad\qquad R_{12} = \frac{R_1 R_2 + R_2 R_3 + R_3 R_1}{R_3}$$

$$R_2 = \frac{R_{23}R_{12}}{\Sigma\, R} \qquad\qquad R_{23} = \frac{R_1 R_2 + R_2 R_3 + R_3 R_1}{R_1}$$

$$R_3 = \frac{R_{31}R_{23}}{\Sigma\, R} \qquad\qquad R_{31} = \frac{R_1 R_2 + R_2 R_3 + R_3 R_1}{R_2}$$

and a combination of resistances in parallel, where applicable, provides the much simpler network of Fig. 2.28c.

At this point, ladder network ideas can be used to finish the example. However, it can be noted that there is another balanced tee in the network of Fig. 2.28c. This can be converted to a balanced pi, as shown in Fig. 2.28d; and a further combination of resistances in parallel yields the network of Fig. 2.28e.

It is readily verified that the resistance presented to the 32-V ideal source in Fig. 2.22e is $974/103\ \Omega$. A simple Ohm's law calculation gives the current i.

$$i = \frac{32}{974/103} = 3.384 \text{ A}$$

(a) A rather complicated network, (b) a reduction based on the identification of a pair of balanced-tee networks, (c) a further simplification after some combination of resistances in parallel, (d) another reduction based on the identification of another balanced tee network and (e) the final network

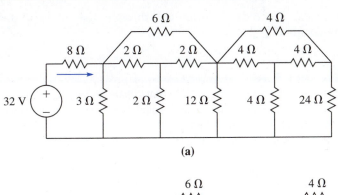

(a)

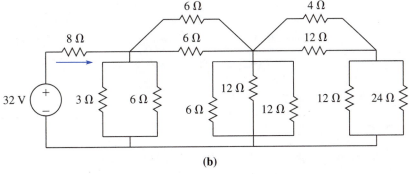

(b)

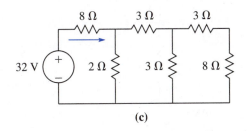

(c)

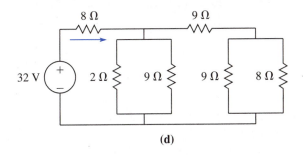

(d)

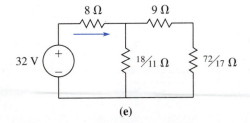

(e)

FIGURE 2.28

SECTION 2.12   SPICE EXAMPLES

Reading: Before the reader proceeds, Sections C.1 through C.4 in Appendix C should be read and understood.

EXAMPLE S2.1

In Section 2.8, the ladder network shown in Fig. 2.18 (repeated here as Fig. S2.1) was analyzed. All branch voltages and branch currents were determined. This example shows that an identical result can be obtained using PSPICE.

The ladder network *made ready* for the PSPICE analysis is shown in Fig. S2.2. The input file is shown in Fig. S2.3, and pertinent extracts from the output file are shown in Fig. S2.4.

Figure S2.1   The ladder network of Fig. 2.18

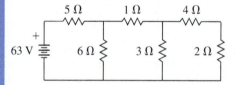

Figure S2.2   The ladder network of Fig. S2.1 made ready for SPICE analysis

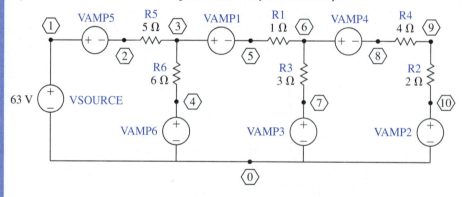

Figure S2.3   SPICE input file for Example S2.1

```
SPICE PROBLEM - CHAPTER 2 - NUMBER 1
R1         5         6         1
R2         9         10        2
R3         6         7         3
R4         8         9         4
R5         2         3         5
R6         3         4         6
VSOURCE              1         0         DC        63
*******************************************************
*HERE ARE THE VOLTAGE SOURCES USED TO MEASURE CURRENT
VAMP1    3         5         DC        0
VAMP2    10        0         DC        0
VAMP3    7         0         DC        0
VAMP4    6         8         DC        0
VAMP5    1         2         DC        0
VAMP6    4         0         DC        0
*******************************************************
*HERE IS THE SINGLE CONTROL STATEMENT
.END
```

Figure S2.4 SPICE output file for Example S2.1

```
SPICE PROBLEM - CHAPTER 2 - NUMBER 1


****      SMALL SIGNAL BIAS SOLUTION        TEMPERATURE =   27.000 DEG C


***********************************************************************


   NODE    VOLTAGE      NODE    VOLTAGE      NODE    VOLTAGE      NODE    VOLTAGE


   (   1)   63.0000   (    2)   63.0000   (    3)   18.0000   (    4)    0.0000

   (   5)   18.0000   (    6)   12.0000   (    7)    0.0000   (    8)   12.0000

   (   9)    4.0000   (   10)    0.0000



        VOLTAGE SOURCE CURRENTS
        NAME            CURRENT

        VSOURCE       -9.000E+00
        VAMP1          6.000E+00
        VAMP2          2.000E+00
        VAMP3          4.000E+00
        VAMP4          2.000E+00
        VAMP5          9.000E+00
        VAMP6          3.000E+00

        TOTAL POWER DISSIPATION    5.67E+02  WATTS

            JOB CONCLUDED
```

Discussion: It may be observed that all branch currents agree with those obtained in Section 2.9. However, the only branch voltages that have been obtained are the voltages between the nodes and the ground node. No branch voltages (voltages between the nodes) have been printed in the output.

Example S2.2 will show how to obtain such branch voltages.

EXAMPLE S2.2

In Exercise 2.5, the network shown in Fig. 2.24 (repeated here as Fig. S2.5) was analyzed. The branch voltages for the 4- and 8-Ω 1 resistors and the branch voltages for the 2- and 5-Ω resistors were determined. This example shows that an identical result can be obtained by using PSPICE.

Figure S2.5 The network of Fig. 2.24

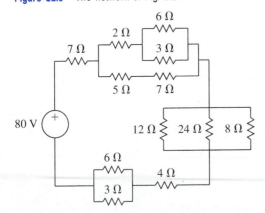

(continues)

Example S2.2 (continued)

Figure S2.6   The network of Fig. S2.5 made ready for SPICE analysis

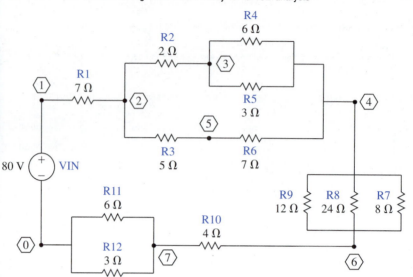

Figure S2.7   SPICE input file for Example S2.2

```
SPICE PROBLEM - CHAPTER 2 - NUMBER 2
R1         1         2         7
R2         2         3         2
R3         2         5         5
R4         3         4         6
R5         3         4         3
R6         5         4         7
R7         4         6         8
R8         4         6         24
R9         4         6         12
R10        6         7         4
R11        7         0         6
R12        7         0         3
*THE LINES BETWEEN THE ASTERISKS PERMIT THE PRINTING OF THE
*BRANCH VOLTAGES. NOTICE HOW THE VALUE OF THE SOURCE IS 0 V,
*BUT THAT THE 80 V VALUE IS ACCOMMODATED IN THE .DC STATEMENT.
****************************************************************
VIN        1         0         DC        0
.DC        VIN       80        80        1
.PRINT  DC      V(2,5)   V(2,3)   I(R10)   I(R7)
****************************************************************
*DON'T FORGET THE .END STATEMENT
.END
```

Figure S2.8   SPICE output file for Example S2.2

```
 SPICE PROBLEM - CHAPTER 2 - NUMBER 2


 ****       DC TRANSFER CURVES                    TEMPERATURE =    27.000 DEG C


 **************************************************************************


   VIN           V(2,5)        V(2,3)       I(R10)       I(R7)

   8.000E+01    5.000E+00    6.000E+00    4.000E+00    2.000E+00



       JOB CONCLUDED
```

The network *made ready* for the PSPICE analysis is shown in Fig. S2.6. The input file is shown in Fig. S2.7, and the output file is shown in Fig. S2.8.

Discussion: It may be observed that the branch currents and voltages agree with those obtained in Section 2.10. Notice how the branch voltages were obtained by using a .DC and .PRINT statement.

CHAPTER 2

SUMMARY

- The Kirchhoff current law: The algebraic sum of the currents flowing into a point (node) in a network at any instant is equal to zero.

- The Kirchhoff voltage law: The algebraic sum of the voltages around a closed path or loop at any instant is equal to zero.

- n resistors in series:

$$R_{eq} = \sum_{k=1}^{n} R_k \qquad (k = 1, 2, 3, \ldots, n)$$

- n inductors in series:

$$L_{eq} = \sum_{k=1}^{n} L_k \qquad (k = 1, 2, 3, \ldots, n)$$

- n capacitors in series:

$$C_{eq} = \frac{1}{\sum_{k=1}^{n} 1/C_k} \qquad (k = 1, 2, 3, \ldots, n)$$

- n resistors in parallel:

$$R_{eq} = \frac{1}{\sum_{k=1}^{n} 1/R_k} \qquad (k = 1, 2, 3, \ldots, n)$$

or

$$G_{eq} = \sum_{k=1}^{n} G_k \qquad (k = 1, 2, 3, \ldots, n)$$

- n inductors in parallel

$$L_{eq} = \frac{1}{\sum_{k=1}^{n} 1/L_k} \qquad (k = 1, 2, 3, \ldots, n)$$

- n capacitors in parallel

$$C_{eq} = \sum_{k=1}^{n} C_k \qquad (k = 1, 2, 3, \ldots, n)$$

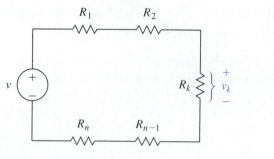
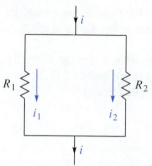

- Voltage division:

$$v_k = \frac{R_k}{\Sigma R} v$$

- Current division:

$$i_1 = \frac{R_2}{R_1 + R_2} i$$

or

$$i_1 = \frac{G_1}{G_1 + G_2} i$$

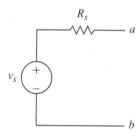
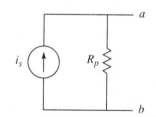

- Source transformations:

$$R_p = R_s = R$$
$$v_s = R i_s$$

and

$$i_s = \frac{v_s}{R}$$

- If an ideal voltage source is to be equal to zero, it is replaced by a short circuit (a voltage source of zero value is a short circuit). If an ideal current source is to be equal to zero, it is replaced by an open circuit (a current source of zero value is an open circuit).

- Ladder networks are analyzed by employing any or all of the following fundamental principles:

—The addition of resistances in series

—The combination of resistances in parallel

—The Kirchhoff voltage law

—The Kirchhoff current law

—The voltage division principle

—The current division principle

—The voltage–to–current source transformation

—The current–to–voltage source transformation

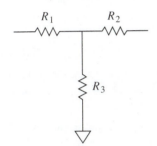 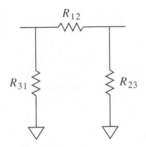

- The tee network with resistive elements R_1, R_2, and R_3 can be transformed to a pi network with elements R_{12}, R_{23}, and R_{31} by

$$R_{12} = \frac{R_1 R_2 + R_1 R_3 + R_2 R_3}{R_3}$$

$$R_{23} = \frac{R_1 R_2 + R_1 R_3 + R_2 R_3}{R_1}$$

and

$$R_{31} = \frac{R_1 R_2 + R_1 R_3 + R_2 R_3}{R_2}$$

- The pi network with elements R_{12}, R_{23}, and R_{31} can be transformed to a tee network with elements R_1, R_2, and R_3 by

$$R_1 = \frac{R_{12} R_{31}}{R_{12} + R_{23} + R_{31}}$$

$$R_2 = \frac{R_{23} R_{12}}{R_{12} + R_{23} + R_{31}}$$

and

$$R_3 = \frac{R_{31} R_{13}}{R_{12} + R_{23} + R_{31}}$$

Additional Readings

Blackwell, W.A., and L.L. Grigsby. *Introductory Network Theory.* Boston: PWS Engineering, 1985, pp. 42–46, 82–88, 116–128, 204–207.

Bobrow, L.S. *Elementary Linear Circuit Analysis.* 2d ed. New York: Holt, Rinehart and Winston, 1987, pp. 11–27, 98, 114–121, 191–194.

Del Toro, V. *Engineering Circuits*. Englewood Cliffs, N.J.: Prentice-Hall, 1987, pp. 14–17, 70–82, 107–109, 145–151, 163–166.

Dorf, R.C. *Introduction to Electric Circuits*. New York: Wiley, 1989, pp. 51–71, 124–129, 175–178, 187–189.

Hayt, W.H., Jr., and J.E. Kemmerly. *Engineering Circuit Analysis*. 4th ed. New York: McGraw-Hill 1986, pp. 27–45, 77–83, 230–235.

Irwin, J.D. *Basic Engineering Circuit Analysis*. 3d ed. New York: Macmillan, 1989, pp. 25–71, 192–197.

Johnson, D.E., J.L. Hilburn, and J.R. Johnson. *Basic Electric Circuit Analysis*. 4th ed. Englewood Cliffs, N.J.: Prentice-Hall, 1989, pp. 25–48, 118, 119, 135–139, 176–179, 186–189.

Karni, S. *Applied Circuit Analysis*. New York: Wiley, 1988, pp. 30–35, 74–85, 120, 293, 304.

Madhu, S. *Linear Circuit Analysis*. Englewood Cliffs, N.J.: Prentice-Hall, 1988, pp. 23–30, 38–40, 56–78, 226.

Nilsson, J.W. *Electric Circuits*. 3d ed. Reading, Mass.: Addison-Wesley, 1990, pp. 25–28, 36–44, 53–55, 93–96, 192–196.

Paul, C.R. *Analysis of Linear Circuits*. New York: McGraw-Hill, 1989, pp. 11–18, 29–31, 39–52, 57–59, 75–80.

PROBLEMS   CHAPTER 2

Section 2.2

2.1 Determine i_1, i_2, and i_3 in the network of Fig. P2.1.

Figure P2.1

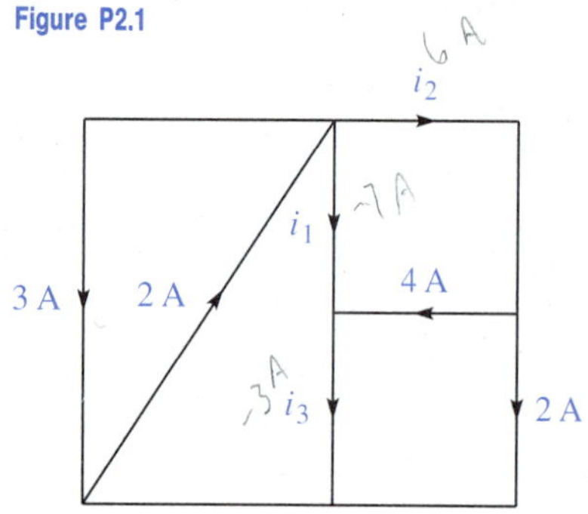

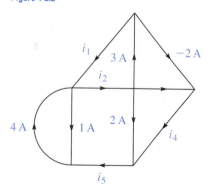

2.2 Determine i_1, i_2, i_3, and i_4 in the network of Fig. P2.2.

Figure P2.2

2.3 Determine i_1, i_2, i_3, and i_4 in the network of Fig. P2.3.

Figure P2.3

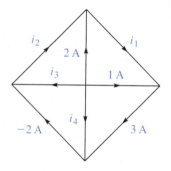

2.4 Determine v_1 through v_5 in the network of Fig. P2.4.

Figure P2.4

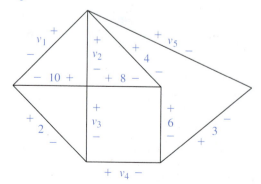

2.5 Determine v_1 through v_7 in the network of Fig. P2.5.

Figure P2.5

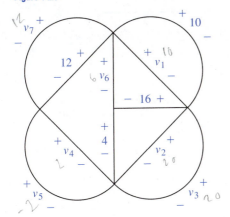

2.6 Determine v_1 through v_6 in the network of Fig. P2.6.

Figure P2.6

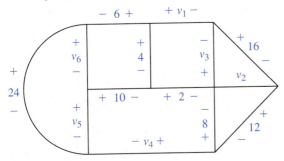

Section 2.5

2.7 In the series network shown in Fig. P2.7, $R_{eq} = 10\ \Omega$, $v_2 = 24$ V, and $p_3 = 16$ W. What are the values of R_1, R_2, and R_3?

Figure P2.7

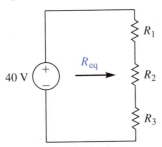

2.8 What is the equivalent capacitance of four capacitors connected in series having capacitance values of $C_1 = 40\ \mu F$, $C_2 = 20\ \mu F$, $C_3 = 10\ \mu F$, and $C_4 = 12\ \mu F$?

2.9 Use voltage division to find the voltage across the 2- and 3-Ω resistors in Fig. P2.8.

Figure P2.8

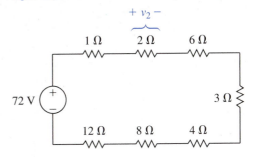

Section 2.6

2.10 Two resistors R_1 and R_2 are connected in parallel. What is the value of the equivalent resistance if (a) $R_1 = 20\ \Omega$ and $R_2 = 20\ \Omega$, (b) $R_1 = 20\ \Omega$ and $R_2 = 80\ \Omega$, (c) $R_1 = 20\ \Omega$ and $R_2 = \infty\ \Omega$, (d) $R_1 = 20\ \Omega$ and $R_2 = 0\ \Omega$, and (e) $R_1 = 20\ \Omega$ and $R_2 = 50\ \Omega$?

2.11 One of the resistance values in the network shown in Fig. P2.9 is blurred due to coffee spillage. If the equivalent resistance of the network is $R_{eq} = 20\ \Omega$, what is the value of the blurred resistor?

Figure P2.9

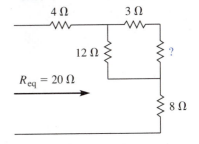

2.12 In Fig. P2.10, what is the value of R to make $i = 2$ A?

Figure P2.10

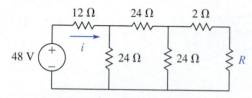

2.13 What is the equivalent resistance looking into terminals a–b of the network shown in Fig. P2.11?

Figure P2.11

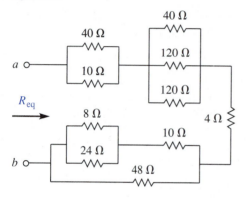

2.14 What is the equivalent resistance looking into terminals a–b of the network shown in Fig. P2.12?

Figure P2.12

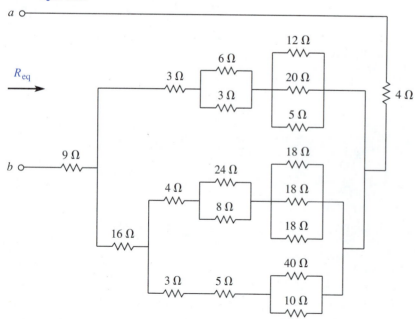

2.15 Select R in Fig. P2.13 so that $i = 2$ A.

Figure P2.13

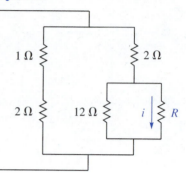

2.16 What is the equivalent resistance looking into terminals a–b of the network shown in Fig. P2.14?

Figure P2.14

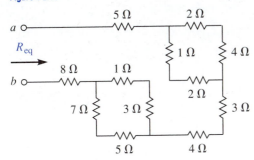

2.17 In the network of Fig. P2.15, what value of R will make the equivalent resistance R_{eq} at terminals a–b equal to R?

Figure P2.15

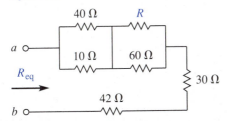

2.18 What is the equivalent resistance looking into terminals a–b of the network shown in Fig. P2.16?

Figure P2.16

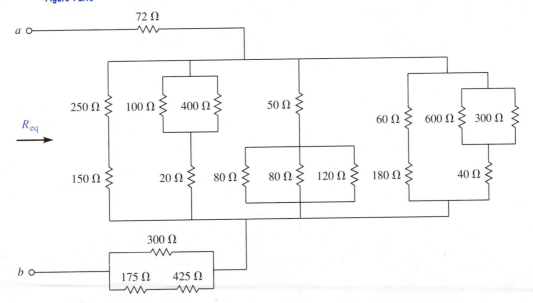

2.19 Two inductors L_1 and L_2 are connected in parallel. What is the value of the equivalent inductance if (a) $L_1 = 6$ H and $L_2 = 3$ H, (b) $L_1 = 8$ H and $L_2 = 8$ H, (c) $L_1 = 12$ mH and $L_2 = 4$ mH, (d) $L_1 = 10$ mH and $L_2 = 120$ mH, and (e) $L_1 = 1$ mH and $L_2 = 99$ mH?

2.20 Use current division to find the currents through the 12- and 24-Ω resistors in Fig. P2.17.

Figure P2.17

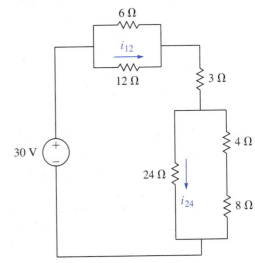

2.21 Use current division to find the current through the 12-Ω resistor in Fig. P2.18.

Figure P2.18

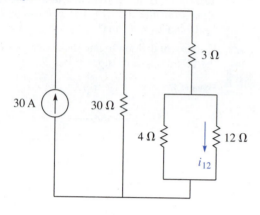

Section 2.7

2.22 Transform the voltage source in the network of Fig. P2.19 to a current source, and use current division to find i_2.

Figure P2.19

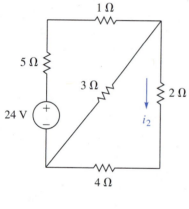

2.23 Transform the current source in the network of Fig. P2.20 to a voltage source, and use voltage division to find v_4.

Figure P2.20

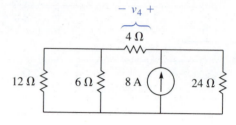

2.24 Convert the configuration of sources in the network of Fig. P2.21 to a single equivalent current source.

Figure P2.21

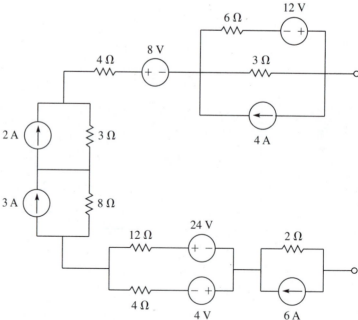

2.25 Convert the configuration of sources in the network of Fig. P2.22 to a single equivalent voltage source.

Figure P2.22

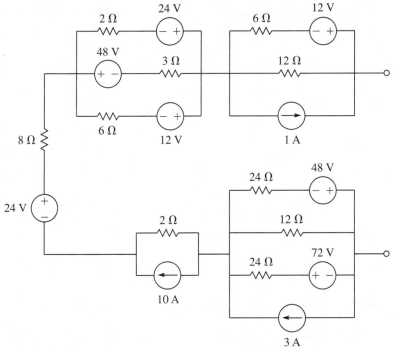

Section 2.9

2.26 Find i_9, i_4, and v_7 in the ladder network shown in Fig. P2.23.

Figure P2.23

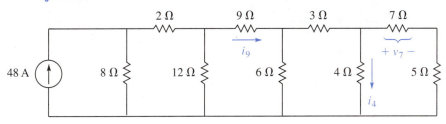

2.27 Find i_3, i_4, v_{12}, and v_{14} in the ladder network shown in Fig. P2.24.

Figure P2.24

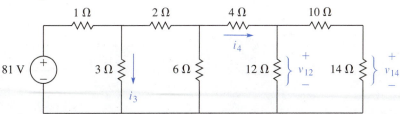

2.28 Find i_6, i_{10}, and v_2 in the ladder network shown in Fig. P2.25.

Figure P2.25

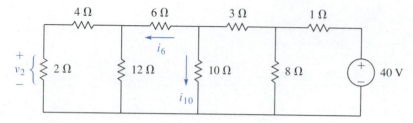

2.29 Find i_9, i_{40}, and v_7 in the ladder network shown in Fig. P2.26.

Figure P2.26

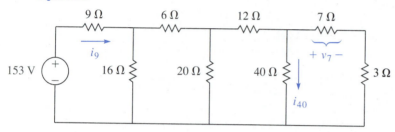

2.30 Find i_{13}, i_{14}, v_6, and v_7 in the ladder network shown in Fig. P2.27.

Figure P2.27

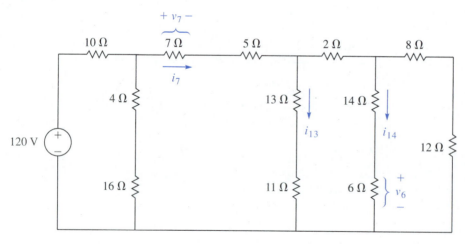

Section 2.10

2.31 Find i_5, v_{20}, and the power dissipated by the 24 Ω resistor in the network shown in Fig. P2.28.

Figure P2.28

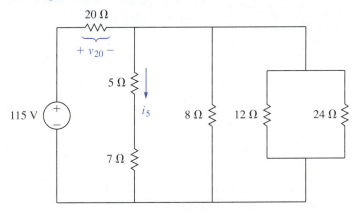

2.32 Find i_5, v_3, and the power dissipated by the 12 Ω resistor in the network shown in Fig. P2.29.

Figure P2.29

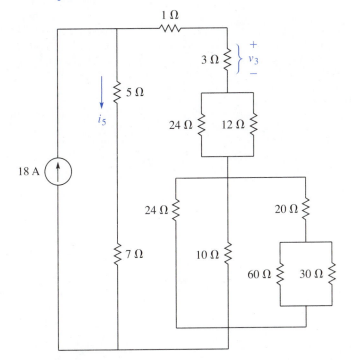

2.33   Find i_6, i_8, v_{12}, and the power dissipated by the 24 Ω resistor in the network shown in Fig. P2.30.

Figure P2.30

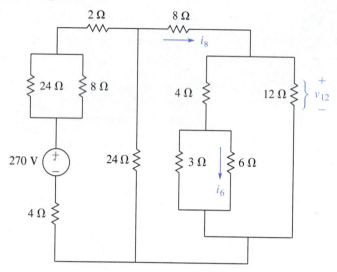

Section 2.11

2.34   Reduce the network shown in Fig. P2.31 to a network of three resistors.

Figure P2.31

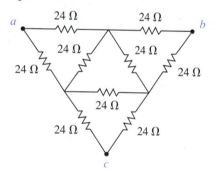

2.35   Reduce the network shown in Fig. P2.32 to a network of three resistors.

Figure P2.32

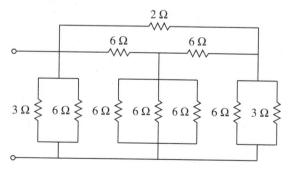

2.36 What current flows in the 4-Ω resistance in the network shown in Fig. P2.33?

Figure P2.33

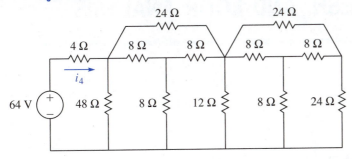

2.37 Find the resistance looking into terminals a–b in the network shown in Fig. P2.34.

Figure P2.34

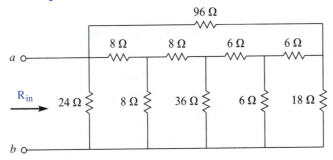

2.38 Find the resistance looking into terminals a–b in the network shown in Fig. P2.35.

Figure P2.35

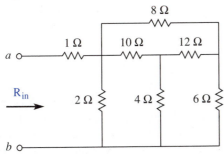

3

NODE, MESH, AND LOOP ANALYSIS

OBJECTIVES

The objectives of this chapter are to:

- Provide a detailed discussion of the elements of network topology.

- Develop a procedure for node, mesh, and loop analysis.

- Consider the fundamental theorem of network topology and its use as a guide for the selection of a method of analysis.

- Consider the computation of the power distribution in general resistive networks.

SECTION 3.1 **INTRODUCTION**

The ladder network was discussed in detail in Chapter 2, where it was shown that it is a configuration that can readily be analyzed. The procedure for the determination of every branch current and every branch voltage involved a simple application of the two Kirchhoff laws, addition of resistors in series and combination of resistors in parallel, voltage and current division, and source transformations. The subject of this chapter is the study of three methods for the analysis of more general networks: the methods of node, mesh, and loop analysis.

The inevitable result in most cases pertaining to a steady-state analysis of more general networks is a set of simultaneous equations in the network variables. The number of equations required depends on the number of network variables needed to yield the values of the currents through and the voltages across each of the network elements. For example, in a node analysis, the formulation of the n linearly independent[1] node equations in n node-to-datum voltages is obtained from a systematic application of the Kirchhoff current law. When these equations are solved, the individual branch voltages can be obtained from the node-to-datum voltages, and it is then a simple matter to obtain the branch currents from the elemental equations that pertain to each branch of the network.

---

[1] The subject of linear independence and the solution of simultaneous equations is discussed in Appendix B.

There is a considerable amount of elegance in network analysis, and the reader can rest assured that there is more to the subject than the "math problem" solution of simultaneous equations. Node, mesh, and loop analyses are standard techniques that can be applied when the network is not a ladder, when the ladder contains more than one source, or when the network cannot be reduced to a ladder. The formulations and solutions of node, mesh, and loop equations in the methods of node, mesh, and loop analysis are considered in this chapter, and a consideration of several network analysis theorems is provided in Chapter 4.

It will be continuously observed that there are several analysis methods available. These range from the ultrasimple for the most elementary to the comprehensive for the more advanced networks. Frequently, the form of the network itself suggests the possibility of alternative methods of analysis, and it is hoped that after the study of this chapter and Chapter 4 has been completed, the reader will have a feeling for the most direct method of analysis for a particular network.

ELEMENTS OF NETWORK TOPOLOGY

SECTION 3.2

Network topology derives from a branch of mathematics that permits the determination of network properties on the basis of structure, geometry, or interconnection of the network. It serves as a basis for a general method of network analysis; it may be used as an aid in the selection of a solution strategy, or it may be used as a guide to writing the independent equations needed to solve for the required variables or quantities. The use of network topology does not depend on the actual network elements that constitute the network.

Every network component of R, L, or C may be replaced by a single line called a *branch*, which has two ends called *nodes*. The terminology of *branch* and *node* will be used in this book, although most treatises on graph theory refer to branches as *edges* and to nodes as *vertices*. Branches that touch a node are said to be *incident* on that node.

As shown in Fig. 3.1, a branch will contain a network component (R, L, or C) and may contain ideal voltage and/or current sources. Notice that the branch currents and branch voltages are designated by j and v, respectively, with the subscript s used to indicate sources. The letters i and e are reserved for mesh or loop currents and node voltages. The discussion that follows in this chapter and in the chapters pertaining to ac analysis will be based on the terminology and the current and voltage orientations in Fig. 3.1.

Note that when ideal sources are present, the branch containing them must also contain a network component, in series with an ideal voltage source and/or in parallel with an ideal current source. There are many reasons why this book has adopted this convention, and here it is only necessary to point out that the ability to perform voltage-to-current and current-to-voltage source transformations[2] on a branch-by-branch basis makes this convention more convenient. It will be shown in Section 3.4.4 that this convention will provide no obstacle to conventional node, mesh, or loop analysis.

A *linear graph* is a collection of branches and nodes that is arranged to exactly represent the geometry of a network, and a *subgraph* is a subset of the nodes and branches of the graph. Figure 3.2a shows a network with several elements, two current sources, and a voltage source; its graph where nodes are designated by numerals

---

[2] See Section 2.7.

FIGURE 3.1   Two alternative representations of a general branch

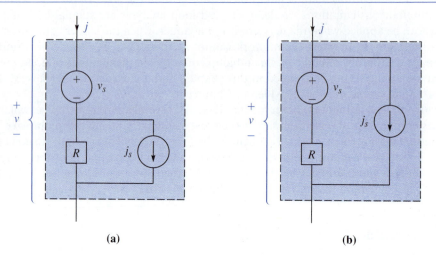

(a)   **(b)**

FIGURE 3.2   (a) A network, (b) its graph, (c) a subgraph, and (d) an oriented graph

Notice in (b) that branches containing ideal sources also contain a network element.

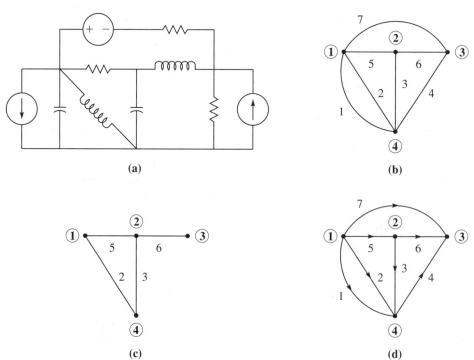

(a)   **(b)**

(c)   **(d)**

within circles, and branches are indicated by numerals is shown in Fig. 3.2b. A subgraph is shown in Fig. 3.2c. When each branch of a graph carries an arrow to indicate its orientation (the branch current direction) as in Fig. 3.2d, the graph is said to be an *oriented graph.*

Orientation of branches is only partially discretionary. Branches that contain voltage sources must show the orientation arrow in the direction of the voltage drop (plus to minus through the voltage source). When current sources are involved, the orientation arrow must be in the direction of the current leaving the source. Figure 3.1 is consistent with these requirements, and in Fig. 3.2d, the oriented graph of the network in Fig. 3.2a, clearly shows that the orientation of branches 1, 4, and 7 is in accordance with the convention established in Fig. 3.1.

A *path* is a subgraph that describes an ordered sequence of branches (without regard to orientation) that satisfies three requirements:

1. The sequence of branches must form a proper subgraph.
2. At all but two nodes in the path (*interior* nodes), only two branches in the path may be incident.
3. At each of the remaining two nodes (*exterior* nodes), only one branch in the path may be incident.

In Fig. 3.2b, the sequence of branches 5, 7, 4 forms a path between nodes 2 and 4. Notice that more than two branches are incident at all nodes in the path, but that no more than two branches that are included in the path are incident at nodes 1 and 3. In addition, only one path branch is incident at nodes 2 and 4. The branches that are accentuated in Fig. 3.3a compose the path, and they do indeed fulfill all of the requirements for a path.

A *connected graph* is a graph in which there is at least one path between any two of the nodes. The graph in Fig. 3.2b is a connected graph. The graph of the network containing a device known as a *transformer* might look like the graph shown in Fig. 3.4b. This graph is not a connected graph.

A *loop*, as the alert reader must have deduced, is a closed path. It is a particular subgraph that is an ordered sequence of branches through nodes at which only two

(a) A network graph with a path shown accentuated and (b) a network graph with two loops shown accentuated

FIGURE 3.3

The dashed accentuated lines form a mesh.

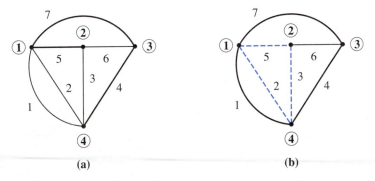

(a)   (b)

FIGURE 3.4
(a) A network containing a pair of inductors that are used to represent a transformer and (b) its unconnected graph

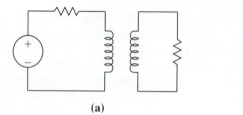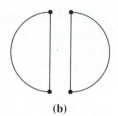

(a)   (b)

branches in the loop may be incident. Figure 3.3b is the graph of the network in Fig. 3.2a. Two loops are indicated by accentuated lines. The ordered sequence of branches 1, 4, and 7 and branches 2, 3, and 5 are both loops. However, the loop composed of branches 2, 3, and 5 does not contain any loops or branches in its interior. This special type of loop is called a *mesh*. Observe that meshes are always loops but that loops are not always meshes. At this point, the reader is alerted to the lack of discussion pertaining to loop or mesh orientation, which is deliberate.

A *tree* of a graph is a connected subgraph of the graph that contains all nodes and no loops. Figure 3.5a is the oriented graph of Fig. 3.2a. Trees of this graph are shown by the oriented accentuated lines in Figs. 3.5b and 3.5c. These two trees show

FIGURE 3.5
(a) An oriented graph, (b) and (c) trees of the oriented graph, and (d) the co-tree of the tree in (b)

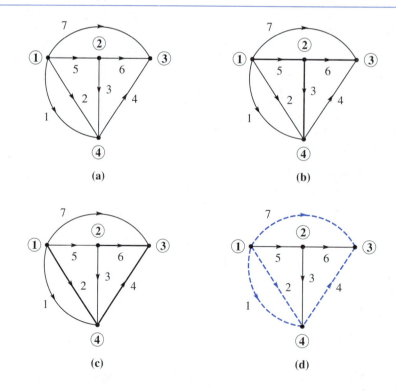

(a) Oriented graph showing an accentuated tree and (b), (c), (d), and (e) loops

FIGURE 3.6

When each link in (a) is reinserted in turn, a loop is formed. Notice that all but one of the loops formed from the tree in (a) are meshes.

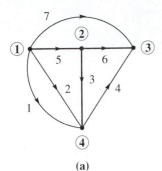

(a)

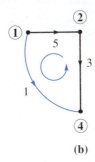

(b)

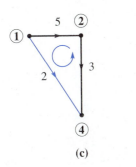

(c)

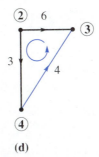

(d)

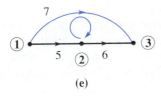

(e)

that more than one tree may be formed from a graph. In both of these trees, the accentuated branches that form the tree are called *tree branches* (or *twigs*). The non-accentuated branches in each tree are called *links* (or *chords*). The links of a tree form a subgraph called a *co-tree*, and the accentuated dashed branches in Fig. 3.5d show the co-tree of the tree in Fig. 3.5b.

Each link, when replaced in the tree one at a time, forms a unique loop that has an orientation in the same direction as the link orientation. For the tree in Fig. 3.5b, repeated in Fig. 3.6a, branches 3, 5, and 6 constitute the tree, and branches 1, 2, 4, and 7 are the links. Replacement of the links, one at a time, forms the loops shown in Figs. 3.6b through 3.6e. Observe that the orientation of each loop is in the direction of the link used to form the loop.

FIGURE 3.7   (a) A graph, (b) a cutset, (c) the same graph, and (d) another cutset

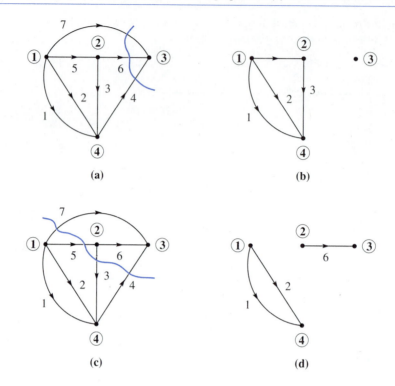

A *cutset* is a set of branches of a graph whose removal causes the graph to be disconnected into exactly two subgraphs, with the further stipulation that the replacement of any one of the branches causes the graph to be once again connected. Figure 3.7b is a cutset of the graph of Fig. 3.7a, and it is seen that the two subgraphs are formed, one of which consists of a single node. In this case, branches 4, 6, and 7 form the cutset. Another cutset consisting of branches 3, 4, 5, and 7 yields the two subgraphs shown in Fig. 3.7d.

A *planar network* is a network that can be drawn on a plane such that no branches cross each other. A *nonplanar network* is, of course, any network that is not planar. Figure 3.8a shows a network that looks as if it is nonplanar but, as Fig. 3.8b clearly shows, it is indeed planar. However, the network displayed in Fig. 3.8c is nonplanar.

SECTION 3.3   **NODE ANALYSIS**

Node analysis involves a systematic application of KCL to all but one node in the network. The one node to which KCL is not applied is called the *datum* or *ground node*. All node voltages that are determined from the analysis are voltages with respect to this datum node. This is the reason the terminology *node-to-datum analysis* rather than just *node analysis* is sometimes used.

FIGURE 3.8

(a) A planar network that looks nonplanar, (b) the same network, and (c) a nonplanar network

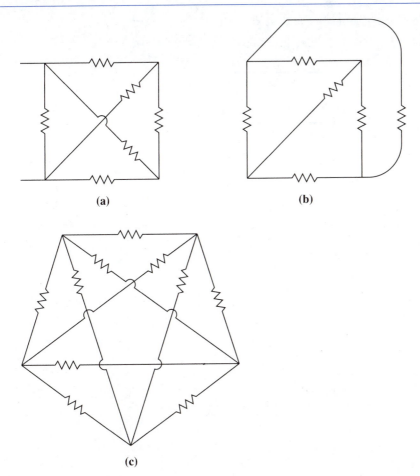

(a)

(b)

(c)

The node voltages are designated with the letter e; and once they are determined, it is a simple matter to compute the individual branch voltages, designated by the letter v, and then the individual branch currents, designated by the letter j. Although node analysis can be conducted in the presence of voltage sources, the method is somewhat more conducive to networks driven by current sources (whether independent or dependent) only. The procedure for conducting a node analysis will now be illustrated.

Consider the network of Fig. 3.9a, which shows three nodes (one of which, node 0, bears the "ground" symbol) and three current sources. The first step is systematic application of the Kirchhoff current law, in turn, to nodes 1 and 2 so that a pair of node equations in terms of the two node voltages will result. With the current flowing *away* from the node in *each* branch taken as positive, one may write $i = \Delta e/R$ expressions for the current in each branch. The same convention must be adopted for each node, and this can be done mentally or as shown in Fig. 3.9b. With e_0 designated as

(a) A network to be subjected to node analysis and (b) the network showing the
current flow convention adopted for each node

The datum or ground node is set at $e_0 = 0$ V.

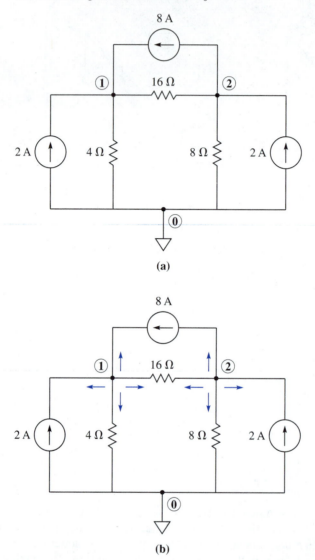

(a)

(b)

the potential of the ground node, the two KCL statements are

$$\frac{1}{4}(e_1 - e_0) + \frac{1}{16}(e_1 - e_2) - 8 - 2 = 0$$

$$\frac{1}{16}(e_2 - e_1) + \frac{1}{8}(e_2 - e_0) + 8 - 2 = 0$$

(3.1)

However, in node analysis, the ground node is the reference node. Thus, the value of e_0 may be taken as zero. With $e_0 = 0$, algebraic adjustment of eqs. (3.1) gives

$$\frac{5}{16} e_1 - \frac{1}{16} e_2 = 10$$

$$-\frac{1}{16} e_1 + \frac{3}{16} e_2 = -6$$

(3.2)

or in the matrix form $\mathbf{GE} = \mathbf{I_s}$,

$$\begin{bmatrix} \frac{5}{16} & -\frac{1}{16} \\ -\frac{1}{16} & \frac{3}{16} \end{bmatrix} \begin{bmatrix} e_1 \\ e_2 \end{bmatrix} = \begin{bmatrix} 10 \\ -6 \end{bmatrix}$$

(3.3)

where $\mathbf{G}$ is a 2×2 coefficient matrix of conductances, $\mathbf{E}$ is a 2×1 column vector of node voltages, and $\mathbf{I_s}$ is a 2×1 current source column vector.

A Cramer's rule solution[3] of eqs. (3.2) gives

$$e_1 = \frac{\begin{vmatrix} 10 & -\frac{1}{16} \\ -6 & \frac{3}{16} \end{vmatrix}}{\begin{vmatrix} \frac{5}{16} & -\frac{1}{16} \\ -\frac{1}{16} & \frac{3}{16} \end{vmatrix}} = \frac{\frac{30}{16} - \frac{6}{16}}{\frac{14}{256}} = \frac{192}{7} \ \text{V}$$

$$e_2 = \frac{\begin{vmatrix} \frac{5}{16} & 10 \\ -\frac{1}{16} & -6 \end{vmatrix}}{\begin{vmatrix} \frac{5}{16} & -\frac{1}{16} \\ -\frac{1}{16} & \frac{3}{16} \end{vmatrix}} = \frac{-\frac{20}{16}}{\frac{14}{256}} = -\frac{160}{7} \ \text{V}$$

Application of a matrix inversion to eqs. (3.3) gives an identical result:

$$\begin{bmatrix} e_1 \\ e_2 \end{bmatrix} = \begin{bmatrix} \frac{5}{16} & -\frac{1}{16} \\ -\frac{1}{16} & \frac{3}{16} \end{bmatrix}^{-1} \begin{bmatrix} 10 \\ -6 \end{bmatrix} = \begin{bmatrix} \frac{24}{7} & \frac{8}{7} \\ \frac{8}{7} & \frac{40}{7} \end{bmatrix} \begin{bmatrix} 10 \\ -6 \end{bmatrix} = \begin{bmatrix} \frac{192}{7} \\ -\frac{160}{7} \end{bmatrix} \ \text{V}$$

In Fig. 3.10a, four nodes can be identified: three bearing numerals contained in circles and one designated by a ground symbol. This time, a voltage source is present, and it may be transformed to a current source, as indicated in Fig. 3.10b.

Here, too, the first step is to adopt a convention for the application of KCL, in turn, to each of the three nodes of interest. Recall that the same convention must be applied to each node. This leads to the second step, which is to write a KCL statement for each node so that a set of n linearly independent node equations in terms of the n node voltages will result. The final step is to solve these equations for the node voltages.

In Fig. 3.10b, let the KCL statements for each node be based on the summation of all currents flowing away from the node, which is equal to zero. With e_0 designated

---

[3] See Appendix B for a discussion of Cramer's rule.

FIGURE 3.10   (a) A network to be subjected to node analysis and (b) the network showing the voltage source transformed to a current source and the current flow convention adopted for each node

The datum or ground node is set at $e_0 = 0$ V.

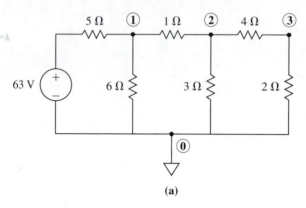

(a)

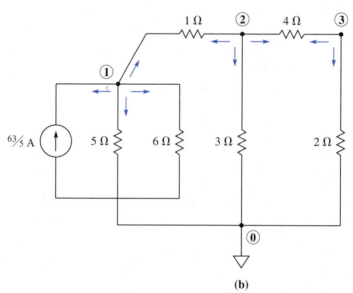

(b)

as the potential of the ground node, the three KCL statements are

$$\frac{1}{5}(e_1 - e_0) + \frac{1}{6}(e_1 - e_0) + \frac{1}{1}(e_1 - e_2) - \frac{63}{5} = 0$$

$$\frac{1}{1}(e_2 - e_1) + \frac{1}{3}(e_2 - e_0) + \frac{1}{4}(e_2 - e_3) = 0 \qquad (3.4)$$

$$\frac{1}{4}(e_3 - e_2) + \frac{1}{2}(e_3 - e_0) = 0$$

Recall that in node analysis, the ground node is the reference node. Thus, the value of e_0 may be taken as zero. With $e_0 = 0$, algebraic adjustment of eqs. (3.1) gives

$$\frac{41}{30}e_1 \quad - e_2 \qquad \qquad = \frac{63}{5}$$

$$-e_1 + \frac{19}{12}e_2 - \frac{1}{4}e_3 = 0 \qquad \qquad \text{(3.5)}$$

$$-\frac{1}{4}e_2 + \frac{3}{4}e_3 = 0$$

or in the matrix form $\mathbf{GE} = \mathbf{I_s}$,

$$\begin{bmatrix} \frac{41}{30} & -1 & 0 \\ -1 & \frac{19}{12} & -\frac{1}{4} \\ 0 & -\frac{1}{4} & \frac{3}{4} \end{bmatrix} \begin{bmatrix} e_1 \\ e_2 \\ e_3 \end{bmatrix} = \begin{bmatrix} \frac{63}{5} \\ 0 \\ 0 \end{bmatrix} \qquad \qquad \text{(3.6)}$$

A Cramer's rule solution for eqs. (3.5) gives

$$e_1 = \frac{\begin{vmatrix} \frac{63}{5} & -1 & 0 \\ 0 & \frac{19}{12} & -\frac{1}{4} \\ 0 & -\frac{1}{4} & \frac{3}{4} \end{vmatrix}}{\begin{vmatrix} \frac{41}{30} & -1 & 0 \\ -1 & \frac{19}{12} & -\frac{1}{4} \\ 0 & -\frac{1}{4} & \frac{3}{4} \end{vmatrix}} = \frac{\frac{567}{40}}{\frac{189}{240}} = 18 \text{ V}$$

$$e_2 = \frac{\begin{vmatrix} \frac{41}{30} & \frac{63}{5} & 0 \\ -1 & 0 & -\frac{1}{4} \\ 0 & 0 & \frac{3}{4} \end{vmatrix}}{\begin{vmatrix} \frac{41}{30} & -1 & 0 \\ -1 & \frac{19}{12} & -\frac{1}{4} \\ 0 & -\frac{1}{4} & \frac{3}{4} \end{vmatrix}} = \frac{\frac{189}{20}}{\frac{189}{240}} = 12 \text{ V}$$

$$e_3 = \frac{\begin{vmatrix} \frac{41}{30} & -1 & \frac{63}{5} \\ -1 & \frac{19}{12} & 0 \\ 0 & -\frac{1}{4} & 0 \end{vmatrix}}{\begin{vmatrix} \frac{41}{30} & -1 & 0 \\ -1 & \frac{19}{12} & -\frac{1}{4} \\ 0 & -\frac{1}{4} & \frac{3}{4} \end{vmatrix}} = \frac{\frac{63}{20}}{\frac{189}{240}} = 4 \text{ V}$$

Application of a matrix inversion to eqs. (3.6) gives an identical result:

$$\begin{bmatrix} e_1 \\ e_2 \\ e_3 \end{bmatrix} = \begin{bmatrix} \frac{41}{30} & -1 & 0 \\ -1 & \frac{19}{12} & -\frac{1}{4} \\ 0 & -\frac{1}{4} & \frac{3}{4} \end{bmatrix}^{-1} \begin{bmatrix} \frac{63}{5} \\ 0 \\ 0 \end{bmatrix}$$

$$= \begin{bmatrix} \frac{10}{7} & \frac{20}{21} & \frac{20}{63} \\ \frac{20}{21} & \frac{82}{63} & \frac{82}{189} \\ \frac{20}{63} & \frac{82}{189} & \frac{838}{567} \end{bmatrix} \begin{bmatrix} \frac{63}{5} \\ 0 \\ 0 \end{bmatrix} = \begin{bmatrix} 18 \\ 12 \\ 4 \end{bmatrix} \text{ V}$$

The three node voltages have been determined. Because the branch voltages bear subscripts that correspond to the branch resistance values, these correspond to $v_6 = 18$ V, $v_3 = 12$ V, and $v_2 = 4$ V in Section 2.9, where the network of Fig. 3.10a was treated as a ladder network driven by a voltage source.

3.3.1 The Symmetry of the Coefficient Matrix

The *characteristic determinant* is the determinant of the coefficient matrix in eqs. (3.6):

$$\det \begin{bmatrix} g_{11} & g_{12} & g_{13} \\ g_{21} & g_{22} & g_{23} \\ g_{31} & g_{32} & g_{33} \end{bmatrix} = \det \begin{bmatrix} \frac{41}{30} & -1 & 0 \\ -1 & \frac{19}{12} & -\frac{1}{4} \\ 0 & -\frac{1}{4} & \frac{3}{4} \end{bmatrix}$$

Its formulation was seen to depend on the premise that the KCL statements were written for individual currents flowing away from all nodes in the network. With this convention, it is possible to write the coefficient matrix from an inspection of the network.

Observe that all elements on the principal diagonal represent the sum of all of the individual conductances that emanate from each node. Refer to Fig. 3.10b and confirm that

$$g_{11} = \frac{1}{5} + \frac{1}{6} + 1 = \frac{41}{30} \qquad g_{22} = 1 + \frac{1}{3} + \frac{1}{4} = \frac{19}{12} \qquad g_{33} = \frac{1}{4} + \frac{1}{2} = \frac{3}{4}$$

and that these are indeed the elements along the principal diagonal.

The off-diagonal elements are all symmetrically nonpositive, and they represent the coupling between the nodes. Thus, node 1 must be coupled to node 2 in the same manner as node 2 is coupled to node 1. This accounts for the value of -1 ℧ in both the g_{12} and g_{21} positions in the coefficient matrix. Similarly, there is a zero in the g_{13} and the g_{31} positions because node 1 is not coupled to node 3. The entries of $-\frac{1}{4}$ ℧ in both the g_{23} and the g_{32} positions show that node 2 is coupled to node 3 (and node 3 is coupled to node 2) by a conductance of $\frac{1}{4}$ ℧.

Current source terms in the original KCL statements, which assume that the currents flowing away from the node are added, will be negative if their direction is into a node. When they are transferred to the other side of the KCL equation, they become positive. This is evident in the first of eqs. (3.4). In the node equations, all current sources are placed on the right-hand side of the individual node equations, positive if their currents enter the node and negative if they leave the node. In the event that there are multiple current sources incident at a node, the net effect is obtained by algebraically adding all currents entering the node.

The symmetry of the coefficient matrix has nothing whatsoever to do with the sources unless controlled sources are present, in which case the coefficient matrix is most likely to exhibit no symmetry. The presence of symmetry is a valuable analytical aid, and the absence of symmetry shows that an error has been made in the writing of the node equations. The error, of course, must be found and corrected before proceeding with the analysis.

A network with five nodes and a datum node

FIGURE 3.11

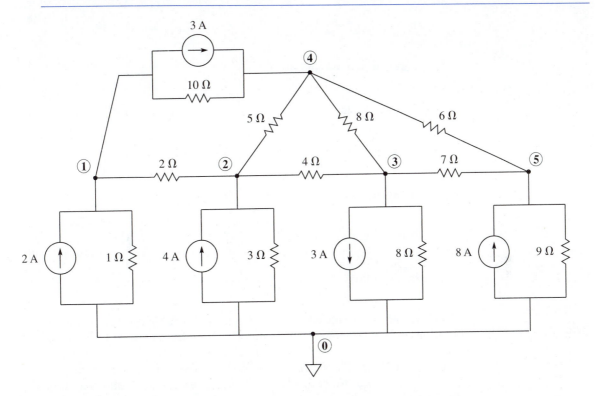

■ **EXAMPLE 3.1**

Write the node equations for the network in Fig. 3.11 but do not attempt to solve them.

Solution   The network contains six nodes, but one of them is the datum or ground node. Five node equations may be written by inspection. The elements of the principal diagonal of the coefficient matrix derive from the sum of the conductances emanating from the individual nodes. With all elements in mhos, the equations are

$$g_{11} = \frac{1}{1} + \frac{1}{2} + \frac{1}{10} = \frac{16}{10} = \frac{8}{5}$$

$$g_{22} = \frac{1}{2} + \frac{1}{3} + \frac{1}{5} + \frac{1}{4} = \frac{77}{60}$$

$$g_{33} = \frac{1}{4} + \frac{1}{8} + \frac{1}{7} + \frac{1}{8} = \frac{36}{56} = \frac{9}{14}$$

$$g_{44} = \frac{1}{10} + \frac{1}{6} + \frac{1}{8} + \frac{1}{5} = \frac{71}{120}$$

$$g_{55} = \frac{1}{6} + \frac{1}{7} + \frac{1}{9} = \frac{53}{126}$$

The off-diagonal elements are established next:

g_{12} between nodes 1 and 2: $-\dfrac{1}{2}\ \mho$ $(g_{12} = g_{21})$

g_{14} between nodes 1 and 4: $-\dfrac{1}{10}\ \mho$ $(g_{14} = g_{41})$

g_{23} between nodes 2 and 3: $-\dfrac{1}{4}\ \mho$ $(g_{23} = g_{32})$

g_{24} between nodes 2 and 4: $-\dfrac{1}{5}\ \mho$ $(g_{24} = g_{42})$

g_{34} between nodes 1 and 2: $-\dfrac{1}{8}\ \mho$ $(g_{34} = g_{43})$

g_{35} between nodes 3 and 5: $-\dfrac{1}{7}\ \mho$ $(g_{35} = g_{53})$

g_{45} between nodes 4 and 5: $-\dfrac{1}{6}\ \mho$ $(g_{45} = g_{54})$

There are five source terms and all appear on the right side of the node equations or in a current source vector. The 2-A source at node 1 is positive because its current enters the node. But the 3-A source leaves node 1 and is negative because its current leaves node 1. The net effect at node 1 is $2 - 3 = -1$ A, negative because the net effect is an energy withdrawal from the network at node 1. A similar argument can be made for the other nodes, and the node equations are

$$\frac{8}{5}e_1 \ -\frac{1}{2}e_2 \qquad\qquad -\frac{1}{10}e_4 \qquad\qquad = 2 - 3 = -1$$

$$-\frac{1}{2}e_1 + \frac{77}{60}e_2 \ -\frac{1}{4}e_3 \ -\frac{1}{5}e_4 \qquad\qquad = 4$$

$$-\frac{1}{4}e_2 + \frac{9}{14}e_3 \ -\frac{1}{8}e_4 \ -\frac{1}{7}e_5 = -3$$

$$-\frac{1}{10}e_1 \ -\frac{1}{5}e_2 \ -\frac{1}{8}e_3 + \frac{71}{120}e_4 \ -\frac{1}{6}e_5 = 3$$

$$-\frac{1}{7}e_3 \ -\frac{1}{6}e_4 + \frac{53}{126}e_5 = 8$$

In matrix form, they are represented by

$$
\begin{bmatrix}
\frac{8}{5} & -\frac{1}{2} & 0 & -\frac{1}{10} & 0 \\
-\frac{1}{2} & \frac{77}{60} & -\frac{1}{4} & -\frac{1}{5} & 0 \\
0 & -\frac{1}{4} & \frac{9}{14} & -\frac{1}{8} & -\frac{1}{7} \\
-\frac{1}{10} & -\frac{1}{5} & -\frac{1}{8} & \frac{71}{120} & -\frac{1}{6} \\
0 & 0 & -\frac{1}{7} & -\frac{1}{6} & \frac{53}{126}
\end{bmatrix}
\begin{bmatrix}
e_1 \\ e_2 \\ e_3 \\ e_4 \\ e_5
\end{bmatrix}
=
\begin{bmatrix}
-1 \\ 4 \\ -3 \\ 3 \\ 8
\end{bmatrix}
$$

3.3.2 Branch Voltages from Node Voltages

With $e_1 = 18$ V, $e_2 = 12$ V, and $e_3 = 4$ V, and with branch voltages designated in accordance with branch resistance values, reference to Fig. 3.10b shows that

$$v_1 = e_1 - e_2 = 18 - 12 = 6 \text{ V} \qquad v_4 = e_2 - e_3 = 12 - 4 = 8 \text{ V}$$

$$v_2 = e_3 = 4 \text{ V} \qquad\qquad\qquad v_6 = e_1 = 18 \text{ V}$$

$$v_3 = e_2 = 12 \text{ V}$$

and v_5, the voltage across the current source, is

$$v_5 = e_1 = 18 \text{ V}$$

This means that the current delivered by the nonideal current source is

$$i_s = \frac{63}{5} - \frac{18}{5} = \frac{45}{5} = 9 \text{ A}$$

The one-to-one correspondence between these values and those in Section 2.9, which considered the same network as a ladder network, may be noted.

3.3.3 Node Analysis in the Presence of Voltage Sources

The network of Fig. 3.12a can be subjected to a node analysis by converting the two voltage sources[4] to current sources in the three-node network of Fig. 3.12b or by considering the five-node problem in which two of the nodes are *dummy nodes*, which possess known node voltages, as shown in Fig. 3.12c.

Consider Fig. 3.12b, and write three node equations by inspection:

$$\left(\frac{1}{8} + \frac{1}{4} + \frac{1}{12}\right)e_1 \qquad -\frac{1}{8}e_2 \qquad -\frac{1}{4}e_3 = 2 + 4 - \frac{5}{2}$$

$$-\frac{1}{8}e_1 + \left(\frac{1}{8} + \frac{1}{6}\right)e_2 \qquad -\frac{1}{6}e_3 = \frac{5}{2}$$

$$-\frac{1}{4}e_1 \qquad -\frac{1}{6}e_2 + \left(\frac{1}{6} + \frac{1}{4} + \frac{1}{2}\right)e_3 = -4$$

or

$$\frac{11}{24}e_1 \quad -\frac{1}{8}e_2 \quad -\frac{1}{4}e_3 = \frac{7}{2}$$

$$-\frac{1}{8}e_1 + \frac{7}{24}e_2 \quad -\frac{1}{6}e_3 = \frac{5}{2} \qquad\qquad\qquad (3.7)$$

$$-\frac{1}{4}e_1 \quad -\frac{1}{6}e_2 + \frac{11}{12}e_3 = -4$$

---

[4] The 20-V ideal source is considered to be in a branch with the 8-Ω resistor.

FIGURE 3.12

(a) A network containing two voltage sources and a current source, (b) the same network with the voltage sources transformed to current sources and ready for a node analysis, and (c) the same network with two dummy nodes

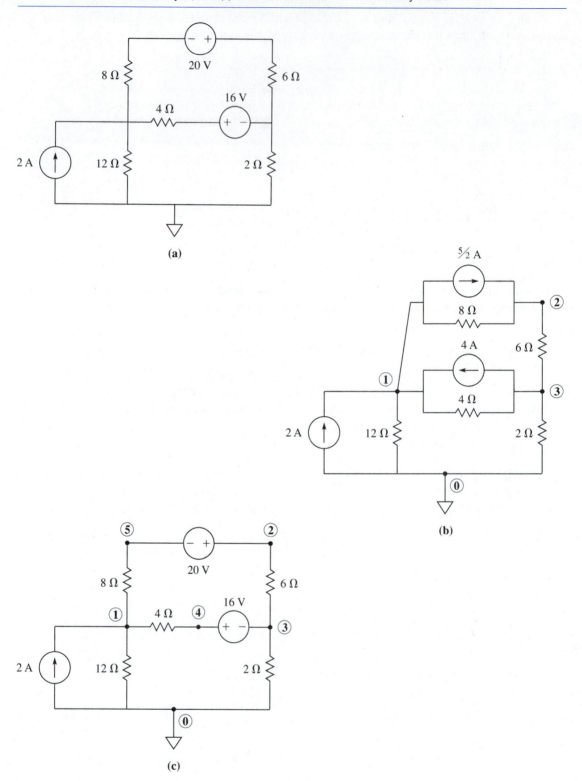

(a)

(b)

(c)

These may be put into matrix form,

$$
\begin{bmatrix} \frac{11}{24} & -\frac{1}{8} & -\frac{1}{4} \\ -\frac{1}{8} & \frac{7}{24} & -\frac{1}{6} \\ -\frac{1}{4} & -\frac{1}{6} & \frac{11}{12} \end{bmatrix}
\begin{bmatrix} e_1 \\ e_2 \\ e_3 \end{bmatrix} =
\begin{bmatrix} \frac{7}{2} \\ \frac{5}{2} \\ -4 \end{bmatrix}
$$

and solved for all node voltages by employing the inverse (and carrying enough significant figures to avoid the phenomena of round-off and what is referred to as catastrophic cancellation):

$$
\begin{bmatrix} e_1 \\ e_2 \\ e_3 \end{bmatrix} =
\begin{bmatrix} 3.5844 & 2.3377 & 1.4026 \\ 2.3377 & 5.3506 & 1.6104 \\ 1.4026 & 1.6104 & 1.7762 \end{bmatrix}
\begin{bmatrix} 3.5000 \\ 2.5000 \\ -4.0000 \end{bmatrix} =
\begin{bmatrix} 12.7792 \\ 15.1169 \\ 1.8701 \end{bmatrix} V \qquad (3.8)
$$

Five nodes are apparent in Fig. 3.12c. Observe that $e_5 = e_2 - 20$ V and $e_4 = e_3 + 16$ V. Nodes 4 and 5 are dummy nodes because their node voltages are functions of the other node voltages. In addition, the current leaving node 2 through the 20-V source is equal to $\frac{1}{8}(e_5 - e_1)$ amperes, and the current leaving node 3 through the 16-V source is equal to $\frac{1}{4}(e_4 - e_1)$ amperes. With these facts, three node equations may be written as

$$
\left(\frac{1}{8} + \frac{1}{4} + \frac{1}{12} \right) e_1 \qquad\qquad\qquad -\frac{1}{4} e_4 - \frac{1}{8} e_5 = 2
$$

$$
-\frac{1}{8} e_1 + \frac{1}{6} e_2 \qquad -\frac{1}{6} e_3 \qquad +\frac{1}{8} e_5 = 0
$$

$$
-\frac{1}{4} e_1 - \frac{1}{6} e_2 + \left(\frac{1}{6} + \frac{1}{2} \right) e_3 + \frac{1}{4} e_4 \qquad = 0
$$

and two more equations involving the dummy nodes, which are KVL statements, are

$$
e_4 = e_3 + 16
$$

$$
e_5 = e_2 - 20
$$

With the fourth and fifth equations put into the first three, the result is

$$
\frac{11}{24} e_1 \;\; -\frac{1}{8} e_2 \;\; -\frac{1}{4} e_3 = \;\; \frac{7}{2}
$$

$$
-\frac{1}{8} e_1 + \frac{7}{24} e_2 \;\; -\frac{1}{6} e_3 = \;\; \frac{5}{2}
$$

$$
-\frac{1}{4} e_1 \;\; -\frac{1}{6} e_2 + \frac{11}{12} e_3 = -4
$$

a set identical to the set of eqs. (3.7) for which eqs. (3.8) are the solution. Notice that the use of dummy nodes is really equivalent to the execution of the voltage-to-current source transformation.

3.3.4 Node Analysis in the Presence of Controlled Sources

The network shown in Fig. 3.13a contains a current source, a voltage source, a voltage-controlled voltage source (VCVS), and a current-controlled current source (ICIS). Three nodes and a datum node are indicated, and it is desired to perform a node analysis.

FIGURE 3.13   (a) A network with controlled sources and (b) the network ready for node analysis, with the controlled sources in terms of node voltages

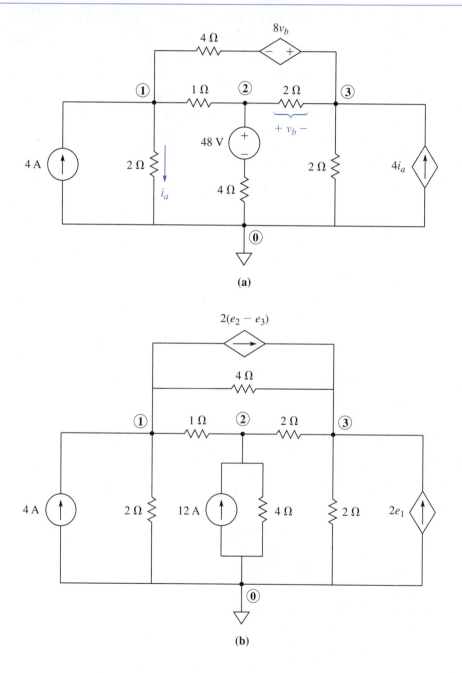

(a)

(b)

Observe that the VCVS can be transformed to a voltage-controlled current source (VCIS) of current magnitude $8v_b/4 = 2v_b$. However, v_b can be expressed in terms of the node voltages e_2 and e_3: $v_b = e_2 - e_3$. Thus, the strength of the VCIS is $2(e_2 - e_3)$.

The ICIS at the right need not be transformed, but its strength must be stated in terms of the node voltages. It is seen that $i_a = e_1/2$, so that the strength of the ICIS is $4i_a = 2e_1$.

Figure 3.13b displays the network with the controlled sources transformed and adjusted. Notice that the 48-V source with the 4-Ω resistor in series has also been transformed to a 12-A current source with the 4-Ω resistor in parallel. The node analysis is now ready to proceed, with Fig. 3.13b as its basis. It may be wise to write the node equations by inspection in two stages. First, establish the coefficients of the node voltages on the left side of the intended set of node equations,

$$\left(\frac{1}{2} + 1 + \frac{1}{4}\right)e_1 \qquad - e_2 \qquad - \frac{1}{4}e_3 =$$

$$-e_1 + \left(1 + \frac{1}{2} + \frac{1}{4}\right)e_2 \qquad - \frac{1}{2}e_3 =$$

$$-\frac{1}{4}e_1 \qquad - \frac{1}{2}e_2 + \left(\frac{1}{2} + \frac{1}{2} + \frac{1}{4}\right)e_3 =$$

and then add the sources, recognizing that the only responsibility here is to take cognizance of whether the current leaves or enters the source:

$$\left(\frac{1}{2} + 1 + \frac{1}{4}\right)e_1 \qquad - e_2 \qquad - \frac{1}{4}e_3 = 4 - 2(e_2 - e_3)$$

$$-e_1 + \left(1 + \frac{1}{2} + \frac{1}{4}\right)e_2 \qquad - \frac{1}{2}e_3 = 12$$

$$-\frac{1}{4}e_1 \qquad - \frac{1}{2}e_2 + \left(\frac{1}{2} + \frac{1}{2} + \frac{1}{4}\right)e_3 = 2e_1 + 2(e_2 - e_3)$$

These may be algebraically adjusted to

$$\frac{7}{4}e_1 \quad + e_2 \quad - \frac{9}{4}e_3 = 4$$

$$-e_1 + \frac{7}{4}e_2 \quad - \frac{1}{2}e_3 = 12 \tag{3.9}$$

$$-\frac{9}{4}e_1 - \frac{5}{2}e_2 + \frac{13}{4}e_3 = 0$$

or in matrix form,

$$\begin{bmatrix} \frac{7}{4} & 1 & -\frac{9}{4} \\ -1 & \frac{7}{4} & -\frac{1}{2} \\ -\frac{9}{4} & -\frac{5}{2} & \frac{13}{4} \end{bmatrix} \begin{bmatrix} e_1 \\ e_2 \\ e_3 \end{bmatrix} = \begin{bmatrix} 4 \\ 12 \\ 0 \end{bmatrix} \tag{3.10}$$

Here, the lack of symmetry in the coefficient matrix is quite evident.

FIGURE 3.14   Network (Exercise 3.1)

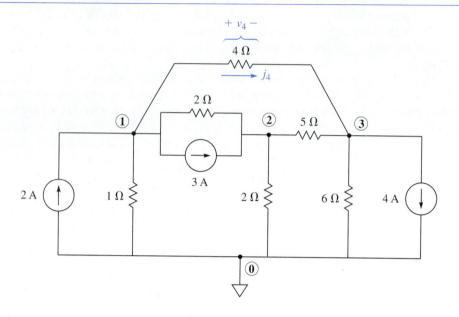

A Cramer's rule solution for e_2 is

$$
e_2 = \frac{\begin{vmatrix} \frac{7}{4} & 4 & -\frac{9}{4} \\ -1 & 12 & -\frac{1}{2} \\ -\frac{9}{4} & 0 & \frac{13}{4} \end{vmatrix}}{\begin{vmatrix} \frac{7}{4} & 1 & -\frac{9}{4} \\ -1 & \frac{7}{4} & -\frac{1}{2} \\ -\frac{9}{4} & -\frac{5}{2} & \frac{13}{4} \end{vmatrix}} = \frac{25}{-75/32} = -10.667 \text{ V}
$$

and the reader may wish to verify that $e_1 = -19.733$ V and $e_2 = -21.867$ V.

EXERCISE 3.1

Use a node analysis to determine the branch voltage v_4 and the branch current j_4 in the branch containing the 4-Ω resistor in Fig. 3.14.

Answer $v_4 = 5.455$ V, $j_4 = 1.361$ A.

EXERCISE 3.2

Use a node analysis to determine the voltage v_b and the current j_b in the branch containing the 2-Ω resistor in Fig. 3.15.

Answer $v_b = \dfrac{56}{93}$ V, $j_b = \dfrac{28}{93}$ A.

Network (Exercise 3.2)

FIGURE 3.15

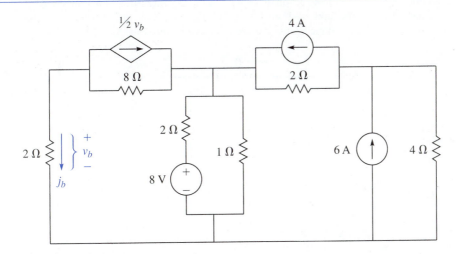

MESH ANALYSIS

The network displayed in Fig. 3.16a is a ladder network containing three meshes. It is a planar network and is identical to the network shown in Figs. 3.10 and 2.14a. An oriented graph is shown in Fig. 3.16b. One particular tree of this graph is shown in Fig. 3.16c, and if the links are replaced one at a time, three loops result. These are shown in Fig. 3.16d. Notice that the three meshes indicated in Fig. 3.16a are also loops, but that two of the three loops in Fig. 3.16d are not meshes. For all three loops to be identical to the three meshes, the tree must be judiciously chosen. Such a tree is shown in Fig. 3.16e. It should be noted that some networks are nonplanar and that mesh equations cannot be written for such networks.

The first step in the method of mesh analysis is to eyeball the meshes and insert clockwise mesh currents. Observe that it is not necessary to draw a tree in order to determine which loops are the meshes. The meshes are quickly identified by noting that meshes are loops with no interior loops or branches. And when the mesh currents are inserted, it will be observed that at least one mesh current passes through each branch of the network. The reason for the use of clockwise mesh currents will be apparent as this study proceeds, and the eventual calculation of a negative mesh current will cause no concern. A negative mesh current does not disturb the validity of the analysis and merely means that the mesh current is actually flowing in the counter-clockwise direction. Figure 3.16a is repeated here as Fig. 3.17, where it may be observed that in this dc analysis, all current variables are designated by a lowercase letter i for the mesh currents and j for the branch currents. This is the customary convention and it is used to avoid confusion between mesh and branch currents.

The second step is a rigorous and systematic application of KVL to each mesh. If the meshes are selected properly, a set of n linear, algebraic, and linearly independent mesh equations will result. Look at mesh 1 and take the algebraic sum of the individual voltages around the closed path:

$$-63 + 5i_1 + 6(i_1 - i_2) = 0$$

FIGURE 3.16

(a) A ladder network showing three meshes, (b) the oriented graph of the network, (c) one possible tree of the graph, (d) the network showing three loops obtained by inserting the tree links back into the graph one at a time, and (e) the tree of the graph that will yield three loops that are the same as the three meshes

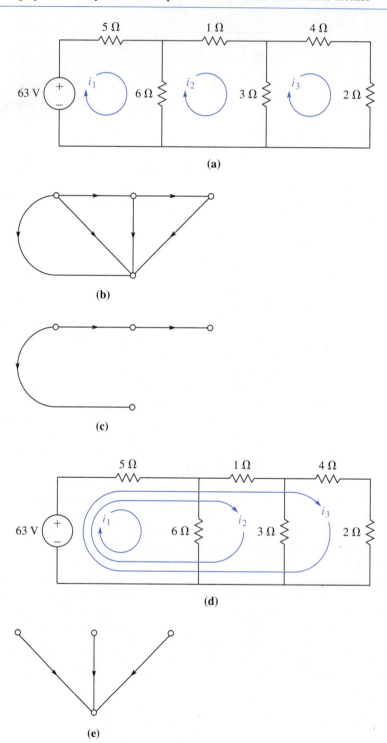

FIGURE 3.17

The network of Fig. 3.16a showing mesh and branch currents

For this dc network, confusion is avoided by designating mesh currents by the letter i and branch currents by the letter j. Observe, however, that the orientation of branch 5 and the direction of branch current j_5 is dictated by the polarity of the ideal 63-V source.

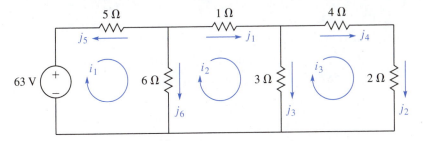

Here, one may note that the source appears as a negative number because it is a voltage rise in the algebraic sum of the voltage drop "scheme of things." Also, the voltage across the 6-Ω resistor in the assumed positive orientation of mesh 1 is in the direction of i_1 and is $6(i_1 - i_2)$. The complete set of three mesh equations is

$$-63 + 5i_1 + 6(i_1 - i_2) = 0$$
$$6(i_2 - i_1) + i_2 + 3(i_2 - i_3) = 0 \qquad \text{(3.11)}$$
$$3(i_3 - i_2) + 4i_3 + 2i_3 = 0$$

Algebraic adjustment gives

$$11i_1 - 6i_2 \qquad = 63$$
$$-6i_1 + 10i_2 - 3i_3 = 0 \qquad \text{(3.12)}$$
$$-3i_2 + 9i_3 = 0$$

and in the matrix form $\mathbf{RI} = \mathbf{V_s}$,

$$\begin{bmatrix} 11 & -6 & 0 \\ -6 & 10 & -3 \\ 0 & -3 & 9 \end{bmatrix} \begin{bmatrix} i_1 \\ i_2 \\ i_3 \end{bmatrix} = \begin{bmatrix} 63 \\ 0 \\ 0 \end{bmatrix} \qquad \text{(3.13)}$$

The mesh currents i_1, i_2, and i_3 may be determined by solving eqs. (3.12). For example, a Cramer's rule solution using determinants may be employed. The *characteristic determinant* is the determinant of the *resistive coefficient matrix* in eqs. (3.13):

$$\begin{vmatrix} 11 & -6 & 0 \\ -6 & 10 & -3 \\ 0 & -3 & 9 \end{vmatrix} = 11(10)(9) - (-3)(-3)(11) - (-6)(-6)(9) = 567$$

Then,

$$i_1 = \frac{\begin{vmatrix} 63 & -6 & 0 \\ 0 & 10 & -3 \\ 0 & -3 & 9 \end{vmatrix}}{567} = \frac{63(10)(9) - (-3)(-3)(63)}{567} = \frac{5103}{567} = 9 \text{ A}$$

$$i_2 = \frac{\begin{vmatrix} 11 & 63 & 0 \\ -6 & 0 & -3 \\ 0 & 0 & 9 \end{vmatrix}}{567} = \frac{-(9)(-6)(63)}{567} = \frac{3402}{567} = 6 \text{ A}$$

$$i_3 = \frac{\begin{vmatrix} 11 & -6 & 63 \\ -6 & 10 & 0 \\ 0 & -3 & 0 \end{vmatrix}}{567} = \frac{63(-6)(-3)}{567} = \frac{1134}{567} = 2 \text{ A}$$

Of course, all three mesh currents are determined at the same time when the coefficient matrix in eq. (3.13) is inverted:

$$\begin{bmatrix} i_1 \\ i_2 \\ i_3 \end{bmatrix} = \begin{bmatrix} 11 & -6 & 0 \\ -6 & 10 & -3 \\ 0 & -3 & 9 \end{bmatrix}^{-1} \begin{bmatrix} 63 \\ 0 \\ 0 \end{bmatrix} = \frac{1}{567} \begin{bmatrix} 81 & 54 & 18 \\ 54 & 99 & 33 \\ 18 & 33 & 74 \end{bmatrix} \begin{bmatrix} 63 \\ 0 \\ 0 \end{bmatrix} = \begin{bmatrix} 9 \\ 6 \\ 2 \end{bmatrix} \text{ A}$$

3.4.1 Formulating the Mesh Equations by Inspection

The characteristic determinant is the determinant of the coefficient matrix. For example, in eqs. (3.13):

$$\det \begin{bmatrix} r_{11} & r_{12} & r_{13} \\ r_{21} & r_{22} & r_{23} \\ r_{31} & r_{32} & r_{33} \end{bmatrix} = \det \begin{bmatrix} 11 & -6 & 0 \\ -6 & 10 & -3 \\ 0 & -3 & 9 \end{bmatrix} \qquad (3.14)$$

The mesh equations from which this coefficient matrix is derived may be formulated directly from an inspection of the network, provided that all mesh currents are inserted so they are oriented in the same direction (clockwise or counterclockwise) and provided that there are no controlled sources present.

In eq. (3.14), observe that the elements of the principal diagonal are all positive and represent the sum of all resistances in each mesh shown in Fig. 3.17. Also, observe that all off-diagonal elements (the elements of R are designated $r_{k\ell}$) are not only nonpositive but also symmetrically nonpositive:

$$r_{k\ell} = r_{\ell k} \qquad (k \neq \ell)$$

This is a consequence of the deliberate selection of the direction of all mesh currents in the clockwise direction. The off-diagonal elements represent the coupling between the meshes; mesh 1 must be coupled to mesh 2 in exactly the same manner as mesh 2 is coupled to mesh 1. Thus, in mesh 1, the coefficient of the second mesh current must be -6, and this is shown by the second term in the first mesh equation [the entry in the first row and second column in eq. (3.14)]. In the second mesh equation, the coefficient of the first mesh current must also be -6 [look at the entry in the second row and first column of eq. (3.14)]. Observe that mesh 1 is not coupled to mesh 3, so that a zero appears in the upper right-hand (mesh 1, column 3; or row 1, column 3) and lower left-hand (mesh 3, column 1; or row 3, column 1) corners of the coefficient matrix.

Source terms in the original KVL statements that equate all voltage drops to zero may be negative when they represent a voltage rise. Thus, when these sources are transferred to the other side of the equation, they become positive. This is clearly evident in the example at hand. The KVL mesh equation for the first mesh is

$$-63 + 5i_1 + 6(i_1 - i_2) = 0$$

and it is then only a matter of algebra to obtain

$$11i_1 - 6i_2 = 63$$

Thus, source terms must be placed to the right of the equal sign—positive if the mesh current leaves the source and the source delivers energy to the network, and non-positive if the mesh current enters the source and the source absorbs energy from the network.

This is a valuable analysis advantage. Before one begins the actual solution of the equations, the symmetry of the coefficient matrix should be observed. If the symmetry is not apparent, an error has been made, and it must be corrected before proceeding. But recall and note well: For the symmetry to be apparent, *all* mesh currents must be in the same (say clockwise) direction, and no controlled sources may be present.

■ **EXAMPLE 3.2**

Write the mesh equations for the network in Fig. 3.18, but do not attempt to solve them.

Solution   The network contains five meshes, and the five mesh equations may be written by inspection. The elements of the principal diagonal of the coefficient matrix derive from the sum of the resistances in the individual meshes. With all elements in ohms, the equations are

$$r_{11} = 1 + 2 = 3 \qquad r_{44} = 3 + 4 + 4 = 11$$

$$r_{22} = 2 + 4 + 1 = 7 \qquad r_{55} = 4 + 2 + 1 + 2 = 9$$

$$r_{33} = 1 + 2 + 3 = 6$$

The off-diagonal elements are established next:

r_{12} between meshes 1 and 2: -2 $(r_{12} = r_{21})$

r_{23} between meshes 2 and 3: -1 $(r_{23} = r_{32})$

r_{24} between meshes 2 and 4: -4 $(r_{24} = r_{42})$

r_{35} between meshes 3 and 5: -2 $(r_{35} = r_{53})$

r_{45} between meshes 4 and 5: -4 $(r_{45} = r_{54})$

FIGURE 3.18   A network with five meshes

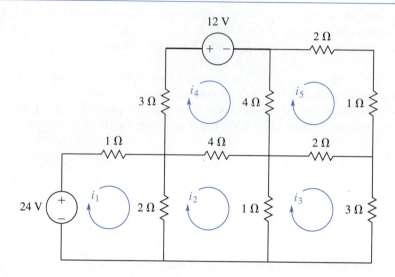

There are two source terms, and both appear on the right side of the equations or in a voltage source vector. The 24-V source in mesh 1 is positive because mesh current 1 leaves this source and the 24 V is delivering energy to the network. The 12-V source in mesh 4 is negative because mesh current 4 enters the source and the source is absorbing energy from the network.

The mesh equations are

$$
\begin{aligned}
3i_1 - 2i_2 &= 24 \\
-2i_1 + 7i_2 - i_3 - 4i_4 &= 0 \\
-i_2 + 6i_3 - 2i_5 &= 0 \\
-4i_2 + 11i_4 - 4i_5 &= -12 \\
-2i_3 - 4i_4 + 9i_5 &= 0
\end{aligned}
$$

and in matrix form, they are represented by

$$
\begin{bmatrix}
3 & -2 & 0 & 0 & 0 \\
-2 & 7 & -1 & -4 & 0 \\
0 & -1 & 6 & 0 & -2 \\
0 & -4 & 0 & 11 & -4 \\
0 & 0 & -2 & -4 & 9
\end{bmatrix}
\begin{bmatrix}
i_1 \\ i_2 \\ i_3 \\ i_4 \\ i_5
\end{bmatrix}
=
\begin{bmatrix}
24 \\ 0 \\ 0 \\ -12 \\ 0
\end{bmatrix}
$$

■

3.4.2 Branch Currents and Branch Voltages from Mesh Currents

The mesh current vector obtained from a solution of eqs. (3.12) or (3.13) is

$$
\mathbf{I} = \begin{bmatrix} 9 \\ 6 \\ 2 \end{bmatrix} A
$$

Reference to Fig. 3.17 shows that

$$j_1 = i_2 = 6 \text{ A} \qquad\qquad j_4 = i_3 = 2 \text{ A}$$
$$j_2 = i_3 = 2 \text{ A} \qquad\qquad j_5 = -i_1 = -9 \text{ A}$$
$$j_3 = i_2 - i_3 = 6 - 2 = 4 \text{ A} \qquad j_6 = i_1 - i_2 = 9 - 6 = 3 \text{ A}$$

which means that j_5 is actually in the opposite direction than that indicated in Fig. 3.17.

Voltages across the individual branches may be obtained from Ohm's law. In this case, the branch voltages, in general, are obtained from

$$v_k = r_k j_k \qquad (k = 1, 2, 3, \ldots)$$

so that with v indicating the branch voltages,

$$v_1 = 1(6) = 6 \text{ V} \qquad v_4 = 4(2) = 8 \text{ V}$$
$$v_2 = 2(2) = 4 \text{ V} \qquad v_6 = 6(3) = 18 \text{ V}$$
$$v_3 = 3(4) = 12 \text{ V}$$

Branch 5, however, is a special case because it includes the 63-V ideal source. With j_5 as indicated in Fig. 3.17,

$$v_5 = 63 + 5j_5 = 63 + 5(-9) = 63 - 45 = 18 \text{ V}$$

and the nonideal voltage source composed of the 63-V ideal source *and* the 5-Ω resistor puts 18 V across the network. All of the foregoing agrees with the solution generated for the ladder network in Section 2.9.

3.4.3 Mesh Analysis in the Presence of Current Sources

Figure 3.19a shows a network with two voltage sources and a single current source; it is the same network shown in Fig. 3.12a. A mesh analysis may be employed in this case, but because of the current source, a slight modification of the previous procedure is necessary. The current source must either be converted to a voltage source by the standard current–to–voltage source transformation, as shown with clockwise mesh currents in Fig. 3.19b, or the network may be treated as a three-mesh, with the third mesh treated as a *dummy mesh* because its current is known, as in Fig. 3.19c.

For Fig. 3.19b, the mesh equations are

$$18i_1 - 4i_2 = 24 - 16 = 8$$
$$-4i_1 + 18i_2 = 20 + 16 = 36 \tag{3.15}$$

or

$$\begin{bmatrix} 18 & -4 \\ -4 & 18 \end{bmatrix} \begin{bmatrix} i_1 \\ i_2 \end{bmatrix} = \begin{bmatrix} 8 \\ 36 \end{bmatrix} \tag{3.16}$$

FIGURE 3.19

(a) A network with a current source, (b) the current source transformed into a voltage source, and (c) the network with a dummy mesh current through the current source

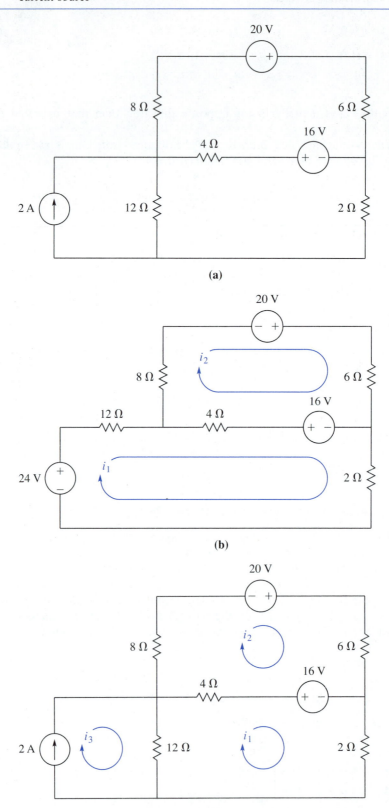

(a)

(b)

(c)

A Cramer's rule solution of eqs. (3.15) gives

$$i_1 = \frac{\begin{vmatrix} 8 & -4 \\ 36 & 18 \end{vmatrix}}{\begin{vmatrix} 18 & -4 \\ -4 & 18 \end{vmatrix}} = \frac{288}{308} = 0.935 \text{ A} \qquad i_2 = \frac{\begin{vmatrix} 18 & 8 \\ -4 & 36 \end{vmatrix}}{308} = \frac{680}{308} = 2.208 \text{ A}$$

A matrix inversion of the system of eq. (3.16) yields both mesh currents in the same operation:

$$\begin{bmatrix} i_1 \\ i_2 \end{bmatrix} = \begin{bmatrix} 18 & -4 \\ -4 & 18 \end{bmatrix}^{-1} \begin{bmatrix} 8 \\ 36 \end{bmatrix} = \frac{1}{308} \begin{bmatrix} 18 & 4 \\ 4 & 18 \end{bmatrix} \begin{bmatrix} 8 \\ 36 \end{bmatrix} = \begin{bmatrix} 0.935 \\ 2.208 \end{bmatrix} \text{ A}$$

For the network shown in Fig. 3.19c, there are three meshes and three mesh currents. Two mesh equations (for meshes 1 and 2) derive from KVL statements. The third mesh current is known (actually from a KCL consideration), and this third mesh is the dummy mesh with the *dummy mesh current* i_3. Thus, the two mesh equations are

$$18i_1 - 4i_2 - 12i_3 = -16$$
$$-4i_1 + 18i_2 \qquad = \quad 36$$

and the third equation is the equation for the dummy mesh whose mesh current is known:

$$i_3 = 2$$

If $i_3 = 2$ A is put into the first mesh equation, the result is

$$18i_1 - 4i_2 = 8 \qquad \text{and} \qquad -4i_1 + 18i_2 = 36$$

which is the set of eqs. (3.15). The solution, of course, is as previously executed; and it can be observed that the adjustment of the mesh equations using the dummy mesh current has the same effect as the use of a current–to–voltage source transformation.

3.4.4 Mesh Analysis in the Presence of Controlled Sources

Mesh analysis may be conducted when the network contains controlled sources and sources unaccompanied by a network element. When such sources are present, the symmetry of the coefficient matrix will be upset and the analysis must begin (at least until some experience is gained) with a KVL statement for each mesh. Figure 3.20a shows a network containing two controlled voltage sources; one is voltage-controlled (by v_a) and one is current-controlled (by i_b). There are three meshes in this network, and the meshes are shown in Fig. 3.20b. Notice that the mesh current i_3 is equal to i_b, so that $2v_a = 2(3i_3) = 6i_3$ and $16i_b = 16i_3$.

KVL statements for each mesh can be written by taking the algebraic sum of the voltages around the closed path in the direction of the assumed clockwise mesh currents:

$$-24 + 4(i_1 - i_3) + 16i_3 + 2(i_1 - i_2) = 0$$
$$-16i_3 + (i_2 - i_3) + 6i_3 + 5i_2 + 2(i_2 - i_1) = 0$$
$$8 + 3i_3 - 6i_3 + (i_3 - i_2) + 4(i_3 - i_1) = 0$$

FIGURE 3.20   (a) A network containing two controlled voltage sources and (b) designation of the meshes for the analysis of the network in (a)

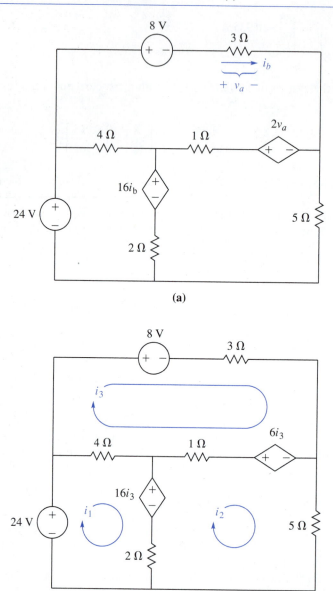

or with some algebra,

$$6i_1 - 2i_2 + 12i_3 = 24$$
$$-2i_1 + 8i_2 - 11i_3 = 0 \tag{3.17}$$
$$-4i_1 - i_2 + 2i_3 = -8$$

The mesh equations may also be formulated by inspection, and as in the case of node analysis, the solution may require a two-step procedure. First, form the coefficients,

$$6i_1 - 2i_2 - 4i_3 =$$
$$-2i_1 + 8i_2 - i_3 =$$
$$-4i_1 - i_2 + 8i_3 =$$

and then, add the source terms by merely noting whether the mesh currents enter or leave the individual sources:

$$6i_1 - 2i_2 - 4i_3 = 24 - 16i_3$$
$$-2i_1 + 8i_2 - i_3 = 16i_3 - 6i_3$$
$$-4i_1 - i_2 + 8i_3 = -8 + 6i_3$$

Algebraic adjustment then gives the equations obtained from the individual KVL statements, and these may also be written in matrix form,

$$\begin{bmatrix} 6 & -2 & 12 \\ -2 & 8 & -11 \\ -4 & -1 & 2 \end{bmatrix} \begin{bmatrix} i_1 \\ i_2 \\ i_3 \end{bmatrix} = \begin{bmatrix} 24 \\ 0 \\ -8 \end{bmatrix} \tag{3.18}$$

and the almost total lack of symmetry can be observed.

Although the formulation of the mesh equations has been demonstrated to be somewhat more complex when controlled sources are present, the solution of the mesh equations is still straightforward. By a matrix inversion process on eqs. (3.18), it is seen that

$$\begin{bmatrix} i_1 \\ i_2 \\ i_3 \end{bmatrix} = \begin{bmatrix} 6 & -2 & 12 \\ -2 & 8 & -11 \\ -4 & -1 & 2 \end{bmatrix}^{-1} \begin{bmatrix} 24 \\ 0 \\ -8 \end{bmatrix}$$

$$= \frac{1}{342} \begin{bmatrix} 5 & -8 & -74 \\ 48 & 60 & 42 \\ 34 & 14 & 44 \end{bmatrix} \begin{bmatrix} 24 \\ 0 \\ -8 \end{bmatrix} = \begin{bmatrix} 2.082 \\ 2.386 \\ 1.357 \end{bmatrix} A$$

In spite of the admonition that "naked" sources should not be placed in a network without an accompanying resistor, little difficulty was encountered in writing the mesh equations shown, in matrix form, in eqs. (3.18). Thus, mesh (or node or loop) analysis of networks containing unaccompanied sources is not at all precluded in most cases.

Sometimes, however, there can be analysis difficulties, and a general network method can be proposed in which *all* ideal sources in a network branch *must* be accompanied by an R, L, or C. Use of this general network method requires unaccompanied sources to be adjusted by techniques known as "pushing a voltage source through a node" and "shifting current sources." All of these techniques, however, are beyond the scope of this book.

The presence of controlled current sources, whether current-controlled or voltage-controlled, should create no problem. They may be converted to controlled voltage sources by using the current–to–voltage source transformation, or they may be used

FIGURE 3.21   (a) A network with two controlled current sources and (b) the network ready for mesh analysis

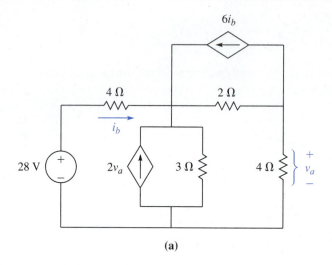

(a)

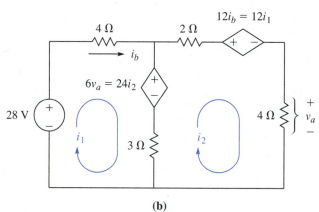

(b)

in dummy meshes. Figure 3.21a presents a network with two controlled current sources. When these controlled current sources are converted or transformed to controlled voltage sources, the network of Fig. 3.21b results.

EXERCISE 3.3

Determine the branch current j_4 and the branch voltage v_4 in the branch containing the 4-Ω resistor in Fig. 3.22.

Answer $j_4 = 0.459$ A, $v_4 = 1.836$ V.

EXERCISE 3.4

Determine the branch current j_8 and the branch voltage v_8 in the branch containing the 8-Ω resistor in Fig. 3.23.

Answer $j_8 = -8.591$ A, $v_8 = -68.728$ V.

FIGURE 3.22

Network (Exercise 3.3)

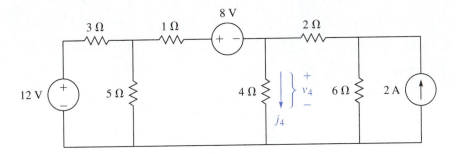

FIGURE 3.23

Network (Exercise 3.4)

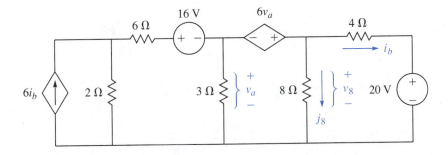

THE FUNDAMENTAL THEOREM OF NETWORK TOPOLOGY

Consider a network with b branches, n_t total nodes, and ℓ loops[5] that are formed from the links in a tree of the network graph. Begin with one branch and note that the branch must have a node at each end, as shown in Fig. 3.24a. The first line in Table 3.1 shows that $b = 1$, $n_t = 2$, and $\ell = 0$. A second branch may be added, but as shown in Fig. 3.24b, no new loop is created. This is summarized in the second line of Table 3.1.

Now observe in the balance of Fig. 3.24 and Table 3.1 two important facts: The addition of a branch without forming a loop adds exactly one more node, and the addition of a branch that forms a loop adds no additional nodes. With these facts, it is possible to conclude by induction that the number of tree branches is

$$b = n_t - 1$$

The number of links is equal to the number of loops, so that the number of branches is the total:

$$b = \ell + n_t - 1 \tag{3.19}$$

---

[5] In a planar network, the number of independent loops is equal to the number of meshes. The proof of this is beyond the scope of this book.

FIGURE 3.24

Steps in the development of the fundamental theorem of network topology by the induction process

The figure should be used in conjunction with Table 3.1.

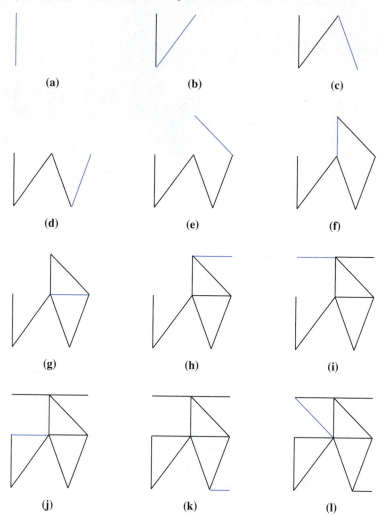

TABLE 3.1

Steps in the development of the fundamental theorem of network topology by induction

Fig. 3.19	b	n_t	ℓ	$\ell = b - n_t + 1$?
(a)	1	2	0	Yes
(b)	2	3	0	Yes
(c)	3	4	0	Yes
(d)	4	5	0	Yes
(e)	5	6	0	Yes
(f)	6	6	1	Yes
(g)	7	6	2	Yes
(h)	8	7	2	Yes
(i)	9	8	2	Yes
(j)	10	8	3	Yes
(k)	11	9	3	Yes
(l)	12	9	4	Yes

Network (Exercise 3.5)

FIGURE 3.25

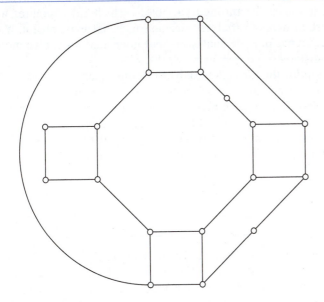

This is the *fundamental theorem of network topology* for a connected graph. It may be used to determine the number of loops after the number of branches and the total number of nodes are counted in an inspection of the given network.

EXERCISE 3.5

Obtain the number of branches, total nodes, and loops in the network whose graph is shown in Fig. 3.25.

Answer $n_t = 18$, $\ell = 8$, $b = 25$.

CHOOSING MESH ANALYSIS OR NODE ANALYSIS

SECTION 3.6

The choice of whether to perform a mesh (or loop[6]) or node analysis for a planar network is easily made on the basis of the number of simultaneous equations that result. Many students who are beginning the network sequence seem reluctant to perform a node analysis, probably because they feel that it is easier to use Ri products than Gv models. In addition, with integer values of resistances, the methods of mesh or loop analysis do not involve fractions or decimals in the coefficient matrix.

    This tendency toward the preference of a mesh analysis seems counterproductive for several reasons. One reason is that in the age of the computer, large-scale analyses are rarely performed by hand, and it behooves the student to be familiar with any and all methods of analysis. Another reason is that attention is most often focused on the across variable, because it is the across variable that is the stress in the classical stress-strain problem. One finds, for example, that analysts are more often interested in the determination of stress or pressure, temperature, and voltages than they are in

[6] Loop analysis is considered in Section 3.7.

the determination of deflections, fluid, heat, and current flows. Because of this, node analysis provides a more straightforward path to the eventual solution of interest, and a procedure that determines the through variables—which then requires an additional step to convert to across variables—seems somewhat extravagant. Yet another reason for the preference of node analysis is that node analysis can be more readily employed in the analysis of nonplanar networks.

However, if the solution for the through variables is more expeditious because a smaller system of simultaneous equations is easier to solve, there is merit in solving for the through variables first. In any given network, it is easy to count the branches and the total number of nodes. Thus, the size of the node analysis problem is quickly established. The fundamental theorem of network topology can then be employed to determine the number of meshes or loops. (*Hint:* Before counting, null the independent sources).

SECTION 3.7   **LOOP ANALYSIS**

In the mesh analysis of planar networks, the meshes may be quickly determined by eyeballing the *window panes* or *fish nets* within the network. Then, the actual mesh analysis is conducted by a systematic application of KVL to each mesh. The inevitable result is a set of linear, simultaneous algebraic equations that can be solved for the mesh currents. Branch currents may be obtained from the mesh currents, and branch voltages can then be evaluated, in a dc analysis containing resistors, from a simple application of Ohm's law. The procedure for loop analysis is somewhat similar.

The network of Fig. 3.16 (not including Fig. 3.16e) is repeated here for the reader's convenience as Fig. 3.26. This network, with loop currents obtained from the tree shown in Fig. 3.26c, will be the network used to demonstrate the method of loop analysis. The network illustrating the three loops is shown in Fig. 3.26d. Observe that the loop currents are all in the clockwise direction. This is not arbitrary; the loop current orientations derive from the orientation of the links associated with the graph and tree of Figs. 3.26b and 3.26c.

A systematic application of KVL, with the algebraic sum of the voltages equal to zero, will yield three loop equations,

$$-63 + 5(i_1 + i_2 + i_3) + 6i_1 = 0$$

$$-63 + 5(i_1 + i_2 + i_3) + (i_2 + i_3) + 3i_2 = 0$$

$$-63 + 5(i_1 + i_2 + i_3) + (i_2 + i_3) + 4i_3 + 2i_3 = 0$$

and a little algebraic adjustment yields

$$11i_1 + 5i_2 + 5i_3 = 63$$

$$5i_1 + 9i_2 + 6i_3 = 63 \tag{3.20}$$

$$5i_1 + 6i_2 + 12i_3 = 63$$

or the matrix representation

$$\begin{bmatrix} 11 & 5 & 5 \\ 5 & 9 & 6 \\ 5 & 6 & 12 \end{bmatrix} \begin{bmatrix} i_1 \\ i_2 \\ i_3 \end{bmatrix} = \begin{bmatrix} 63 \\ 63 \\ 63 \end{bmatrix} \tag{3.21}$$

(a) A ladder network showing three meshes, (b) the oriented graph of the network, (c) one possible tree of the graph, and (d) the network showing three loops obtained by inserting the tree links back into the graph one at a time

FIGURE 3.26

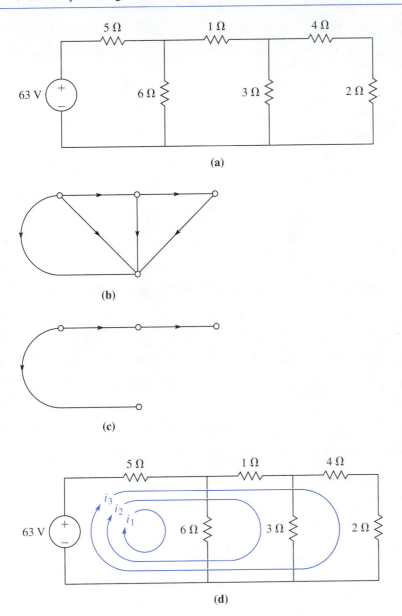

(a)

(b)

(c)

(d)

Here, too, there is a certain symmetry. The elements on the principal diagonal of the coefficient matrix do indeed represent the total resistance of the particular loop, but the off-diagonal elements are not easily identified. However, the solution is quite straightforward. From eqs. (3.20), a Cramer's rule solution for i_3, which should be

FIGURE 3.27   Another tree of the graph of Fig. 3.26b (Exercise 3.6)

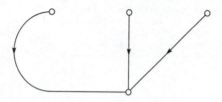

equal to 2 A, is

$$i_3 = \frac{\begin{vmatrix} 11 & 5 & 63 \\ 5 & 9 & 63 \\ 5 & 6 & 63 \end{vmatrix}}{\begin{vmatrix} 11 & 5 & 5 \\ 5 & 9 & 6 \\ 5 & 6 & 12 \end{vmatrix}} = \frac{1134}{567} = 2 \text{ A}$$

A matrix inversion of the system of eqs. (3.21) will provide all three loop currents:

$$\begin{bmatrix} i_1 \\ i_2 \\ i_3 \end{bmatrix} = \begin{bmatrix} 11 & 5 & 5 \\ 5 & 9 & 6 \\ 5 & 6 & 12 \end{bmatrix}^{-1} \begin{bmatrix} 63 \\ 63 \\ 63 \end{bmatrix} = \frac{1}{567} \begin{bmatrix} 72 & -30 & -15 \\ -30 & 107 & -41 \\ -15 & -41 & 74 \end{bmatrix} \begin{bmatrix} 63 \\ 63 \\ 63 \end{bmatrix} = \begin{bmatrix} 3 \\ 4 \\ 2 \end{bmatrix} \text{ A}$$

In Fig. 3.17, the correspondence of

$$j_2 = i_3 = 2 \text{ A} \qquad j_3 = i_2 = 4 \text{ A} \qquad j_6 = i_1 = 3 \text{ A}$$

as well as all other branch currents, is noted.

EXERCISE 3.6

For the network of Fig. 3.26a, use the tree shown in Fig. 3.27 to formulate the loop equations. Then, solve the equations to confirm that the branch currents have the same value as obtained by the method of mesh analysis.

Answer $i_1 = 3$ A, $i_2 = 6$ A, and $i_3 = 2$ A; the branch currents do have the same values.

SECTION 3.8   POWER IN RESISTIVE NETWORKS

The conservation of energy principle—which states that energy can neither be created nor destroyed—can be invoked to quickly point out that because power is the rate of energy transformation or the rate of doing work, the power dissipated by a resistive network is equal to the power delivered to the network. The power dissipated by a

FIGURE 3.28

The network of Fig. 3.10 repeated

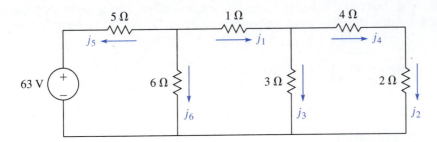

resistive branch will always be equal to the product of the branch current and the branch voltage, $p = vj$. The power supplied by any source is, of course, p_s, and the conservation of energy principle dictates that

$$\sum p_s = \sum vj$$

or

$$\sum p_s - \sum vj = 0 \qquad\qquad (3.22a)$$

Alternatively,

$$\sum p_s - \sum j^2 R = 0 \qquad\qquad (3.22b)$$

and

$$\sum p_s - \sum \frac{v^2}{R} = 0 \qquad\qquad (3.22c)$$

Equations (3.22) apply to individual elements and to the equivalents that may be obtained from the individual elements. For example, in the network shown in Fig. 3.10a, repeated here as Fig. 3.28 with branch current designators inserted, it was found in Section 3.3 that the branch currents and branch voltages are

$$j_1 = 6 \text{ A} \qquad v_1 = 6 \text{ V}$$
$$j_2 = 2 \text{ A} \qquad v_2 = 4 \text{ V}$$
$$j_3 = 4 \text{ A} \qquad v_3 = 12 \text{ V}$$
$$j_4 = 2 \text{ A} \qquad v_4 = 8 \text{ V}$$
$$j_6 = 3 \text{ A} \qquad v_6 = 18 \text{ V}$$

For the 5-Ω resistor that is included with the 63-V source in branch 5, the current is 9 A and the voltage drop is 45 V. The source voltage (63 V) delivers 9 A to the resistive network.

Conservation of energy dictates that in accordance with eq. (3.22a),

$$63(9) - 6(6) - 4(2) - 12(4) - 8(2) - 45(9) - 18(3) = 0$$

which is easily verified. Moreover, the resistance presented to the 63-V ideal source was shown in Section 2.7 to be 7 Ω. In this case, conservation of energy must also apply to the network represented by this equivalent or driving-point impedance. Here, 9 A enters this equivalent resistance, and by eq. (3.22b), the power delivered by the source is equal to the power dissipated by the network:

$$63(9) - (9)^2(7) = 567 - 567 = 0$$

SECTION 3.9   SPICE EXAMPLES

Reading: In addition to reading Sections C.1 through C.4, which were recommended for reading in Chapter 2, the reader should understand Section C.5 before proceeding to the SPICE examples that follow.

EXAMPLE S3.1

This example is a sensitivity analysis that shows the effect of variations in the magnitude of the component resistances on the current i in the network in Fig. S3.1. The network made ready for the PSPICE analysis is shown in Fig. S3.2. The input file is shown in Fig. S3.3, and pertinent extracts from the output file are shown in Fig. S3.4.

Figure S3.1   Network used to demonstrate a sensitivity analysis

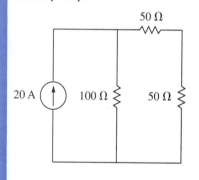

Figure S3.2   Network of Fig. S3.1 made ready for SPICE analysis

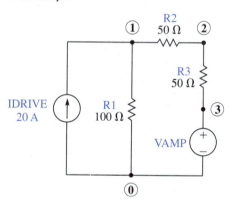

Figure S3.3   SPICE input file for Example S3.1

```
SPICE PROBLEM - CHAPTER 3 - NUMBER 1 - SENSITIVITY ANALYSIS
***************************************************************
*FIRST, THE ELEMENTS AND THE SINGLE CURRENT SOURCE.
IDRIVE      0      1      DC      20
R1          1      0      100
R2          1      2      50
R3          2      3      50
***************************************************************
*THEN THE VOLTMETER USED TO MEASURE THE CURRENT
VAMP        3      0      DC      0
***************************************************************
*FINALLY, THE TWO CONTROL STATEMENTS
.SENS I(VAMP)
.END
```

Figure S3.4   SPICE output file for Example S3.1

```
SPICE PROBLEM - CHAPTER 3 - NUMBER 1 - SENSITIVITY ANALYSIS


****      DC SENSITIVITY ANALYSIS            TEMPERATURE =   27.000 DEG C

*****************************************************************************


DC SENSITIVITIES OF OUTPUT I(VAMP)

         ELEMENT          ELEMENT         ELEMENT         NORMALIZED
          NAME             VALUE        SENSITIVITY       SENSITIVITY
                                        (AMPS/UNIT)     (AMPS/PERCENT)

          R1            1.000E+02         5.000E-02        5.000E-02
          R2            5.000E+01        -5.000E-02       -2.500E-02
          R3            5.000E+01        -5.000E-02       -2.500E-02
          VAMP          0.000E+00        -5.000E-03        0.000E+00
          IDRIVE        2.000E+01         5.000E-01        1.000E-01


          JOB  CONCLUDED
```

EXAMPLE S3.2

In Section 3.4.4, the network shown in Fig. 3.20 (repeated here as Fig. S3.5) was analyzed. The mesh currents were obtained and were displayed in the mesh current vector,

$$I = \begin{bmatrix} 2.082 \\ 2.386 \\ 1.357 \end{bmatrix} A$$

This example shows that an identical result can be obtained by using PSPICE.

The network made ready for the PSPICE analysis is shown in Fig. S3.6. The input file is shown in Fig. S3.7, and the output file is displayed in Fig. S3.8.

Figure S3.5   Network used to demonstrate an analysis containing controlled sources

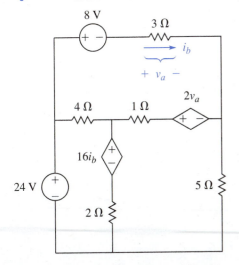

(continues)

Example S3.2 (*continued*)

Figure S3.6   Network of Fig. S3.5 made ready for SPICE analysis

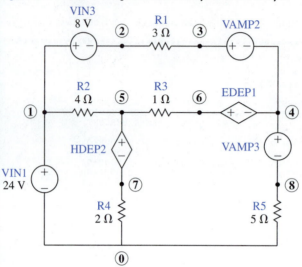

Figure S3.7   SPICE input file for Example S3.2

```
SPICE PROBLEM - CHAPTER 3 - NUMBER 2 - CONTROLLED SOURCE PROBLEM
*THE PROBLEM IS MORE ADVANCED BECAUSE OF THE
*TWO CONTROLLED SOURCES.
*****************************************************************
*FIRST, THE SOURCES AND THE RESISTORS
VIN1    1    0    DC    24
VIN3    1    2    DC     8
R1      2    3    3
R2      1    5    4
R3      5    6    1
R4      7    0    2
R5      8    0    5
*****************************************************************
*NEXT, THE CONTROLLED SOURCES
EDEP1   6    4    2    3    2
HDEP2   5    7    VAMP2    16
*****************************************************************
*THEN, THE VOLTAGE SOURCES THAT MEASURE CURRENT.
*THERE IS NO NEED TO MEASURE I1 AS VIN1 MEASURES
*I1 IN THE REVERSE DIRECTION.
VAMP2   3    4    DC    0
VAMP3   4    8    DC    0
*****************************************************************
*DON'T FORGET THE .END STATEMENT
.END
```

Figure S3.8 SPICE output file for Example S3.2

```
SPICE PROBLEM - CHAPTER 3 - NUMBER 2 - CONTROLLED SOURCE PROBLEM

****      SMALL SIGNAL BIAS SOLUTION        TEMPERATURE =    27.000 DEG C


********************************************************************


NODE    VOLTAGE      NODE    VOLTAGE      NODE    VOLTAGE      NODE    VOLTAGE


(    1)     24.0000  (    2)     16.0000  (    3)     11.9300  (    4)     11.9300

(    5)     21.0990  (    6)     20.0700  (    7)     -.6082   (    8)     11.9300

VOLTAGE SOURCE CURRENTS
NAME            CURRENT

VIN1          -2.082E+00
VIN2           1.357E+00
VAMP2          1.357E+00
VAMP3          2.386E+00

TOTAL POWER DISSIPATION    3.91E+01   WATTS
```

CHAPTER 3

SUMMARY

- Node analysis depends upon a repeated application of KCL to form n node equations in n unknown node voltages of the form

$$\mathbf{GE} = \mathbf{I_s}$$

- Every independent voltage source that is included in a node analysis creates a dummy node.

- Mesh analysis depends upon a repeated application of KVL to form ℓ mesh equations in ℓ unknown mesh currents of the form

$$\mathbf{RI} = \mathbf{V_s}$$

- Every current source that is included in a mesh analysis creates a dummy mesh.

- The fundamental theorem of network topology is

$$b = \ell + n_t - 1$$

where b is the number of branches, n_t is the total number of nodes, and ℓ is the number of loops.

- Loop analysis is conducted in the same manner as mesh analysis. Loop analysis also depends upon a repeated application of KVL to form ℓ loop equations in ℓ unknown loop currents of the form

$$\mathbf{RI} = \mathbf{V_s}$$

However, a linearly independent set of ℓ loop equations in the ℓ loop currents is guaranteed only if the loops are formed from a tree of the oriented network graph.

Additional Readings

Blackwell, W.A., and L.L. Grigsby. *Introductory Network Theory*. Boston: PWS Engineering, 1985, pp. 77–82, 88–106, 174–186.

Bobrow, L.S. *Elementary Linear Circuit Analysis*. 2d ed. New York: Holt, Rinehart and Winston, 1987, pp. 51–93.

Del Toro, V. *Engineering Circuits*. Englewood Cliffs, N.J.: Prentice-Hall, 1987, pp. 85–101, 110–116, 159–163.

Dorf, R.C. *Introduction to Electric Circuits*. New York: Wiley, 1989, pp. 86–107.

Hayt, W.H., Jr., and J.E. Kemmerly. *Engineering Circuit Analysis*. 4th ed. New York: McGraw-Hill, 1986, pp. 58–73, 92–103.

Irwin, J.D. *Basic Engineering Circuit Analysis*. 3rd ed. New York: Macmillan, 1989, pp. 83–120, 879–889.

Johnson, D.E., and J.R. Johnson. *Graph Theory with Engineering Applications*. New York: Ronald Press, pp. 3–33.

Johnson, D.E., J.L. Hilburn, and J.R. Johnson. *Basic Electric Circuit Analysis*. 4th ed. Englewood Cliffs, N.J.: Prentice-Hall, 1989, pp. 78–98, 152–164.

Karni, S. *Applied Circuit Analysis*. New York: Wiley, 1988, pp. 35–42, 50–61, 108–114.

Madhu, S. *Linear Circuit Analysis*. Englewood Cliffs, N.J.: Prentice-Hall, 1988, pp. 104–139, 168–213.

Nilsson, J.W. *Electric Circuits*. 3d ed. Reading, Mass.: Addison-Wesley, 1990, pp. 66–92, 121–142.

Paul, C.R. *Analysis of Linear Circuits*. New York: McGraw-Hill, 1989, pp. 80–86, 145–183.

PROBLEMS   CHAPTER 3

Section 3.2

3.1 For the network in Fig. P3.1, draw the oriented graph and draw any tree, and for the tree, show all loops and all cutsets.

Figure P3.1

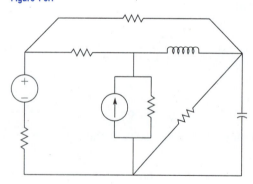

3.2 Consider the graph of a network and a particular set of loops, as indicated in Fig. P3.2. Identify the tree from which the loops were formed.

Figure P3.2

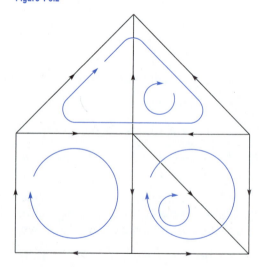

3.3 In the graph whose branches and nodes are indicated in Fig. P3.3, identify two paths from node 1 to node 4 and two loops that include branch 3.

3.4 Identify all meshes in the graph of Fig. P3.3

Figure P3.3

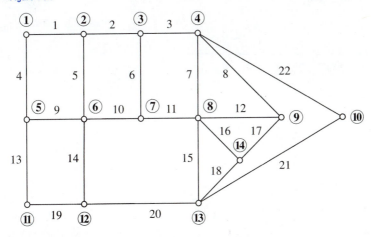

Section 3.3

3.5 Write and solve the equations for the node volt-ages in the network of Fig. P3.4. Then, find the branch current j_6 and the branch voltage v_3.

Figure P3.4

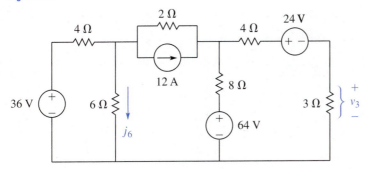

3.6 Write and solve the equations for the node volt-ages in the network of Fig. P3.5. Then, find the branch current j_4 and the branch voltage v_8.

Figure P3.5

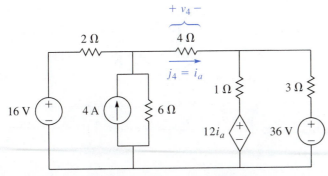

3.7   Write and solve the equations for the node voltages in the network of Fig. P3.6. Then, find the branch current j_8 and the branch voltage v_3.

Figure P3.6

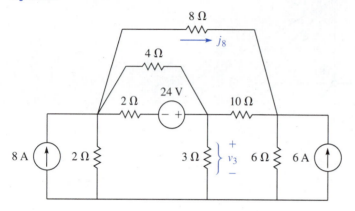

3.8   Write and solve the equations for the node voltages in the network of Fig. P3.7. Then, find the branch current j_{16} and the branch voltage v_8.

Figure P3.7

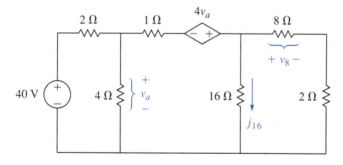

3.9   In the network of Fig. P3.8, what value of E will make the branch current j_6 numerically equal to zero?

3.10   In the network of Fig. P3.9, what is the value of i_a?

Figure P3.9

Figure P3.8

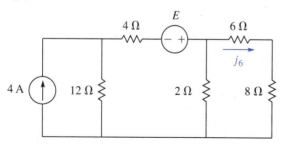

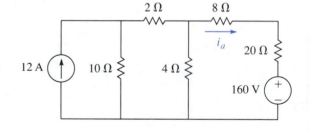

3.11 Write a set of node equations (but do not attempt to solve them) for the network with node designations in Fig. P3.10.

Figure P3.10

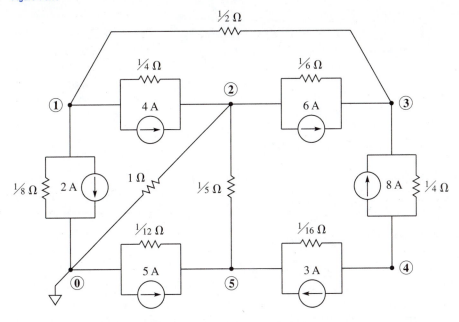

3.12 In Fig. P3.11, what is the value of i_s to produce a node voltage of 36 V at node 2?

Figure P3.11

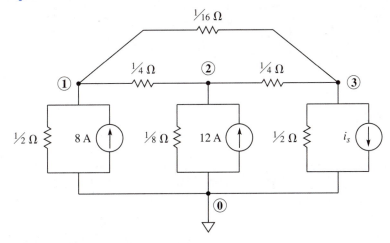

3.13 In Fig. P3.12, the resistor R is variable. If 100.6 mA is to flow through it, determine its value.

Figure P3.12

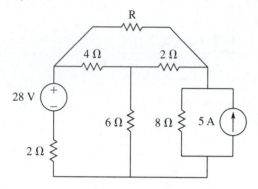

Figure P3.14

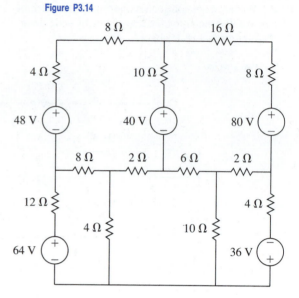

3.14 In the network of Fig. P3.13, what is the value of the branch current j_{40}?

Figure P3.13

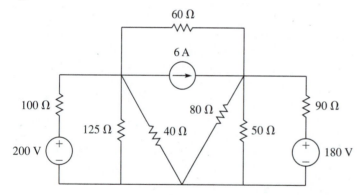

Section 3.4

3.15 Repeat Problem 3.6 by using mesh analysis.

3.16 Repeat Problem 3.8 by using mesh analysis.

3.17 Repeat Problem 3.9 by using mesh analysis.

3.18 Repeat Problem 3.10 by using mesh analysis.

3.19 Write a set of mesh equations (but do not attempt to solve them) for the network in Fig. P3.14.

Section 3.5

3.20 From the network graph displayed in Fig. P3.15, determine the total number of nodes, branches, and loops.

Figure P3.15

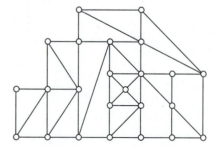

3.21 From the network graph displayed in Fig. P3.16, determine the total number of nodes, branches, and loops.

Figure P3.16

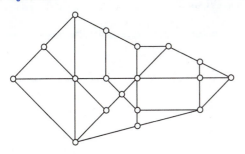

3.22 From the network graph displayed in Fig. P3.17, determine the total number of nodes, branches, and loops.

Figure P3.17

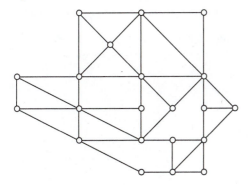

Section 3.7

3.23 Repeat Problem 3.5 by using loop analysis.

3.24 Repeat Problem 3.6 by using loop analysis.

3.25 Repeat Problem 3.8 by using loop analysis.

3.26 Repeat Problem 3.9 by using loop analysis.

3.27 Repeat Problem 3.10 by using loop analysis.

Section 3.8

3.28 Determine the total power drawn by the network in Fig. P3.4.

3.29 Determine the total power drawn by the network in Fig. P3.18.

Figure P3.18

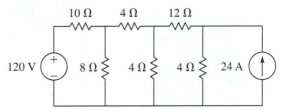

3.30 Determine the total power drawn by the network in Fig. P3.19.

Figure P3.19

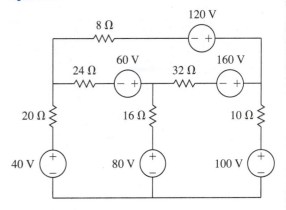

4

NETWORK THEOREMS

OBJECTIVES

The objectives of this chapter are to:

- Show how the superposition principle can be used in the analysis of networks.

- Provide a detailed discussion of the network theorems of Thévenin and Norton for resistive networks

- Reconsider power in resistive networks, introduce Tellegen's theorem, and show how Tellegen's theorem is used.

- Discuss the reciprocity principle and derive the general reciprocity relationship.

- Introduce the maximum power transfer theorem.

SECTION 4.1   INTRODUCTION

In Section 3.1, it was mentioned that there is a considerable amount of elegance in network analysis and that there is more to the subject than the "math problem" solution of simultaneous equations that derive from the methods of node, mesh, and loop analysis discussed in Chapter 3. The *Thévenin* and *Norton theorems* are considered in this chapter, as well as the concepts of *linearity* and *superposition*. The latter may well be one of the more important fundamental concepts at the disposal of the engineer. The principle of *reciprocity* is also important, as are the conditions for maximum power transfer; these concepts, too, are treated in this chapter.

It will be continuously observed that there are several analysis methods available. These range from the ultrasimple for the most elementary to the comprehensive for the more advanced networks. Frequently, the form of the network itself suggests the possibility of alternative methods of analysis, and it is hoped that after the study of this chapter has been completed, the reader will be motivated to search for the most direct method of analysis for a particular given network.

THE SUPERPOSITION THEOREM

SECTION 4.2

The homogeneity and superposition conditions for linearity were cited in Section 1.5 in connection with the discussion of linear network elements. Although both are necessary for linearity, the superposition condition, in itself, provides the basis for the superposition theorem:

> **Superposition Theorem:** If cause and effect are linearly related, the total effect due to several causes acting simultaneously is equal to the sum of the individual effects due to each of the causes acting one at a time.

An application of the superposition theorem for a linear system can be obtained from a consideration of a set of n mesh equations in n unknown mesh currents,

$$\mathbf{RI} = \mathbf{V_s} \tag{4.1}$$

where $\mathbf{I}$ is an $n \times 1$ column vector of mesh currents, $\mathbf{R}$ is an $n \times n$ resistance coefficient matrix,

$$\mathbf{R} = \begin{bmatrix} r_{11} & r_{12} & r_{13} & \cdots & r_{1n} \\ r_{21} & r_{22} & r_{23} & \cdots & r_{2n} \\ r_{31} & r_{32} & r_{33} & \cdots & r_{3n} \\ \cdots & \cdots & \cdots & \cdots & \cdots \\ r_{n1} & r_{n2} & r_{n3} & \cdots & r_{nn} \end{bmatrix}$$

and $\mathbf{V_s}$ is an $n \times 1$ column vector of voltage sources that cannot be null. A matrix inversion process will yield $\mathbf{I}$:

$$\begin{bmatrix} i_1 \\ i_2 \\ i_3 \\ \vdots \\ i_4 \end{bmatrix} = \frac{1}{\det \mathbf{R}} \begin{bmatrix} R_{11} & R_{21} & R_{31} & \cdots & R_{n1} \\ R_{12} & R_{22} & R_{32} & \cdots & R_{n2} \\ R_{13} & R_{23} & R_{33} & \cdots & R_{n3} \\ \cdots & \cdots & \cdots & \cdots & \cdots \\ R_{1n} & R_{2n} & R_{3n} & \cdots & R_{nn} \end{bmatrix} \begin{bmatrix} v_{s1} \\ v_{s2} \\ v_{s3} \\ \vdots \\ v_{sn} \end{bmatrix} \tag{4.2}$$

where $\det \mathbf{R} = |R|$ and the capital R's with double subscripts are the cofactors in the resistive coefficient matrix.

A matrix multiplication will yield the elements of the $\mathbf{I}$ vector. For example, for the element i_k (lowercase i to be consistent with the designation of a matrix by a boldface capital letter and the elements of that matrix by lowercase letters),

$$i_k = \frac{1}{\det \mathbf{R}} (R_{1k}v_{s1} + R_{2k}v_{s2} + R_{3k}v_{s3} + \cdots + R_{nk}v_{sn}) \tag{4.3}$$

or

$$i_k = \sum_{m=1}^{n} \frac{R_{mk}}{\det \mathbf{R}} v_{sm} \tag{4.4}$$

Equation (4.4) clearly shows that the additive or superposition condition holds and proves that systems described by eq. (4.1) are linear. It also demonstrates the

FIGURE 4.1 (a) A network to be solved by employing superposition, (b) the network with the current source nulled, and (c) the network with the voltage source nulled

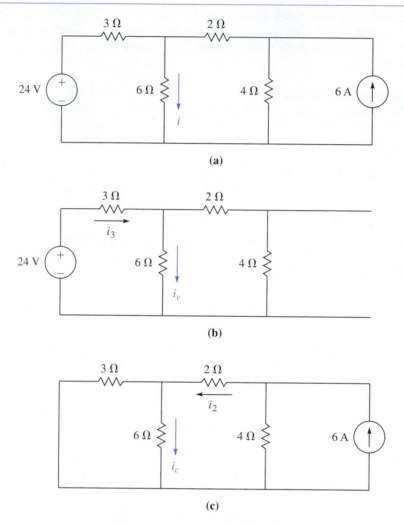

(a)

(b)

(c)

homogeneity condition for linearity; the magnitude of each current component is directly proportional to the magnitude of the source that is producing it.

For the use of the superposition theorem, consider a ladder network driven by a voltage source at the left and a current source at the right, as shown in Fig. 4.1a. The current indicated by i is to be determined. According to the superposition theorem (hereinafter to be invoked by merely stating "by superposition"), the current i will be the sum of the two current components i_v and i_c shown in Figs. 4.1b and 4.1c. The nomenclature is designed to indicate that i_v is the current component due to the voltage source acting alone (with the current source nulled) and i_c is the current component due to the current source acting alone (with the voltage source nulled). Figures 4.1b and 4.1c follow from Section 2.7, where it was observed that voltage sources are nulled in a network by replacing them with short circuits and current sources are nulled in a network by replacing them with open circuits.

The networks displayed in Figs. 4.1b and 4.1c are simple ladder networks, and the strategy is to first determine the equivalent resistances presented to the 24-V and

6-A sources. For Fig. 4.1b,

$$R_{eq} = \frac{(4+2)(6)}{4+2+6} + 3 = \frac{36}{12} + 3 = 3 + 3 = 6 \ \Omega$$

and the current that flows in the 3-Ω resistor is

$$i_3 = \frac{24}{6} = 4 \ A$$

The current i_v can be obtained from a current division:

$$i_v = \left(\frac{4+2}{6+4+2} \right) i_3 = \left(\frac{6}{12} \right)(4) = 2 \ A$$

For Fig. 4.1c, there is an equivalent resistance acting in parallel with the 4-Ω resistor. Its value is

$$R_{eq} = \frac{6(3)}{6+3} + 2 = \frac{18}{9} + 2 = 2 + 2 = 4 \ \Omega$$

The current from the 6-A source must divide equally between $R_{eq} = 4 \ \Omega$ and the 4-Ω source resistor to produce $i_2 = 6/2 = 3 \ A$. The value of i_c can be obtained from another current division:

$$i_c = \left(\frac{3}{6+3} \right) i_2 = \left(\frac{3}{9} \right)(3) = 1 \ A$$

The value of i in Fig. 4.1a must be

$$i = i_v + i_c = 2 + 1 = 3 \ A$$

EXERCISE 4.1

Use superposition to find the voltage across the 5-Ω resistor in Fig. 4.2.

Answer   18 V.

Network (Exercise 4.1)

FIGURE 4.2

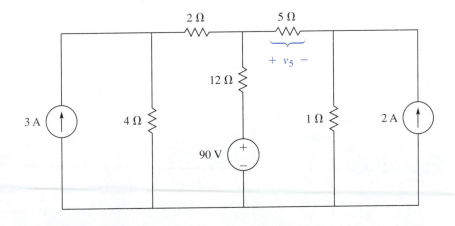

SECTION 4.3 THE NETWORK THEOREMS OF THÉVENIN AND NORTON

If interest is to be focused on the voltages across and the currents through a portion of a network, such as network B in Fig. 4.3a, it is convenient to replace network A, which is complicated and of little interest, by a simple equivalent. The simple equivalent may contain a single, equivalent voltage source in series with an equivalent resistance, as displayed in Fig. 4.3b. In this case, the equivalent is called a *Thévenin equivalent*. Alternatively, the simple equivalent may consist of an equivalent current source in parallel with an equivalent resistance. This equivalent, shown in Fig. 4.3c, is called a *Norton equivalent*. Observe that as long as R_T (subscript T for Thévenin) is equal to R_N (subscript N for Norton), the two equivalents may be obtained from

FIGURE 4.3   (a) Two one-port networks, (b) the Thévenin equivalent for network A, and (c) the Norton equivalent for network A

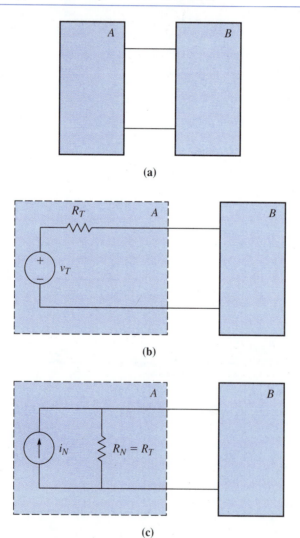

(a)

(b)

(c)

one another by a simple source transformation. This section provides the procedure for the determination of the Thévenin and Norton equivalents.

4.3.1 Conditions of Application

The Thévenin and Norton network equivalents are only valid at the terminals of network A in Fig. 4.3, and they do not extend to its interior. In addition, there are certain restrictions on networks A and B. Network A may contain only linear elements but may contain both independent and dependent sources. Network B, on the other hand, is not restricted to linear elements; it may contain nonlinear or time-varying elements and may also contain both independent and dependent sources. There can be no controlled-source coupling or magnetic coupling between networks A and B.

4.3.2 The Thévenin Theorem

The derivation of the Thévenin theorem is interesting, but its presentation will serve no useful purpose here. Refer to Figs. 4.3a and 4.3b and consider the definition for the case of resistive networks.

> **Thévenin's Theorem:** Insofar as a load that has no magnetic or controlled-source coupling to a one-port is concerned, a network containing linear elements and both independent and controlled sources may be replaced by an ideal voltage source of strength v_T and an equivalent resistance R_T in series with the source. The value of v_T is the open-circuit voltage appearing across the terminals of the network, and R_T is the driving-point resistance at the terminals of the network, obtained with all independent sources set equal to zero.

4.3.3 The Norton Theorem

The Norton theorem involves a current source equivalent. This time, refer to Figs. 4.3a and 4.3c, and consider the statement of the Norton theorem for resistive networks:

> **Norton's Theorem:** Insofar as a load that has no magnetic or controlled-source coupling to a one-port is concerned, the network containing linear elements and both independent and controlled sources may be replaced by an ideal current source of strength i_N and an equivalent resistance R_N in parallel with the source. The value of i_N is the short-circuit current that results when the terminals of the network are shorted, and R_N is the driving-point resistance at the terminals when all independent sources are set equal to zero.

4.3.4 The Equivalent Resistance $R_T = R_N$

Three methods are commonly used for the determination of R_T. All of them are applicable at the analyst's discretion. However, when controlled sources are present, the first method cannot be employed.

The first method involves the direct calculation of $R_{eq} = R_T = R_N$ by series-parallel combination looking into the terminals of the network after all independent sources have been nulled. As indicated in Section 2.7, independent sources are nulled in a network by replacing all independent voltage sources with a short circuit and all independent current sources with an open circuit.

The second method, which may be used when controlled sources are present in the network (and, of course, when independent sources are present), requires the computation of both the Thévenin equivalent voltage (the open-circuit voltage at the terminals of the network) and the Norton equivalent current (the current through the short-circuited terminals of the network). The equivalent resistance is the ratio of these two quantities:

$$R_T = R_N = R_{eq} = \frac{v_T}{i_N} \tag{4.5}$$

The third method can always be used. A test voltage may be placed across the terminals, with a resulting current calculated or measured after all independent sources are nulled. Alternatively, a test current may be injected into the terminals, with a resulting voltage determined after all independent sources are nulled. In either case, the equivalent resistance can be obtained from the absolute value of the ratio of the test voltage, v_o, to the resulting current, i_o, or the ratio of the absolute value of the observed or calculated voltage to the test current:

$$R_T = \left| \frac{v_o}{i_o} \right| \tag{4.6}$$

■ **EXAMPLE 4.1**

For the network shown in Fig. 4.4a, determine the value of the current i through the 10-Ω resistor by using Thévenin's theorem.

Solution   In Fig. 4.4b, there is no current through and no voltage drop across the 2-Ω resistor. No current can flow through it when the 10-Ω resistor carrying i is removed, because an open circuit is formed. This means that $v_{oc} = v_T$ appears across the 8-Ω resistor. The voltage drop across the 8-Ω resistor can be obtained from a mesh analysis using two meshes. Hence, for the two meshes shown in Fig. 4.4b, the mesh equations are

$$18i_1 - 6i_2 = 24$$

$$-6i_1 + 18i_2 = 16$$

A Cramer's rule solution provides

$$i_2 = \frac{\begin{vmatrix} 18 & 24 \\ -6 & 16 \end{vmatrix}}{\begin{vmatrix} 18 & -6 \\ -6 & 18 \end{vmatrix}} = \frac{432}{288} = \frac{3}{2} \text{ A}$$

The value of v_T is $v_T = v_{oc} = 8i_2 = 8(3/2) = 12$ V.

(a) A network in which the current i is to be determined, (b) the network with the 10-Ω resistor carrying i removed, (c) the network used to establish R_T, (d) the Thévenin equivalent with the 10-Ω resistor reconnected, (e) the network used to calculate the Norton equivalent current, and (f) the Norton equivalent with the 10-Ω resistor reconnected

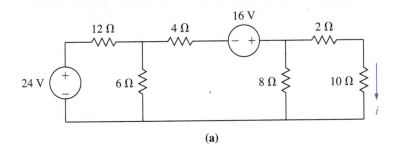

(a)

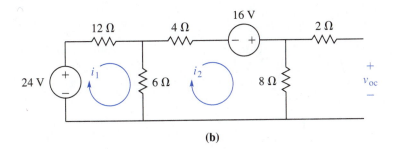

(b)

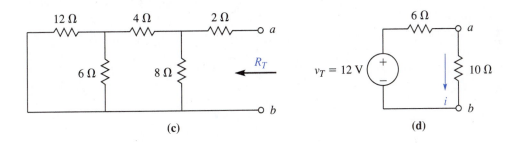

(c)   **(d)**

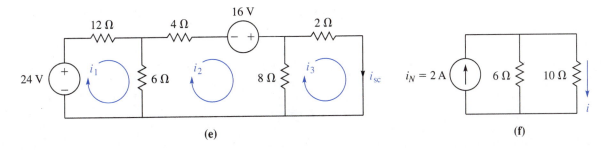

(e)   **(f)**

The value of R_T is obtained from the ladder network shown in Fig. 4.4c with both voltage sources nulled by replacing them with short circuits:

$$R_a = \frac{6(12)}{6+12} = \frac{72}{18} = 4\ \Omega$$

$$R_b = R_a + 4 = 4 + 4 = 8\ \Omega$$

$$R_c = \frac{R_b(8)}{R_b+8} = \frac{8(8)}{8+8} = \frac{64}{16} = 4\ \Omega$$

$$R_T = R_c + 2 = 4 + 2 = 6\ \Omega$$

The Thévenin equivalent with the 10-Ω resistor reconnected across its terminals is shown in Fig. 4.4d. The value of i is

$$i = \frac{v_T}{R_T + 10} = \frac{12}{6+10} = \frac{12}{16} = \frac{3}{4}\ \text{A}$$

Now, for completeness, let a Norton equivalent be established with $R_N = R_T = 6\ \Omega$. The Norton equivalent depends on the short-circuit current, i_{sc}, which is shown in Fig. 4.4e. Here, a three mesh can be used to determine its value. The mesh equations are

$$18i_1 - 6i_2 \qquad\quad = 24$$

$$-6i_1 + 18i_2 - 8i_3 = 16$$

$$-8i_2 + 10i_3 = 0$$

The current $i_{sc} = i_3$ can be determined from an application of Cramer's rule:

$$i_{sc} = i_3 = \frac{\begin{vmatrix} 18 & -6 & 24 \\ -6 & 18 & 16 \\ 0 & -8 & 0 \end{vmatrix}}{\begin{vmatrix} 18 & -6 & 0 \\ -6 & 18 & -8 \\ 0 & -8 & 10 \end{vmatrix}} = \frac{3456}{1728} = 2\ \text{A}$$

The Norton equivalent, with the 10-Ω resistor connected across its terminals, is shown in Fig. 4.4f. By a current division,

$$i = \left(\frac{6}{6+10}\right) i_N = \left(\frac{3}{8}\right)(2) = \frac{3}{4}\ \text{A}$$

which checks the previous result.

Additional confirmation can be obtained from eq. (4.5):

$$R_T = R_N = \frac{v_T}{i_N} = \frac{12}{2} = 6\ \Omega$$

This is essentially a computation of the Thévenin equivalent resistance by the second method, and it agrees with the determination of R_T by the first method. ■

■ **EXAMPLE 4.2**

Use Thévenin's theorem to establish the value of i through the 6-Ω resistor in the network shown in Fig. 4.5a.

Solution In Fig. 4.5a, there are two controlled sources. If the mesh currents are assigned in Fig. 4.5b so that i_1 flows in the mesh at the left and i_2 flows in the mesh in the center, $v_b = 2(i_1 - i_2)$, and the VCVS has strength $4(i_1 - i_2)$ volts. Because $i_a = i_1$, the ICIS may be transformed into an ICVS having a strength $3(4i_1) = 12i_1$ volts. Figure 4.5b shows this version of the controlled sources with an open-circuited pair of terminals and two mesh currents.

The open-circuit voltage across the terminals will be $v_{oc} = v_T = 5i_2$, because no current is carried by the 1-Ω resistor when the 6-Ω resistor is removed from the circuit. The mesh equations are

$$6i_1 \;-\; 2i_2 = 16 - 4(i_1 - i_2)$$

$$-2i_1 + 10i_2 = 4(i_1 - i_2) + 12i_1$$

or

$$10i_1 \;-\; 6i_2 = 16$$

$$-18i_1 + 14i_2 = 0$$

A Cramer's rule solution gives

$$i_2 = \cfrac{\begin{vmatrix} 10 & 16 \\ -18 & 0 \end{vmatrix}}{\begin{vmatrix} 10 & -6 \\ -18 & 14 \end{vmatrix}} = \frac{288}{32} = 9 \text{ A}$$

which means that $v_T = v_{oc} = 5(9) = 45$ V.

Because there are controlled sources, R_T is determined by applying a 1-V test source at the open-circuited terminals after the 16-V independent source is nulled by replacing it with a short circuit. The three mesh of Fig. 4.5c results, and $i_o = -i_3$. The mesh equations are

$$6i_1 \;-\; 2i_2 \qquad\quad = -4(i_1 - i_2)$$

$$-2i_1 + 10i_2 - 5i_3 = 4(i_1 - i_2) + 12i_1$$

$$-5i_2 + 6i_3 = -1$$

or

$$10i_1 \;-\; 6i_2 \qquad\quad = \;\; 0$$

$$-18i_1 + 14i_2 - 5i_3 = \;\; 0$$

$$-5i_2 + 6i_3 = -1$$

FIGURE 4.5   (a) A network with two controlled sources, (b) the network with the 6-Ω resistor
removed and ready for mesh analysis, (c) the network used to establish the Thévenin
equivalent resistance, and (d) the Thévenin equivalent network with the 6-Ω
resistor reconnected

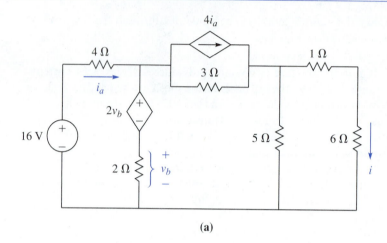

(a)

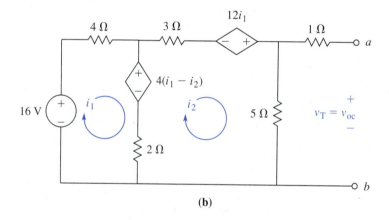

(b)

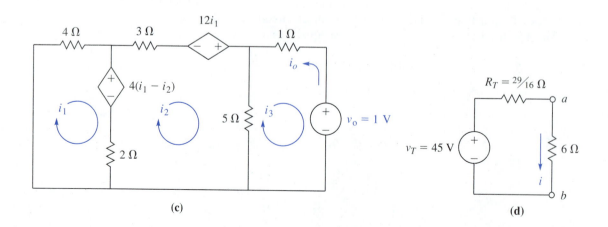

(c)   **(d)**

A Cramer's rule solution for i_3 yields

$$i_3 = \frac{\begin{vmatrix} 10 & -6 & 0 \\ -18 & 14 & 0 \\ 0 & -5 & -1 \end{vmatrix}}{\begin{vmatrix} 10 & -6 & 0 \\ -18 & 14 & -5 \\ 0 & -5 & 6 \end{vmatrix}} = \frac{-32}{-58} = \frac{16}{29} \text{ A}$$

With $i_o = -i_3 = -16/29$ A and $v_o = 1$ V, the value of R_T is the reciprocal of the absolute value of i_o, or

$$R_T = \left| \frac{1}{i_o} \right| = \frac{29}{16} \, \Omega$$

The Thévenin equivalent with the 6-Ω resistor reconnected is shown in Fig. 4.5d. The value of i is

$$i = \frac{45}{\frac{29}{16} + 6} = \frac{45}{\frac{125}{16}} = \frac{144}{25} = 5.760 \text{ A}$$ ∎

EXERCISE 4.2

Use Thévenin's theorem to determine the current i indicated in Fig. 4.6.

Answer $v_T = 14$ V, $R_T = 1.875 \, \Omega$, and $i = 5.6$ A.

Network (Exercise 4.2)

FIGURE 4.6

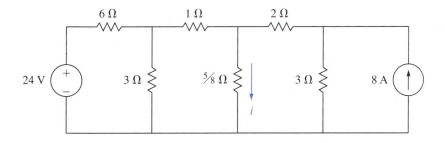

EXERCISE 4.3

Use Thévenin's Theorem to determine the current i indicated in Fig. 4.7.

Answer $v_T = 26$ V, $R_T = \dfrac{130}{34} \, \Omega$, and $i = \dfrac{442}{235}$ A.

FIGURE 4.7   Network (Exercise 4.3)

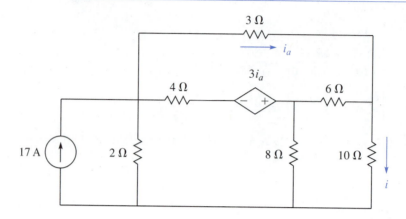

SECTION 4.4   **TELLEGEN'S THEOREM**

Tellegen's theorem, relating branch voltages and currents is as follows:

> **Tellegen's Theorem:** In an arbitrarily lumped network subject to KVL and KCL constraints, with reference directions of the branch currents, and branch voltages associated with the KVL and KCL constraints, the product of all branch currents and branch voltages must equal zero.

Tellegen's theorem may be summarized by the equation

$$\sum_{k=1}^{b} v_k j_k = 0 \tag{4.7}$$

where the lowercase letters v and j represent any branch voltages consistent with KVL and branch currents consistent with KCL, respectively, and where b is the total number of branches. A matrix representation employing the branch current and branch voltage vectors also exists. Because **V** and **J** are column vectors,

$$\sum_{k=1}^{b} v_k j_k = \mathbf{V}^T \mathbf{J} = \mathbf{J}^T \mathbf{V} \tag{4.8}$$

In Fig. 3.17, repeated here as Fig. 4.8, the 63-V source is included with the 5-Ω resistor in branch 5. For the entire branch,

$$j_5 = -9 \text{ A} \quad \text{and} \quad v_5 = 63 - 5(9) = 63 - 45 = 18 \text{ V}$$

where the branch current has been established in a direction in accordance with the orientation recommended in Section 3.2 (Fig. 3.1). For the entire network shown in

A network used to show the validity of Tellegen's theorem

FIGURE 4.8

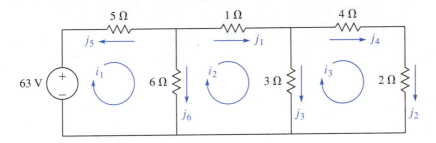

Fig. 4.8, there is a branch current vector **J** and a branch voltage vector **V** as follows:

$$
\mathbf{J} = \begin{bmatrix} 6 \\ 2 \\ 4 \\ 2 \\ -9 \\ 3 \end{bmatrix}
\qquad
\mathbf{V} = \begin{bmatrix} 6 \\ 4 \\ 12 \\ 8 \\ 18 \\ 18 \end{bmatrix}
$$

Then,

$$
\sum_{k=1}^{b} v_k j_k = \mathbf{V}^T \mathbf{J} = \begin{bmatrix} 6 & 4 & 12 & 8 & 18 & 18 \end{bmatrix} \begin{bmatrix} 6 \\ 2 \\ 4 \\ 2 \\ -9 \\ 3 \end{bmatrix} = 0
$$

which is a verification of Tellegen's theorem.

4.4.1 KVL and KCL Constraints

The prerequisite concerning the KVL and KCL constraints in the statement of Tellegen's theorem is of crucial importance and so is the observation that the *v-i* relationship of elements is irrelevant. It is well to consider these constraints at this point through the use of an example, which will further demonstrate the usefulness of Tellegen's theorem.

■ EXAMPLE 4.3

There are ten branches labeled consecutively by numbers and eight nodes labeled by letters within circles in the oriented graph of a particular network shown in Fig. 4.9. In this graph, several branch currents and branch voltages that are known are indicated. Because the type of elements or their values are immaterial to Tellegen, the other branch currents and branch voltages may be evaluated from repeated applications of KCL and KVL. KCL may be used first at the various nodes.

FIGURE 4.9

An oriented graph of a particular network with some known branch currents and branch voltages

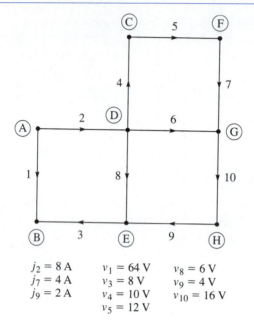

$j_2 = 8$ A $v_1 = 64$ V $v_8 = 6$ V
$j_7 = 4$ A $v_3 = 8$ V $v_9 = 4$ V
$j_9 = 2$ A $v_4 = 10$ V $v_{10} = 16$ V
                         $v_5 = 12$ V

for node a: $j_1 = -j_2 = -8$ A

for node b: $j_3 = -j_1 = -(-8) = 8$ A

for node f: $j_5 = j_7 = 4$ A

for node c: $j_4 = j_5 = 4$ A

for node h: $j_{10} = j_9 = 2$ A

for node g: $j_6 = j_{10} - j_7 = 2 - 4 = -2$ A

for node d: $j_8 = j_2 - j_4 - j_6 = 8 - 4 - (-2) = 6$ A

A check at node e shows that

$$j_8 = j_3 - j_9 = 8 - 2 = 6 \text{ A}$$

which is a verification.

KVL may be employed in several of the observed meshes. For the mesh composed of branches 6, 10, 9, and 8,

$$v_6 + v_{10} + v_9 - v_8 = 0$$

$$v_6 = v_8 - v_9 - v_{10} = 6 - 4 - 16 = -14 \text{ V}$$

For the mesh composed of branches 1, 2, 8, and 3,

$$v_2 + v_8 + v_3 - v_1 = 0$$

$$v_2 = v_1 - v_3 - v_8 = 64 - 8 - 6 = 50 \text{ V}$$

And for the mesh composed of branches 4, 5, 7, and 6,

$$v_4 + v_5 + v_7 - v_6 = 0$$

$$v_7 = v_6 - v_4 - v_5 = -14 - 10 - 12 = -36 \text{ V}$$

The foregoing values of branch currents and branch voltages may be placed into branch current and branch voltage vectors along with the known values listed in Fig. 4.9:

$$\mathbf{V} = \begin{bmatrix} 64 \\ 50 \\ 8 \\ 10 \\ 12 \\ -14 \\ -36 \\ 6 \\ 4 \\ 16 \end{bmatrix} \qquad \mathbf{J} = \begin{bmatrix} -8 \\ 8 \\ 8 \\ 4 \\ 4 \\ -2 \\ 4 \\ 6 \\ 2 \\ 2 \end{bmatrix}$$

Tellegen's theorem then yields

$$\sum_{k=1}^{b} v_k j_k = \mathbf{J}^T \mathbf{V} = \begin{bmatrix} -8 & 8 & 8 & 4 & 4 & -2 & 4 & 6 & 2 & 2 \end{bmatrix} \begin{bmatrix} 64 \\ 50 \\ 8 \\ 10 \\ 12 \\ -14 \\ -36 \\ 6 \\ 4 \\ 16 \end{bmatrix} = 0$$

which the reader may wish to verify.   ■

4.4.2 A Caution Regarding the Use of Superposition in Power Calculations

Superposition is a principle that is based upon linearity. Because power may always be expressed in terms of a product of voltage and current, it is a quadratic quantity that is not linear. Although branch currents and branch voltages may be obtained by using the superposition principle, the addition of powers that are drawn from individual sources is an incorrect procedure.

Look at Fig. 4.10a, which shows a simple network containing two voltage sources and two meshes. The mesh equations are

$$6i_1 - 4i_2 = 8$$

$$-4i_1 + 10i_2 = -10$$

FIGURE 4.10   (a) A network containing two voltage sources and two meshes, (b) the network with the 10-V source removed, and (c) the network with the 8-V source removed

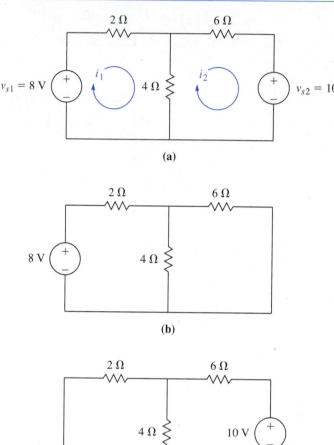

(a)

(b)

(c)

These may be written in matrix form as

$$\mathbf{RI} = \mathbf{V_s} = \begin{bmatrix} 6 & -4 \\ -4 & 10 \end{bmatrix} \begin{bmatrix} i_1 \\ i_2 \end{bmatrix} = \begin{bmatrix} 8 \\ -10 \end{bmatrix}$$

and from an application of the matrix inverse,

$$\mathbf{I} = \begin{bmatrix} i_1 \\ i_2 \end{bmatrix} = \frac{1}{44} \begin{bmatrix} 10 & 4 \\ 4 & 6 \end{bmatrix} \begin{bmatrix} 8 \\ -10 \end{bmatrix} = \begin{bmatrix} \frac{20}{22} \\ -\frac{14}{22} \end{bmatrix} \text{A}$$

which the reader may wish to verify by Cramer's rule.
    The power delivered to the network will be

$$\sum P_s = v_{s1}(i_1) - v_{s2}(i_2) = 8\left(\frac{20}{22}\right) - 10\left(-\frac{14}{22}\right) = \frac{300}{22} = \frac{150}{11} \text{ W}$$

Now, repeat the process by using superposition. For Fig. 4.10b, which considers only the 8-V source (the 10-V source is nulled by replacing it with a short circuit),

$$R_{eq} = \frac{4(6)}{4+6} + 2 = \frac{12}{5} + 2 = \frac{22}{5}\,\Omega$$

The power drawn from the 8-V source is

$$p_8 = \frac{v^2}{R_{eq}} = \frac{(8)^2}{\frac{22}{5}} = \frac{64(5)}{22} = \frac{160}{11}\,W$$

For Fig. 4.10c, which considers only the 10-V source (the 8-V source has been nulled by replacing it with a short circuit),

$$R_{eq} = \frac{2(4)}{2+4} + 6 = \frac{4}{3} + 6 = \frac{22}{3}\,\Omega$$

The power drawn from the 10-V source is

$$p_{10} = \frac{v^2}{R_{eq}} = \frac{(10)^2}{\frac{22}{3}} = \frac{100(3)}{22} = \frac{150}{11}\,W$$

The sum

$$\frac{160}{11} + \frac{150}{11} = \frac{310}{11}\,W$$

is most certainly not equal to the correct value of 150/11 W. This demonstrates that individual power values obtained by an application of the superposition principle may not be added to obtain the total power drawn by a network.

THE RECIPROCITY PRINCIPLE SECTION 4.5

A general set of mesh equations in n unknown mesh currents for any resistive, linear, passive, and bilateral network containing n meshes may be represented by

$$\mathbf{RI} = \mathbf{V_s} \tag{4.9}$$

Here, the matrix $\mathbf{R}$ is the $n \times n$ resistive coefficient matrix, and it is *symmetrical*. The $n \times 1$ column vector $\mathbf{I}$ can be obtained from a matrix inversion:[1]

$$
\begin{bmatrix} i_1 \\ i_2 \\ i_3 \\ \vdots \\ i_N \end{bmatrix}
= \frac{1}{\det \mathbf{R}}
\begin{bmatrix}
R_{11} & R_{21} & R_{31} & \cdots \\
R_{12} & R_{22} & R_{32} & \cdots \\
R_{13} & R_{23} & R_{33} & \cdots \\
\cdots & \cdots & \cdots & \cdots \\
R_{1n} & R_{2n} & R_{3n} & \cdots
\end{bmatrix}
\begin{bmatrix} v_{s1} \\ v_{s2} \\ v_{s3} \\ \vdots \\ v_{sn} \end{bmatrix}
\tag{4.10}
$$

[1] See Appendix B.

The elements of the mesh current vector as well as the elements of the $n \times 1$ voltage source vector are designated with lowercase letters, which is customary. The capital R's in the $n \times n$ matrix represent cofactors of the $n \times n$ coefficient matrix.

Consider the case where all voltage sources are equal to zero except v_{s2}. Then, i_1 can be determined quite easily by expanding eq. (4.10):

$$i_1 = \frac{R_{21}}{\det \mathbf{R}}\, v_{s2}$$

The same procedure applies if all voltage sources are zero except v_{s1}. In this case, i_2 will be

$$i_2 = \frac{R_{12}}{\det \mathbf{R}}\, v_{s1}$$

The two cofactors, R_{12} and R_{21}, are equal, because symmetrical matrices possess symmetrical cofactor matrices. Thus, with $R_{12} = R_{21}$, i_1 will equal i_2 if v_{s1} is equal to v_{s2}. This demonstrates the important *principle of reciprocity*, which is a useful general property that applies to all linear, passive, and bilateral networks. It applies only to cases where current and voltage are involved and points out that the ratio of a single excitation applied at one point to an observed response at another is invariant with respect to an interchange of the points of excitation and observation. The reciprocity principle also applies if the excitation is a current and the observed response is a voltage.

Confirmation of the foregoing can be obtained from Tellegen's theorem. Figure 4.11 represents a network containing n elements and shows two pairs of external voltages and currents. With no specific values of external voltages and currents, all of the voltages across and the currents through the n elements are related by Ohm's law:

$$v_k = R_k i_k \qquad k = 1, 2, 3, \ldots \tag{4.11}$$

Now, consider a different set of currents and voltages in the network that result from different external connections. Denote these by v_k' and i_k', and observe that they satisfy

$$v_k' = R_k i_k' \qquad k = 1, 2, 3, \ldots \tag{4.12}$$

FIGURE 4.11   A network with two excitations

Tellegen's theorem states that

$$v_1' i_1 + v_2' i_2 = \sum_{k=1}^{n} v_k' i_k \tag{4.13}$$

$$v_1 i_1' + v_2 i_2' = \sum_{k=1}^{n} v_k i_k' \tag{4.14}$$

Note that Tellegen's theorem does not require the primed and unprimed variables to exist *simultaneously* in the network.

Now, by eq. (4.12),

$$\sum_{k=1}^{n} v_k' i_k = \sum_{k=1}^{n} R_k i_k' i_k$$

and by eq. (4.11),

$$\sum_{k=1}^{n} v_k i_k' = \sum_{k=1}^{n} R_k i_k i_k'$$

so that

$$\sum_{k=1}^{n} v_k' i_k = \sum_{k=1}^{n} v_k i_k'$$

and

$$v_1' i_1 + v_2' i_2 = v_1 i_1' + v_2 i_2' \tag{4.15}$$

which is the general form of the reciprocity principle.

Reciprocity is a property of a linear, passive, bilateral network and is summarized by:

> In any linear, passive, bilateral network, if the single voltage source, v_{sp} in branch p produces the current response i_q in branch q, then the removal of v_{sp} from branch p and its placement in branch q will produce the current response in branch p.

■ **EXAMPLE 4.4**

Show that the positions of v_s and i in Fig. 4.12a may be interchanged as in Fig. 4.12b without changing the ratio i/v_s.

Solution   In Fig. 4.12a,

$$i_a = \frac{v_s}{3 + [6(12)/(6 + 12)]} = \frac{v_s}{3 + 4} = \frac{v_s}{7}$$

and by current division,

$$i = \frac{6}{6 + 12} i_a = \frac{1}{3} \cdot \frac{v_s}{7} = \frac{v_s}{21}$$

so that $i/v_s = 1/21$.

FIGURE 4.12   Networks used to illustrate the reciprocity principle

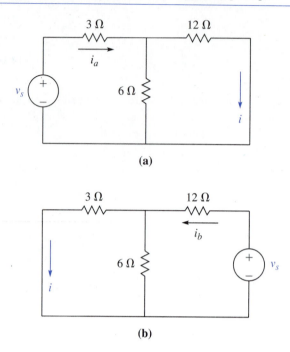

(a)

(b)

In Fig. 4.12b,

$$i_b = \frac{v_s}{12 + [6(3)/(6 + 3)]} = \frac{v_s}{12 + 2} = \frac{v_s}{14}$$

and by current division,

$$i = \frac{6}{3 + 6} i_b = \frac{2}{3} \cdot \frac{v_s}{14} = \frac{v_s}{21}$$

so that $i/v_s = 1/21$. The network is reciprocal. ∎

SECTION 4.6

THE MAXIMUM POWER TRANSFER THEOREM

A question is frequently asked with regard to any network: "What load should be connected to a particular network so that the power transferred to the load is a maximum?" In the case of a resistive network, which possesses a Thévenin equivalent and a resistive load designated as R_o, the network with the load in series can be represented as shown in Fig. 4.13. Observe that the current flowing through the load

A resistive load connected across the Thévenin equivalent of a network

FIGURE 4.13

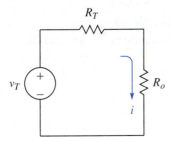

is given by

$$i = \frac{v_T}{R_T + R_o}$$

and the power drawn by the load is

$$p_o = i^2 R_o = \left(\frac{v_T}{R_T + R_o}\right)^2 R_o = v_T^2 R_o (R_T + R_o)^{-2}$$

To find the load R_o that draws maximum power, one employs the standard technique in any maximum or minimum problem: Determine where the derivative $\partial p_o / \partial R_o$ vanishes. For fixed v_T and R_o, the partial derivatives are not required. Thus,

$$\frac{dp_o}{dR_o} = v_T^2 (R_T + R_o)^{-2} - 2v_T^2 R_o (R_T + R_o)^{-3} = 0$$

and a little algebraic adjustment provides

$$R_T + R_o = 2R_o$$

which can only be satisfied when the load resistance R_o matches the Thévenin equivalent resistance R_T.

The maximum power transfer theorem for a resistive network and a resistive load is as follows:

> **Maximum Power Transfer Theorem:** For a resistive network and a resistive load, maximum power will be transferred to the load when the load is equal to the Thévenin equivalent resistance of the network.

The resistive case is, of course, the simplest case, and there are additional implications when one is dealing with alternating-current circuits containing inductors and capacitors. Thus, the subject of maximum power transfer must be revisited when alternating current is considered. However, the simple resistive case shows that the maximum power that can be drawn by a load can never exceed 50% of the power delivered by a source.

■ **EXAMPLE 4.5**

For the network shown in Fig. 4.14, determine the following:

a. Its Thévenin equivalent with respect to terminals a and b
b. The load to be connected to terminals a and b so that the load draws maximum power
c. The power delivered by the source and the power drawn by the load

Solution

a. The open-circuit voltage across terminals a–b in Fig. 4.14a will be, by superposition, the sum of the open-circuit voltages due to the voltage source acting alone and the current source acting alone:

$$v_T = v_{oc} = v_{oc,v} + v_{oc,i}$$

For the voltage source configuration in Fig. 4.14b, a simple voltage division yields

$$v_{oc,v} = \left(\frac{2+4}{6+2+4}\right)(12) = \left(\frac{1}{2}\right)(12) = 6 \text{ V}$$

For the current source configuration in Fig. 4.14c, a current division gives

$$i_a = \left(\frac{4}{6+2+4}\right)(3) = \left(\frac{1}{3}\right)(3) = 1 \text{ A}$$

Then, by Ohm's law,

$$v_{oc,i} = 6i_a = 6(1) = 6 \text{ V}$$

The Thévenin equivalent voltage is then

$$v_T = v_{oc,v} + v_{oc,i} = 6 + 6 = 12 \text{ V}$$

The Thévenin equivalent resistance is easily determined from Fig. 4.14d:

$$R_T = \frac{6(2+4)}{6+2+4} = \frac{36}{12} = 3 \ \Omega$$

and the Thévenin equivalent for the network is shown in Fig. 4.14e.
b. The load required for a maximum power transfer must match the Thévenin equivalent resistance:

$$R_o = R_T = 3 \ \Omega$$

c. The total resistance in the network is $3 + 3 = 6 \ \Omega$, so that the current drawn from the source is

$$i = \frac{V_T}{6} = \frac{12}{6} = 2 \text{ A}$$

(a) A simple network with two sources, (b) the network with the current source removed, (c) the network with the voltage source removed, (d) the network with both sources removed, and (e) the Thévenin equivalent of the network

FIGURE 4.14

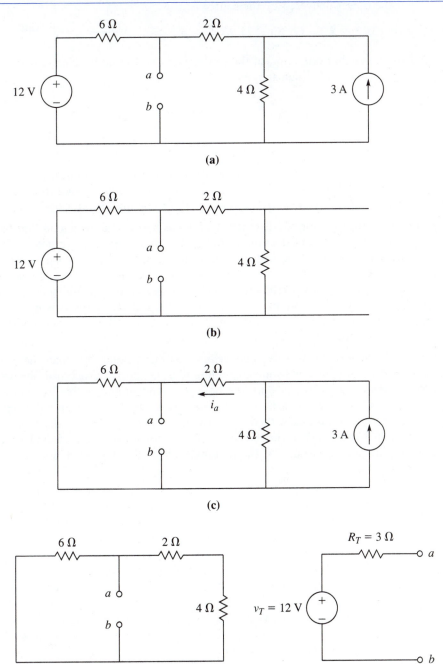

The power dissipated by the load is

$$p_o = i^2 R_o = (2)^2(3) = 12 \text{ W}$$

and the power delivered by the source is

$$p_s = i^2(R_T + R_o) = (2)^2(6) = 24 \text{ W}$$

This shows that only 50% of the power is taken by the load under the most favorable of circumstances. ∎

CHAPTER 4

SUMMARY

- The superposition theorem states that if cause and effect are linearly related, the total effect due to several causes acting simultaneously is equal to the sum of the individual effects due to each of the causes acting one at a time.

- Thévenin's theorem for resistive networks states that insofar as a load that has no magnetic or controlled-source coupling to a one-port is concerned, a network containing linear elements and both independent and controlled sources may be replaced by an ideal voltage source of strength v_T and an equivalent resistance R_T in series with the source. The value of v_T is the open-circuit voltage appearing across the terminals of the network, and R_T is the driving-point resistance at the terminals of the network, obtained with all independent sources set equal to zero.

- Norton's theorem for resistive networks states that insofar as a load that has no magnetic or controlled-source coupling to a one-port is concerned, the network containing linear elements and both independent and controlled sources may be replaced by an ideal current source of strength i_N and an equivalent resistance R_N in parallel with the source. The value of i_N is the short-circuit current that results when the terminals of the network are shorted, and R_N is the driving-point resistance at the terminals when all independent sources are set equal to zero.

- Tellegen's theorem states that, in an arbitrarily lumped network subject to KVL and KCL constraints, with reference directions of the branch currents and branch voltages associated with the KVL and KCL constraints the product of all branch currents and branch voltages must equal zero.

- The reciprocity principle states that in any linear, passive, bilateral network, if the single voltage source v_{sp} in branch p produces the current response i_q in branch q, then the removal of v_{sp} from branch p and its placement in branch q will produce the current response in branch p.

- Maximum power for a resistive network will be transferred from a source network to a load when the magnitude of the load is equal to the Thévenin equivalent resistance of the source network.

Additional Readings

Blackwell, W.A., and L.L. Grigsby. *Introductory Network Theory*. Boston: PWS Engineering, 1985, pp. 71, 107–116, 128–140.

Bobrow, L.S. *Elementary Linear Circuit Analysis*. 2d ed. New York: Holt, Rinehart and Winston, 1987, pp. 121–149, 576.

Del Toro, V. *Engineering Circuits.* Englewood Cliffs, N.J.: Prentice-Hall, 1987, pp. 82–85, 101–109, 166–172.

Dorf, R.C. *Introduction to Electric Circuits.* New York: Wiley, 1989, pp. 29–32, 129–149.

Hayt, W.H., Jr., and J.E. Kemmerly. *Engineering Circuit Analysis.* 4th ed. New York: McGraw-Hill, 1986, pp. 74–91, 468, 469.

Irwin, J.D. *Basic Engineering Circuit Analysis.* 3d ed. New York: Macmillan, 1989, pp. 132–134, 180–191, 197–217.

Johnson, D.E., J.L. Hilburn, and J.R. Johnson. *Basic Electric Circuit Analysis.* 4th ed. Englewood Cliffs, N.J.: Prentice-Hall, 1989, pp. 34, 44, 45, 121–134, 140–142, 377.

Karni, S. *Applied Circuit Analysis.* New York: Wiley, 1988, pp. 42–44, 101–108, 114–124.

Madhu, S. *Linear Circuit Analysis.* Englewood Cliffs, N.J.: Prentice-Hall, 1988, pp. 48, 244–256.

Nilsson, J.W. *Electric Circuits.* 3d ed. Reading, Mass.: Addison-Wesley, 1990, pp. 96–112, 766, 767.

Paul, C.R. *Analysis of Linear Circuits.* New York: McGraw-Hill, 1989, pp. 123–145, 196–210.

CHAPTER 4   **PROBLEMS**

Section 4.2

4.1 Use superposition to find the current i in the network of Fig. P4.1.

Figure P4.1

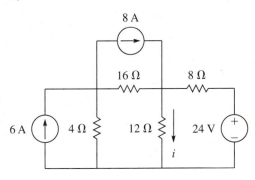

4.3 Use superposition to find the voltage v in the network of Fig. P4.3.

Figure P4.3

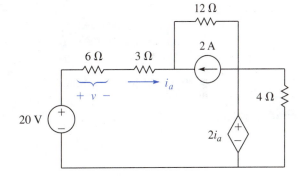

4.2 Use superposition to find the current i in the network of Fig. P4.2.

Figure P4.2

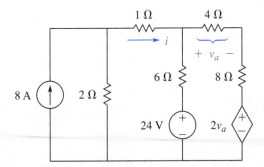

4.4 Use superposition to find the voltage v in the network of Fig. P4.4.

Figure P4.4

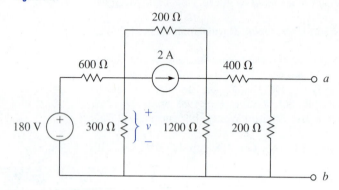

4.5 Use superposition to find the current i in the network of Fig. P4.5.

Figure P4.5

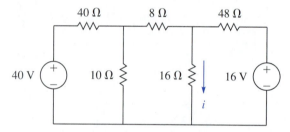

4.6 Use superposition to find the current i in the network of Fig. P4.6.

Figure P4.6

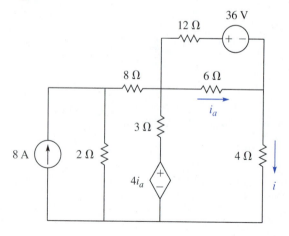

4.7 Use superposition to determine the value of v_s that will cause the current indicated by i_a in Fig.P4.7 to equal 3.00 A.

Figure P4.7

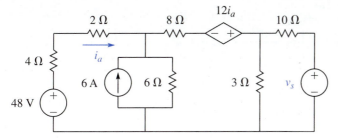

Section 4.3

4.8 Determine the Thévenin equivalent at terminals a–b for the network in Fig. P4.8.

Figure P4.8

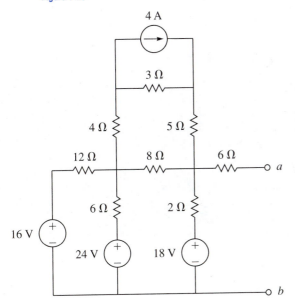

4.9 Use Thévenin's theorem to find the current i in the network of Fig. P4.9.

Figure P4.9

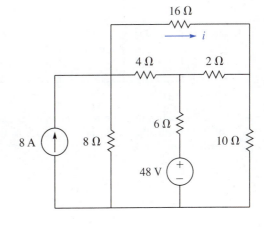

4.10 Use Thévenin's theorem to find the current i in the network of Fig. P4.10.

Figure P4.10

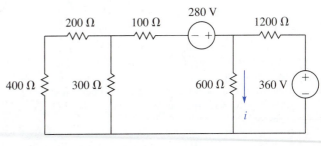

4.11 Determine the Thévenin equivalent at terminals
a–b for the network in Fig. P4.11.

Figure P4.11

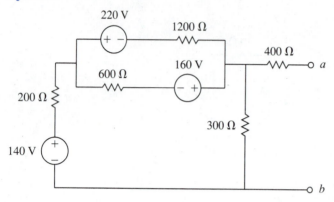

4.12 Use Thévenin's theorem to find the voltage v in
the network of Fig. P4.12.

Figure P4.12

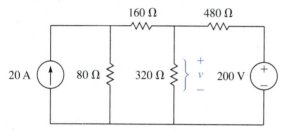

4.13 Use Thévenin's theorem to determine the value
of R that will allow a current $i = 1.00$ A to flow
through the 2 Ω resistor in Fig. P4.13.

Figure P4.13

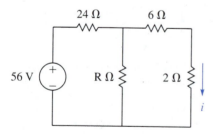

4.14 Find the current through the 3 Ω resistor in
Fig. P4.14. Use superposition to find the Thévenin equivalent voltage, being careful to take
cognizance of its polarity. Then use Thévenin's
theorem to find the current through the 3 Ω
resistor.

Figure P4.14

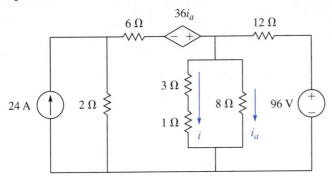

Section 4.4

4.15 Find all of the branch voltages and branch currents in the network of Fig. P4.15, and verify the result by using Tellegen's theorem.

Figure P4.15

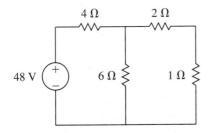

4.16 Find all of the branch voltages and branch currents in the network of Fig. P4.5, and verify the result by using Tellegen's theorem.

4.17 Find all of the branch voltages and branch currents in the network of Fig. P4.16, and verify the result by using Tellegen's theorem.

Figure P4.16

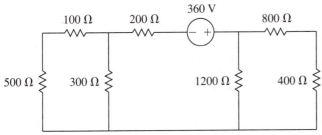

4.18 Find all of the branch voltages and branch currents in the network of Fig. P4.12, and verify the result by using Tellegen's theorem.

Section 4.5

4.19 Use the reciprocity principle to demonstrate that the currents designated by i in Figs. P4.17a and P4.17b are identical.

Figure P4.17

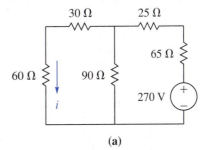

(a)

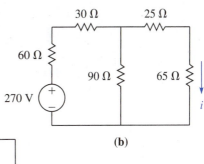

(b)

4.20 Use the reciprocity principle to demonstrate that the voltages designated by v in Figs. P4.18a and P4.18b are identical.

Figure P4.18

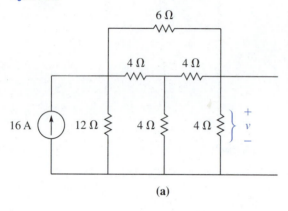

(a)

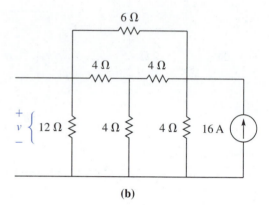

(b)

4.21 Use the reciprocity principle to demonstrate that the voltages designated by v in Figs. P4.19a and P4.19b are identical.

Figure P4.19

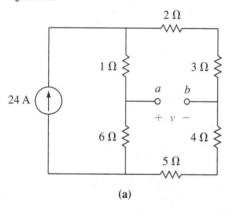

(a)

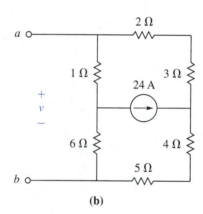

(b)

Section 4.6

4.22 What load should be connected across terminals a–b in Fig. P4.4 in order for the load to absorb maximum power? How much power is delivered to the load?

4.23 What load should be connected across terminals a–b in Fig. P4.8 in order for the load to absorb maximum power? How much power is delivered to the load?

4.24 What load should be connected across terminals a–b in Fig. P4.11 in order for the load to absorb maximum power? How much power is delivered to the load?

4.25 What power is dissipated in the 320 Ω resistor in Fig. P4.12? In an attempt to draw maximum power from the sources, the 320 Ω resistor is to be changed. What is the new value of R?

SIGNALS AND SINGULARITY FUNCTIONS

5

OBJECTIVES

The objectives of this chapter are to:

- Display and discuss three of the most common *singularity* functions that are employed as signals in electric networks. These are the unit step, the unit ramp, and the unit impulse functions.

- Show how to represent waveforms in terms of the singularity functions.

- Learn how to shift waveforms (translate waveforms in time), using the unit step function.

- Study the exponential and sinusoidal functions.

- Develop expressions for the average and effective values of a periodic function.

INTRODUCTION

A signal may be considered as a physical variable that is a function of time. In electric networks, the variables of interest are usually the across and the through variables.

Although these are most often the across variable of voltage and the through variable of current, sometimes interest is directed toward the integrated through variable of charge or the integrated across variable of flux linkage. All of these may be considered as signals.

Input signals may be provided by independent voltage and current sources, and output signals may be any voltage or current produced by the network.

The signals treated in this book are deterministic, continuous time, periodic, and aperiodic, which will be defined shortly. No attempt will be made to consider discrete time signals such as those observed in a *sampled data system*. In addition, *random signals* will not be considered in this book.

A *deterministic signal* is a signal that is completely specified as a function of time. Four examples of deterministic voltage signals are the exponential signal,

$$v(t) = Ve^{-at} \qquad (t \geq 0)$$

the sinusoidal signal,

$$v(t) = V \sin(\omega t + \theta) \qquad \text{(all } t\text{)}$$

the pulse of height or strength V of duration 1 s

$$v(t) = V\Pi(t) = \begin{cases} V, & 0 \le t \le 1 \\ 0, & \text{otherwise} \end{cases} \tag{5.1}$$

and a signal

$$v(t) = \frac{Vt^2}{3 + t}$$

All of these signals are *continuous time signals*. This applies to the pulse, which, although a mathematically discontinuous function, is a function of a continuous time variable.

A signal designated as $f(t)$ is *periodic* if and only if

$$f(t) = f(t + T) \tag{5.2}$$

for all t where the smallest nonzero value of T that causes this equation to be valid is designated as the *fundamental period*, or merely the *period*. Of the foregoing signals, only the sinusoid is periodic. Any signal that does not satisfy eq. (5.2) is termed an *aperiodic* signal.

SECTION 5.2   **THE SINGULARITY FUNCTIONS**

Singularity functions are functions that are discontinuous or have discontinuous derivatives. They are used to account for the action of switches within a network and for analytical representations of shifted or delayed functions. The three most important singularity functions (for the purposes required by the study of electric networks) are the unit step, the unit ramp, and the unit impulse functions. These functions will now be discussed in detail.

5.2.1 The Unit Step Function

The unit step function, designated as $u(t)$, has zero magnitude for all values of time less than $t = 0$ and unit magnitude for all time greater than $t = 0$. It exhibits a discontinuity at exactly $t = 0$, and it is written as

$$u(t) = \begin{cases} 0, & t < 0 \\ 1, & t > 0 \end{cases} \tag{5.3}$$

The unit step function is displayed in Fig. 5.1a.

Notice here that although $u(t)$ is undefined at $t = 0$, its value is known for all points in time arbitrarily close to $t = 0$. For this reason, one may frequently see the use of $t = 0^-$ and $t = 0^+$ in $u(t)$ and the use of $u(t = 0^-)$ and $u(t = 0^+)$ is consistent with the definition of the unit step given by eq. (5.3).

(a) The unit step function, (b) the delayed unit step function, (c) a dc battery and a network with a switch, and (d) the use of the unit step function to indicate that the switch closes at $t = 0$

Observe in (d), which applies when $t > 0$, that the switch is not shown because the switching instant ($t = 0$) is contained in the functional relationship for $u(t)$.

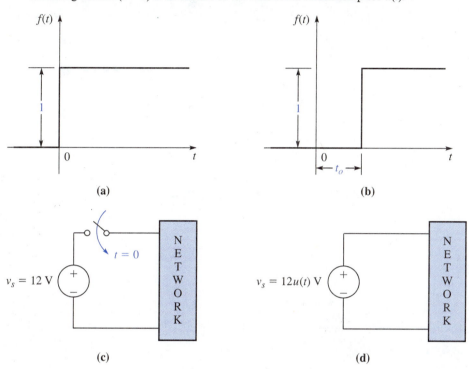

(a)

(b)

(c)

(d)

If it is desired to apply the unit step function at some time other than $t = 0$, such as $t = t_o$, the delayed or shifted unit step function is used. It is defined as

$$u(t - t_o) = \begin{cases} 0, & t < t_o \\ 1, & t > t_o \end{cases} \tag{5.4}$$

and it is displayed in Fig. 5.1b.

The analytical neatness in the use of the unit step or delayed unit step function is clearly evident. Figure 5.1c shows a network, a switch, and a 12-V battery. The switch closes at time $t = 0$ (Fig. 5.1d), and the battery voltage may be written as $12u(t)$ volts. The time of switching has been *built into* the designation of the voltage source. In addition, in $12u(t)$, the 12 carries the units of volts and is termed the strength of the unit step function; the $u(t)$ is dimensionless.

5.2.2 The Unit Ramp Function

The unit ramp function is the integral of the unit step function and is defined by

$$r(t) = \int_{-\infty}^{t} u(t) \, dt = \begin{cases} 0, & t < 0 \\ t, & t > 0 \end{cases} \tag{5.5}$$

and is shown in Fig. 5.2a.

FIGURE 5.2   (a) The unit ramp, (b) the delayed unit ramp with delay t_o, and (c) the ramp of strength 2

The strength requires appropriate units.

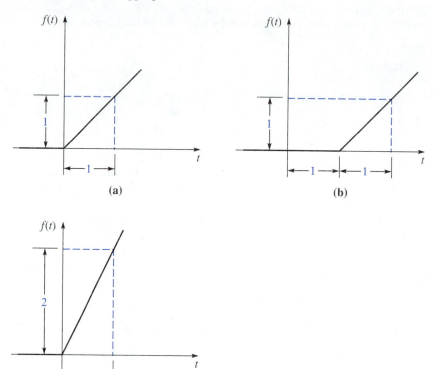

(a)

(b)

(c)

The delayed unit ramp is defined by

$$r(t - t_o) = \begin{cases} 0, & t < t_o \\ t - t_o, & t > t_o \end{cases}$$

(5.6)

and is shown in Fig. 5.2b.

The unit ramp has the dimensions of time and has a slope or strength of unity. Figure 5.2c displays a voltage ramp, $v(t) = 2r(t)$, which has a slope or strength of magnitude 2. In this case, the ramp carries the dimension of volts so that the unit ramp, with the associated strength, carries the dimensions of volts per unit of time.

5.2.3 The Unit Impulse Function

The unit impulse, designated as $\delta(t)$, is a most important singularity function and is the derivative of the unit step function. Figure 5.1a shows the unit step function, and it is evident that the derivative of the unit step function is always zero except at the exact instant where $t = 0$. Indeed, it can be stated that the derivative of the unit step function always has a magnitude equal to zero except for the fleeting infinitesimal instant between $t = 0^-$ and $t = 0^+$.

FIGURE 5.3

(a) An approximation to the unit step function on an expanded time scale, (b) the derivative of the unit step, and (c) the derivative of the unit step for ever-decreasing values of Δ

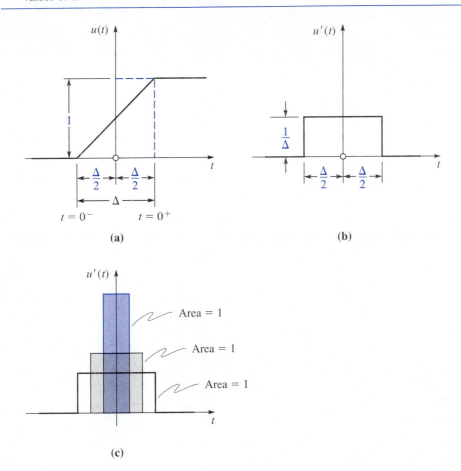

(a)

(b)

(c)

With this in mind, let the unit step function be approximated and plotted on expanded time scale, as shown in Fig. 5.3a. The time interval between $t = 0^-$ and $t = 0^+$ is designated as Δ. The derivative of this approximation is a constant and has a magnitude of $1/\Delta$, as shown in Fig. 5.3b. Here, it may be observed that the area under the derivative curve is always $(1/\Delta)\,\Delta = 1$, regardless of the value of Δ. This is shown in Fig. 5.3c for ever-decreasing values of Δ.

In the limit as $\Delta \to 0$, the area under the derivative curve in Fig. 5.3c remains fixed at unity, but the magnitude of the derivative becomes unbounded. For this reason, the unit impulse function, which can be visualized as a spike of very, very large magnitude that is applied or occurs over an infinitesimal period of time, is defined on the basis of an integral:

$$\int_a^b \delta(t)\, dt = 1 \qquad (a \le 0 \le b) \tag{5.7}$$

The unit impulse function may be considered as an applied or resulting sudden shock, depending on whether it arises as a signal or as a response to a signal. It is

FIGURE 5.4   (a) The unit impulse function and (b) the delayed unit impulse function

Here, the delay is t_o time units.

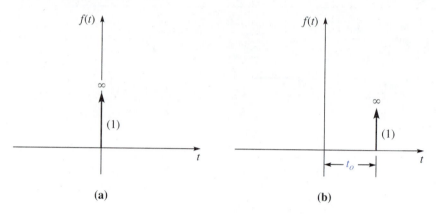

(a)   (b)

a very useful function, as will be seen as this study progresses, and it is frequently referred to as the *Dirac delta function*, which has a two-part definition:

$$\delta(t) = \begin{cases} 0, & t < 0 \\ 0, & t > 0 \end{cases} \qquad\qquad (5.8a)$$

but with

$$\int_{0-}^{0+} \delta(t)\, dt = 1 \qquad\qquad (5.8b)$$

The unit impulse may, of course, be delayed in time and applied at some time t_o ($t_o \neq 0$), and in this case, it is defined by

$$\int_a^b \delta(t - t_o)\, dt = 1 \qquad (a \leq t_o \leq b) \qquad\qquad (5.9)$$

Symbols for unit impulses applied at $t = 0$ and $t = t_o$ are shown in Fig. 5.4. Observe here that the use of the numeral (1) enclosed in parentheses designates an area and not a magnitude.

5.2.4 A Dilemma That Is Easily Resolved

Figure 5.5a shows a voltage source v_s, a switch, and a capacitor connected in series. When the switch closes instantaneously, as indicated in Fig. 5.5b, KVL dictates that the source voltage must appear across the capacitor, $v_C = v_s$. A dilemma arises because the continuity of stored energy principle says, *The voltage across a capacitor may not change instantaneously.* Apparently, KVL and the continuity of stored energy are in direct contradiction.

The dilemma is easily resolved when the source voltage is written, as indicated in Fig. 5.5c, in terms of the unit step, $v_s = 12u(t)$. Then,

$$i = C\frac{dv}{dt} = C\frac{d}{dt}\left[12u(t)\right] = 12C\delta(t)$$

FIGURE 5.5

(a) A battery, a switch, and a capacitor in series, (b) the switch closed
instantaneously, (c) an impulse of current flow, and (d) a network

A network such as the simple *RL* series network in (d) will show a spark across the
switch when it is opened instantaneously.

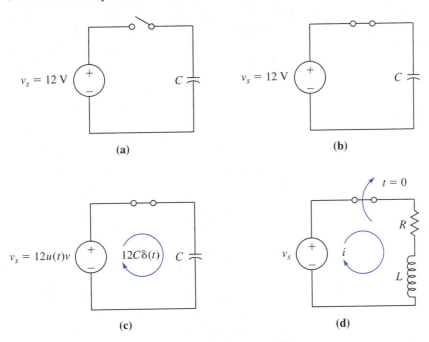

(a) (b) (c) (d)

and when the switch is instantaneously closed at $t = 0^-$, an impulse or spike of current flows to the capacitor. At $t = 0^+$, the voltage across the capacitor, due to the
accumulation of charge from the current impulse acting between $t = 0^-$ and $t = 0^+$,
is $v_C = v_s = 12u(t)$. A little reflection will show that this does not violate the continuity
of stored energy principle because the impulse is defined as occurring in the limit as
$\Delta \to 0$ (Fig. 5.3c).

Thus, the dilemma is resolved from an analytical point of view. From a practical
point of view, opening the switch in the simple network in Fig. 5.5d will result in a
spark at the switch, because the continuity of stored energy principle also says that
currents through inductors may not change instantaneously. This gives the assurance
that what has been discussed analytically can be realized in practice.

THE USE OF RAMPS AND STEPS TO REPRESENT A WAVEFORM   SECTION 5.3

An excellent analytical technique is now available for the representation of some
complex waveforms consisting of ramps and steps of various strengths. Figure 5.6a
displays a piecewise linear waveform, which may be considered as an input voltage
signal to a particular network. Notice that nothing but inclined or horizontal straight
lines are evident. At $t = 0$, the signal begins with a ramp of slope $2/2 = 1$, so that
the signal begins with $r(t)$.

At $t = 2$ s, a unit ramp must be subtracted to level off the effect that began at
$t = 0$. In addition, a negative unit step must be provided to drop the magnitude of

FIGURE 5.6   (a) A voltage signal and (b) through (f) the superposition of step and ramp functions
of various strengths that compose the signal

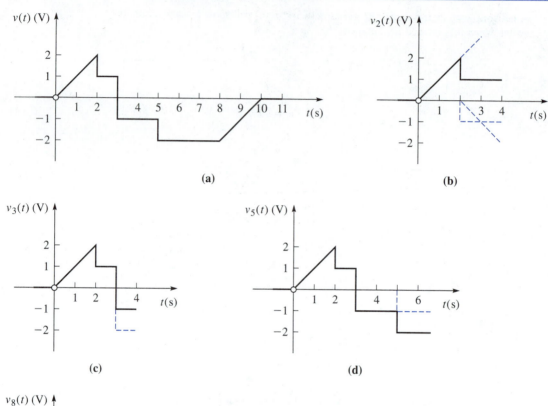

(a)   (b)

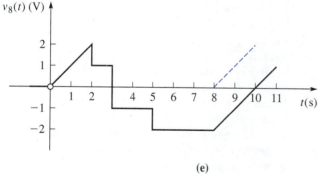

(c)   (d)

(e)

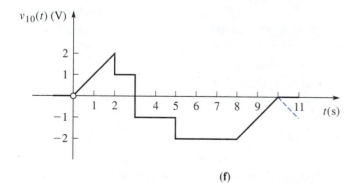

(f)

the signal from 2 to 1 V. Thus at $t = 2$ s, three components are evident: a unit ramp that began at $t = 0$, a negative unit ramp that begins at $t = 2$ s, and a negative unit step that also begins at $t = 2$ s. If a subscript is utilized to mark the beginning of a time interval, this superposition of functions may be written as

$$v_2(t) = r(t) - r(t - 2) - u(t - 2) \text{ volts}$$

and is shown in Fig. 5.6b with the individual components and with $v_2(t)$ shown as a bold, solid line.

If at $t = 3$ s, another negative unit step is provided, this time with a strength of -2 V, the situation is as shown in Fig. 5.6c, with the cumulative effect designated as $v_3(t)$, shown as a solid line:

$$v_3(t) = r(t) - r(t - 2) - u(t - 2) - 2u(t - 3) \text{ volts}$$

This holds over the period that starts at $t = 3$ s and lasts until $t = 5$ s.

At $t = 5$ s, another negative unit step is applied. Thus,

$$v_5(t) = r(t) - r(t - 2) - u(t - 2) - 2u(t - 3) - u(t - 5) \text{ volts}$$

and this lasts until $t = 8$ s. It is displayed in Fig. 5.6d as a solid line.

At $t = 8$ s, a positive unit ramp with slope

$$\frac{0 - (-2)}{10 - 8} = \frac{2}{2} = 1$$

is applied, and this is shown in Fig. 5.6e. Observe that if a negative unit ramp is not applied at $t = 10$ s, the inclined line beginning at $t = 8$ s will continue indefinitely. Thus, in keeping with the previous nomenclature,

$$v_8(t) = r(t) - r(t - 2) - u(t - 2) - 2u(t - 3) - u(t - 5) + r(t - 8) \text{ volts}$$

and

$$v_{10}(t) = r(t) - r(t - 2) - u(t - 2) - 2u(t - 3)$$
$$- u(t - 5) + r(t - 8) - r(t - 10) \text{ volts}$$

and these are depicted respectively by the solid lines in Figs. 5.6e and 5.6f.

The ability to combine step and ramp functions with various strengths is invaluable in network analysis. This will be even more apparent as this study continues.

EXERCISE 5.1

Express the current waveform shown in Fig. 5.7 in terms of ramp and step functions.

Answer

$$i(t) = u(t) + r(t - 2) - r(t - 5) - 5u(t - 5) + 4u(t - 8)$$
$$- \frac{1}{2}r(t - 9) + \frac{1}{2}r(t - 15) \text{ amperes}$$

FIGURE 5.7   Current waveform (Exercise 5.1)

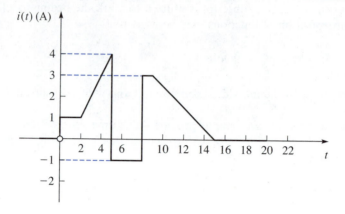

FIGURE 5.8   Four shifted signals

Notice that all are different.

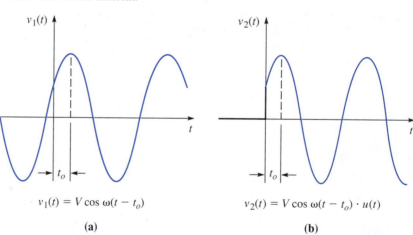

$v_1(t) = V \cos \omega(t - t_o)$

(a)

$v_2(t) = V \cos \omega(t - t_o) \cdot u(t)$

(b)

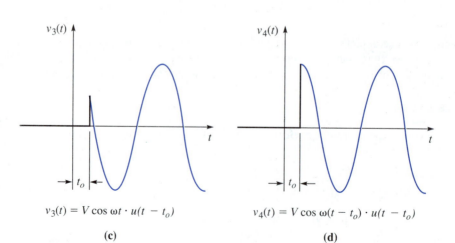

$v_3(t) = V \cos \omega t \cdot u(t - t_o)$

(c)

$v_4(t) = V \cos \omega(t - t_o) \cdot u(t - t_o)$

(d)

THE TIME SHIFTING OF WAVEFORMS

The unit step function is also very useful in the shifting of waveforms. However, caution is necessary, as will now be demonstrated using the voltage signal

$$v(t) = V \cos \omega t$$

Quite naturally, a particular shifted signal is desired, and Figs. 5.8a through 5.8d show that the four signals

$$v_1(t) = V \cos \omega(t - t_o)$$

$$v_2(t) = V \cos \omega(t - t_o) \cdot u(t)$$

$$v_3(t) = V \cos \omega t \cdot u(t - t_o)$$

$$v_4(t) = V \cos \omega(t - t_o) \cdot u(t - t_o)$$

all have different representations.

THE EXPONENTIAL FUNCTION

Consider the decaying exponential, written here in terms of a voltage:

$$v(t) = Ve^{-at} \qquad (5.10)$$

This may also be written as

$$v(t) = Ve^{-t/T} \qquad (5.11)$$

where T is frequently referred to as *the time constant*. Equation (5.10) represents the general case, whereas the form that embraces the time constant usually pertains to the transient or natural response of a first order (RL or RC) network.[1] Similar representations may be written for an exponential current,

$$i(t) = Ie^{-at} \qquad \text{or } i(t) = Ie^{-t/T}$$

or an exponential charge,

$$q(t) = Qe^{-at} \qquad \text{or } q(t) = Qe^{-t/T}$$

Equation (5.10) is shown in Fig. 5.9, where it may be noted that at $t = 0$, $v(t) = V$.

The slope of the exponential is

$$\frac{dv(t)}{dt} = -\frac{V}{T} e^{-t/T}$$

and the initial slope is

$$\frac{dv(t)}{dt}\bigg|_{t=0} = -\frac{V}{T}$$

[1] To be discussed in Chapter 6.

FIGURE 5.9   An exponential voltage

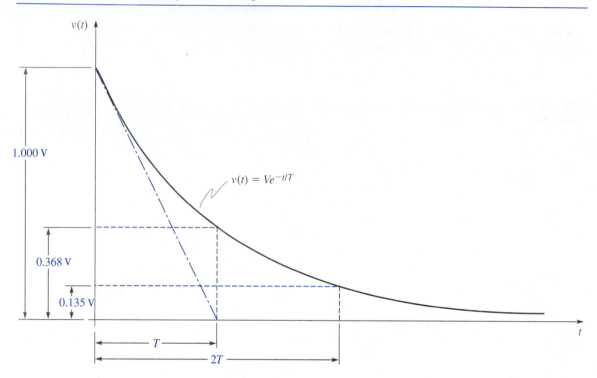

This initial slope is shown as the dashed line in Fig. 5.9, and it may be observed that the initial slope intersects the time axis at $t = T$ time units. At $t = T$, the value of the exponential function is

$$v(t) = Ve^{-t/T} = Ve^{-1} = 0.368 \text{ V}$$

which shows that the exponential function decays to 36.8% of its initial value during an elapsed time equal to one time constant. Moreover, it is seen that no matter what the initial time of reference, the function decays 36.8% during a time period of one time constant. Notice in Fig. 5.9 that at $t = 2T$, the function has decayed to 36.8% of its value at $t = T$ and that it has decayed to $(0.368)(0.368) = 0.135$, or 13.5%, of its initial value.

The importance of the exponential function in the study of network theory cannot be overemphasized.

■ **EXAMPLE 5.1**

Measurements of an exponentially decaying current yield the following data:

Time (s)	Current (mA)
0.250	294.30
0.275	266.30

Determine the initial value of the current and the time constant.

Solution   The statement is made that the current is decaying exponentially. Thus, one writes $i(t) = Ie^{-t/T}$ and uses the data provided:

$$294.30 = Ie^{-0.25/T} \tag{a}$$

$$266.30 = Ie^{-0.275/T} \tag{b}$$

If eq. (a) is divided by eq. (b), the result is

$$1.105 = e^{0.025/T}$$

Use the natural logarithm to obtain

$$0.100 = \frac{0.025}{T}$$

and then solve for T, the time constant:

$$T = \frac{0.025}{0.100} = 0.25 \text{ s}$$

This makes $a = 1/0.25 = 4$ and

$$i(t) = Ie^{-4t} \text{ mA} \tag{c}$$

Use of eq. (c) with the data employed to form eq. (a) gives

$$294.30 = Ie^{-4(0.25)} = Ie^{-1} = 0.368I$$

and hence $I = 294.30/0.368 = 800$. The result is

$$i(t) = 800e^{-4t} \text{ mA}$$

where 800 mA is the initial value of the current.   ■

EXERCISE 5.2

A voltage is known to be exponentially decaying. Measurements indicate the following:

Time (ms)	Voltage (V)
10	53.375
25	25.214

Determine the initial value of the voltage and the time constant.

Answer $V = 88$ V, $T = 20$ ms.

SECTION 5.6   THE SINUSOIDAL FUNCTION

The sinusoidal voltage function,

$$v(t) = V \sin(\omega t + \theta) \tag{5.12}$$

has a period T in seconds and a frequency of f in hertz (Hz). The frequency may be thought of as the number of cycles that occur during a time period, T (usually 1 s). Observe that the *angular frequency* is

$$\omega = 2\pi f = \frac{2\pi}{T} \qquad \text{(radians per second)} \tag{5.13}$$

and θ is the *phase angle* or *phase* of the function, in radians.

The meaning of the angular frequency may be recalled from considerations of simple harmonic motion, a topic in the curriculum of a course in basic or elementary physics. Simple harmonic motion governs the movement of a pendulum of constant length oscillating between two maximum amplitudes X, as shown in Fig. 5.10a. The projection of the oscillation onto the horizontal plane provides a single straight line, as shown in Fig. 5.10b. This line, in turn, may be generated by a rotating line of length equal to the amplitude, X, and rotating at an angular velocity of $\omega = 2\pi f$ radians per second. This is shown in Fig. 5.10c, and the line in Fig. 5.10b is seen to be the horizontal projection of the rotating line. The simple harmonic motion of the pendulum may be, depending on the time of initial release, completely described by either

$$x(t) = X \cos 2\pi f t = X \cos \omega t \tag{5.14a}$$

or

$$x(t) = X \sin 2\pi f t = X \sin \omega t \tag{5.14b}$$

The sinusoidal signal described by eq. (5.12) is periodic because it satisfies eq. (5.2) expressed in terms of voltage:

$$v(t) = v(t + T)$$

To verify this, consider the voltage sinusoid

$$v(t) = V \sin[\omega(t + T) + \theta]$$

FIGURE 5.10   (a) A pendulum in simple harmonic motion, (b) the projection of the motion on the horizontal plane, and (c) the concept of a rotating line segment

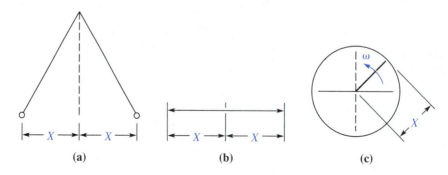

(a)   (b)   (c) and by eq. (5.13), with $T = 2\pi/\omega$,

$$v(t) = V \sin(\omega t + 2\pi + \theta)$$

From the trigonometric identity,

$$\sin(A + B) = \sin A \cos B + \cos A \sin B$$

select $A = \omega t + \theta$ and $B = 2\pi$. Then,

$$V \sin(\omega t + 2\pi + \theta) = V \sin(\omega t + \theta) \cos 2\pi + V \cos(\omega t + \theta) \sin 2\pi$$

and because $\cos 2\pi = 1$ and $\sin 2\pi = 0$,

$$v(t) = V \sin(\omega t + 2\pi + \theta) = V \sin(\omega t + \theta)$$

Thus, the sinusoidal function truly satisfies the conditions for a periodic function.

A further recollection of elementary trigonometry will show that the sine and cosine functions are related by $\pi/2$ radians or by $90°$:

$$\sin \theta = \cos\left(\theta - \frac{\pi}{2}\right) = \cos(\theta - 90°)$$

or

$$\sin \theta = \cos \phi$$

where $\theta = \phi + \pi/2$ or $\theta = \phi + 90°$. Thus, alternative representations for the sinusoidal voltage are

$$v(t) = V \sin(\omega t + \theta) \qquad\qquad \textbf{(5.15)}$$

or

$$v(t) = V \cos(\omega t + \phi) \qquad\qquad \textbf{(5.16)}$$

Equation (5.16) is plotted in Fig. 5.11 with an arbitrary ϕ.

The function $v(t) = V \cos(\omega t + \phi)$

FIGURE 5.11

This function, even though closer to a cosine, is considered to be a sinusoid because sines and cosines are related by $\pi/2$ radians or $90°$.

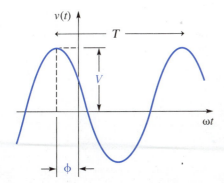

Observe that the period and the maximum amplitude (or merely the amplitude) V are clearly indicated. The amplitude is always taken as positive, and any negative values of $v(t)$ derive from the cosine term, which may fluctuate between $+1$ and -1. As a final point, notice the use here of the uppercase letter V (without subscript) to designate the maximum amplitude. In Chapter 11, where the power in alternating-current networks is treated in detail, it will be necessary to distinguish between maximum amplitude V_m and rms (or root-mean-square) amplitude V.

■ **EXAMPLE 5.2**

Consider a sinusoidal current that passes through a negative maximum of -20 mA at time $t = 2$ ms. It also passes through the next positive maximum at $t = 12$ ms. Using the representation of eq. (5.16), find I, ω, and ϕ.

Solution   Here is eq. (5.16) written for a sinusoidal current:

$$i(t) = I \cos(\omega t + \phi)$$

where I is the amplitude, which in this case is 20 mA, or 0.020 A. Thus,

$$i(t) = 0.020 \cos(\omega t + \phi)$$

and it is observed that any negative values of $i(t)$ will be attributed to the cosine term.

The signal goes from a maximum negative to a maximum positive value in $0.012 - 0.002 = 0.010$ s. This is seen (Fig. 5.11) to be half the period, so that

$$T = 2(0.010) = 0.020 \text{ s}$$

and the frequency is

$$f = \frac{1}{T} = \frac{1}{0.020} = 50 \text{ Hz}$$

This makes the angular frequency

$$\omega = 2\pi f = 2\pi(50) = 100\pi \text{ rad/s}$$

and now

$$i(t) = 0.020 \cos(100\pi t + \phi) \text{ amperes}$$

To evaluate ϕ, use the second data point at $t = 0.012$ s, where the instantaneous value of the current is $i(t = 0.012) = 0.020$ A:

$$0.020 = 0.020 \cos[100\pi(0.012) + \phi]$$
$$1 = \cos(1.2\pi + \phi)$$

The angle whose cosine is unity is 0 rad. Thus,

$$1.2\pi + \phi = 0$$

and

$$\phi = -1.2\pi \qquad (-216°)$$

The sinusoidal current may be represented by any or all of the following:

$$i(t) = 0.020 \cos(100\pi t - 1.2\pi) \text{ amperes}$$

$$i(t) = 0.020 \cos(100\pi t - 216°) \text{ amperes}$$

$$i(t) = 0.020 \cos(100\pi t + 0.8\pi) \text{ amperes}$$

$$i(t) = 0.020 \cos(100\pi t + 144°) \text{ amperes}$$

Notice that if the phase angle is expressed in radians, the radian unit is understood and there is no need to indicate that the angle is in radians. If, however, the angle is given in degrees, the degree sign must always be present. For those with hand-held calculators, it is best to press the radian button before making calculations involving angular frequencies in radians per second, angles in radians, and times in seconds. ■

EXERCISE 5.3

A sinusoidal voltage of the form given in eq. (5.16) has a zero magnitude at 1.25 ms. The next zero magnitude occurs at 6.25 ms. At 8 ms, the amplitude of the voltage is 89.10 V. Determine the amplitude V, the frequency, and the phase ϕ of this voltage signal.

Answer $V = 100$ V, $f = 100$ Hz, and $\phi = \pi/4$ or $45°$.

AVERAGE AND EFFECTIVE VALUES

Signals may take almost any shape or form, and examples of exponential and sinusoidal signals were presented in Sections 5.5 and 5.6. Although the complete specification of these and other signals is desirable, quite often incomplete information about a signal is a useful and worthy substitute. For example, if one is fearful of exceeding the voltage rating of a particular network element or device, just a knowledge of the maximum value of the voltage signal can be quite helpful. Two such useful specifications of a signal are its *average* and its *effective*, or *root-mean-square* (hereinafter referred to as *rms*), values.

5.7.1 The Average Value

The average or mean value of a signal over a certain time interval $T = t_1 - t_0$ is the area under its magnitude curve over the time interval divided by the time interval. The area under the magnitude curve is the integral of the function, and hence,

$$f_{av} = \frac{1}{T} \int_{t_0}^{t_1} f(z)\, dz \tag{5.17}$$

If the function is periodic, as defined by eq. (5.2), then T is the period; and eq. (5.17) may be written as

$$f_{av} = \frac{1}{T} \int_0^T f(t)\, dt \tag{5.18}$$

■ **EXAMPLE 5.3**

Assume that one period of a voltage signal is as shown in Fig. 5.12. What is its average value?

Solution   The period of this voltage signal is $T = 8$ s, and the signal may be completely specified by

$$v(t) = \begin{cases} t, & 0 < t < 2 \\ 1, & 2 < t < 3 \\ -1, & 3 < t < 5 \\ -2, & 5 < t < 8 \end{cases}$$

The average value is determined from eq. (5.18):

$$v_{av} = \frac{1}{8}\left[\int_0^2 t\, dt + \int_2^3 1\, dt + \int_3^5 (-1)\, dt + \int_5^8 (-2)\, dt \right]$$

$$= \frac{1}{8}\left[\frac{1}{2} t^2 \Big|_0^2 + t \Big|_2^3 - t \Big|_3^5 - 2t \Big|_5^8 \right]$$

$$= \frac{1}{8}\left[\frac{1}{2}(4 - 0) + (3 - 2) - (5 - 3) - 2(8 - 5) \right]$$

$$= \frac{1}{8}(2 + 1 - 2 - 6) = \frac{1}{8}(-5) = -0.625 \text{ V}$$

■

FIGURE 5.12   A voltage signal for which the average and rms values are sought

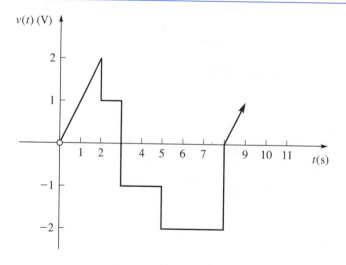

FIGURE 5.13

Voltage waveform (Exercise 5.4)

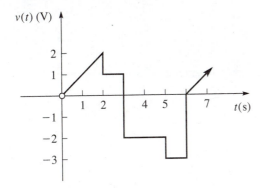

EXAMPLE 5.4

Consider the sinusoidal current $i(t) = I \cos \omega t$. Determine its average value.

Solution Recall that $T = 1/f$; and because $\omega = 2\pi f$, $T = 2\pi/\omega$. The average value will be

$$i(t) = \frac{1}{T} \int_0^T i(t)\, dt$$

or

$$i(t) = \frac{1}{T} \int_0^T I \cos \omega t\, dt = \frac{\omega I}{2\pi \omega} \sin \omega t \Big|_0^{T = 2\pi/\omega}$$

$$= \frac{I}{2\pi} \left[\sin \omega \left(\frac{2\pi}{\omega} \right) - \sin(0) \right] = \frac{I}{2\pi} \left[\sin 2\pi - \sin 0 \right]$$

and hence

$$i_{av} = \frac{I}{2\pi} (0 - 0) = 0$$

This example shows that the average value of a sinusoid is zero. ■

EXERCISE 5.4

What is the average value of the voltage waveform shown in Fig. 5.13?

Answer $v_{av} = -\dfrac{2}{3}$ V.

5.7.2 The Effective or Root-Mean-Square Value

Suppose that a periodic voltage, $v(t)$, is placed across a resistor. The current that flows will be $i(t) = v(t)/R$ and the instantaneous power will be $p(t) = v^2(t)/R$. The total energy delivered to the resistor over one period, T, will be

$$w = \int_0^T p(t)\, dt = \frac{1}{R} \int_0^T [v(t)]^2\, dt$$

If the periodic voltage is replaced by the constant or effective voltage, v_0, with $p = v_0^2/R$, the energy delivered to the resistor over one period will be

$$w = \int_0^T p\,dt = \frac{1}{R}\int_0^T v_0^2\,dt$$

If the energy dissipated by the resistor in each case is identical, then

$$\frac{1}{R}\,v_0^2 t\Big|_0^T = \frac{1}{R}\,v_0^2 T = \frac{1}{R}\int_0^T [v(t)]^2\,dt$$

and the effective, or root-mean-square (rms) value is

$$v_{\text{rms}} = v_0 = \sqrt{\frac{1}{T}\int_0^T [v(t)]^2\,dt}$$

This idea can be extended to a periodic current signal so that, in both cases, the effective or rms value is that constant value that will produce the same energy transfer by the periodic signal to a resistor. Thus, in general, when dealing with a periodic signal, $f(t)$, of period, T, the rms value is

$$f_{\text{rms}} = \sqrt{\frac{1}{T}\int_0^T [f(t)]^2\,dt} \tag{5.19}$$

■ **EXAMPLE 5.5**

For the voltage waveform of Fig. 5.12, find v_{rms}.

Solution Working with the mathematical representation of $v(t)$ provided in Example 5.3 and using eq. (5.19) gives

$$v_{\text{rms}}^2 = \frac{1}{8}\left[\int_0^2 t^2\,dt + \int_2^3 (1)^2\,dt + \int_3^5 (-1)^2\,dt + \int_5^8 (-2)^2\,dt\right]$$

Observe that the functions are squared under the integrals to avoid any inadvertent subtractions, which must not occur in the process of finding the rms value. Now evaluate the integral:

$$v_{\text{rms}}^2 = \frac{1}{8}\left(\frac{1}{3}t^3\Big|_0^2 + t\Big|_2^3 + t\Big|_3^5 + 4t\Big|_5^8\right)$$

$$= \frac{1}{8}\left[\frac{1}{3}(8-0) + (3-2) + (5-3) + 4(8-5)\right]$$

$$= \frac{1}{8}\left[\frac{8 + 3 + 6 + 36}{3}\right] = \frac{1}{8}\left(\frac{53}{3}\right) = \frac{53}{24}$$

This makes

$$v_{\text{rms}} = \sqrt{\frac{53}{24}} = 1.486 \text{ V}$$

The rms value of a signal will always be positive and will always be greater than or equal to the average value. ◼

◼ **EXAMPLE 5.6**

For the current sinusoid $i(t) = I \cos \omega t$, find i_{rms}.

Solution   The period is $T = 2\pi/\omega$, and from eq. (5.19),

$$i_{rms}^2 = \frac{\omega}{2\pi} \int_0^{2\pi/\omega} I^2 \cos^2 \omega t \, dt$$

There is another useful trigonometric identity:

$$\cos^2 \omega t = \frac{1}{2}(1 + \cos 2\omega t)$$

so that

$$
\begin{aligned}
i_{rms}^2 &= \frac{\omega I^2}{4\pi} \left(\int_0^{2\pi/\omega} dt + \int_0^{2\pi/\omega} \cos 2\omega t \, dt \right) \\
&= \frac{\omega I^2}{4\pi} \left(t \Big|_0^{2\pi/\omega} + \frac{1}{2\omega} \sin 2\omega t \Big|_0^{2\pi/\omega} \right) \\
&= \frac{\omega I^2}{4\pi} \left[\left(\frac{2\pi}{\omega} - 0 \right) + \frac{1}{2\omega}(\sin 4\pi - \sin 0) \right]
\end{aligned}
$$

With $\sin 4\pi = \sin 0 = 0$,

$$i_{rms}^2 = \frac{\omega I^2}{4\pi} \left(\frac{2\pi}{\omega} \right) = \frac{I^2}{2}$$

and

$$i_{rms} = \sqrt{\frac{I^2}{2}} = \frac{I}{\sqrt{2}} = \frac{\sqrt{2}}{2} I = 0.707 I \text{ A}$$ ◼

EXERCISE 5.5

What is the effective or rms value of the voltage waveform shown in Fig. 5.13?

Answer $v_{rms} = 1.795$ V.

CHAPTER 5

SUMMARY

▪ Singularity functions are as follows.

—The unit step function is

$$u(t) = \begin{cases} 0, & t < 0 \\ 1, & t > 0 \end{cases}$$

—The unit ramp function is

$$r(t) = \int_{\infty}^{t} u(t)\,dt = \begin{cases} 0, & t < 0 \\ t, & t > 0 \end{cases}$$

—The unit impulse function is

$$\delta(t) = \begin{cases} 0, & t < 0 \\ 0, & t > 0 \end{cases}$$

with

$$\int_{0-}^{0+} \delta(t)\,dt = 1$$

▪ Delayed singularity functions are as follows.
—The delayed unit step function is

$$u(t - t_0) = \begin{cases} 0, & t < t_o \\ 1, & t > t_o \end{cases}$$

—The delayed unit ramp function is

$$r(t - t_0) = \begin{cases} 0, & t < t_o \\ t - t_o, & t > t_o \end{cases}$$

—The delayed unit impulse function is

$$\int_{a}^{b} \delta(t - t_0)\,dt = 1 \qquad (a \le t_o \le b)$$

or

$$\int_{-\infty}^{t} f(z)\,dz = u(t)$$

▪ The exponential function is

$$f(t) = Ae^{-t/T}$$

where T is the time constant.

▪ The general sinusoid is

$$f(t) = A\cos(\omega t + \phi)$$

where A is the amplitude, T is the period in seconds, f is the frequency in hertz, $\omega = 2\pi f$ is the angular frequency in radians per second, and ϕ is the phase angle in radians or degrees.

▪ The average value of $f(t)$ over the time interval from 0 to T is

$$f_{\text{av}} = \frac{1}{T} \int_{0}^{T} f(t)\,dt$$

- The effective value of a periodic function with period T, $f(t)$ is

$$f_{\text{rms}} = \sqrt{\frac{1}{T} \int_0^T [f(t)]^2 \, dt}$$

Additional Readings

Blackwell, W.A., and L.L. Grigsby. *Introductory Network Theory*. Boston: PWS Engineering, 1985, pp. 36–40, 244–245, 257–264.

Bobrow, L.S. *Elementary Linear Circuit Analysis*. 2d ed. New York: Holt, Rinehart and Winston, 1987, pp. 172–191, 348–354, 398, 599.

Del Toro, V. *Engineering Circuits*. Englewood Cliffs, N.J.: Prentice-Hall, 1987, pp. 183–190, 272, 295, 296, 301, 318, 352—354, 356, 357.

Dorf, R.C. *Introduction to Electric Circuits*. New York: Wiley, 1989, pp. 212–215, 250–253, 320–324, 391–394, 509–513.

Hayt, W.H., Jr., and J.E. Kemmerly. *Engineering Circuit Analysis*. 4th ed. New York: McGraw-Hill, 1986, pp. 119, 150–152, 169–172, 233–235, 284–297, 530, 536, 572.

Irwin, J.D. *Basic Engineering Circuit Analysis*. 3d ed. New York: Macmillan, 1989, pp. 257, 260, 279, 280, 353–358, 458–462, 729, 816.

Johnson, D.E., J.L. Hilburn, and J.R. Johnson. *Basic Electric Circuit Analysis*. 4th ed. Englewood Cliffs, N.J.: Prentice-Hall, 1989, pp. 211, 212, 230–234, 237, 304–306, 382–384, 590, 591, 600–604.

Karni, S. *Applied Circuit Analysis*. New York: Wiley, 1988, pp. 180, 183, 203, 249–255, 333–337, 414–428, 451.

Madhu, S. *Linear Circuit Analysis*. Englewood Cliffs, N.J.: Prentice-Hall, 1988, pp. 272, 277, 278, 290, 291, 319–326, 334–340, 705–710.

Nilsson, J.W. *Electric Circuits*. 3d ed. Reading, Mass.: Addison-Wesley, 1990, pp. 205, 206, 313–316, 373–377, 539, 542, 596–606.

Paul, C.R. *Analysis of Linear Circuits*. New York: McGraw-Hill, 1989, pp. 240, 299–303, 413, 425, 428, 433–442, 512, 657.

CHAPTER 5   **PROBLEMS**

Section 5.3

5.1 Express the waveform in Fig. P5.1 in terms of the unit step and the unit ramp functions.

Figure P5.1

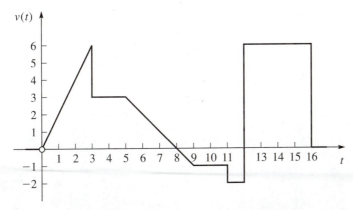

5.2   Express the waveform in Fig. P5.2 in terms of
the unit step and the unit ramp functions.

Figure P5.2

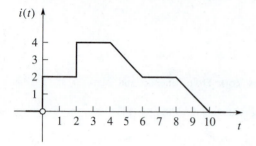

5.3   Express the waveform in Fig. P5.3 in terms of
the unit step and the unit ramp functions.

Figure P5.3

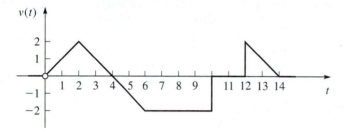

5.4   Express the waveform in Fig. P5.4 in terms of
the unit step and the unit ramp functions.

Figure P5.4

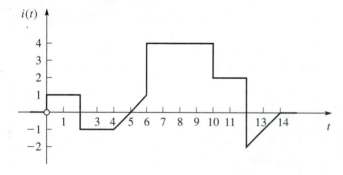

5.5   Express the waveform in Fig. P5.5 in terms of the unit step and the unit ramp functions.

Figure P5.5

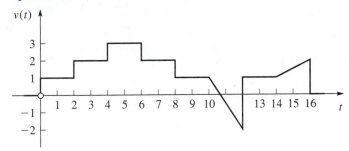

5.6   Sketch the waveform that is represented by

$$v(t) = 2u(t) + 3r(t - 2) - 6r(t - 5)$$
$$+ 2u(t - 5) + r(t - 8) \text{ volts}$$

5.7   Sketch the waveform that is represented by

$$v(t) = 2r(t) + r(t - 2) + 2u(t - 4) - 6u(t - 5)$$
$$- 4r(t - 7) + r(t - 12) - 8u(t - 14) \text{ volts}$$

5.8   Sketch the waveform that is represented by

$$i(t) = r(t) + 2u(t) - u(t - 4) + r(t - 6)$$
$$- 2u(t - 8) - r(t - 10) - 4u(t - 12)$$
$$- r(t - 14) \text{ volts}$$

Section 5.4

5.9   Sketch the functions $4e^{-2t}$ and $4e^{-2(t-2)}$.

5.10   Sketch the functions $4e^{-2t}$ and $4e^{-2(t-2)}u(t)$.

5.11   Sketch the functions $4e^{-2t}$ and $4e^{-2t}u(t - 2)$.

5.12   Sketch the functions $4e^{-2t}$ and $4e^{-2(t-2)}u(t - 2)$.

Section 5.5

5.13   Measurements of the exponentially decaying charge on a sphere provide the following data:

Time (ms)	Charge (μC)
0.40	170.429
0.425	168.733

Determine the initial value of the charge, the time constant, and the charge at 0.625 ms.

5.14   Measurements of an exponentially decaying voltage provide the following data:

Time (s)	Voltage (V)
0.12	658.574
0.15	566.840

Determine the initial value of the voltage, the time constant, and the voltage at 32.5 ms.

5.15   If an exponentially decaying current has a time constant of 100 ms and the current has a value of 12.42 A at $t = 165$ ms, what is the value of the current at $t = 0$, and at what time will the current have a value of 4.571 A?

5.16   An exponentially decaying voltage has an initial value of 1.2 kV and a time constant $T = 400$ ms. What is the value of the voltage at $t = 125$ ms and what is the value of the voltage five time constants later?

Section 5.6

5.17   A sinusoidal voltage that has a frequency of 80 Hz and an amplitude of 120 V passes through 0 V with a positive slope at $t = 0.8$ ms. Find V, ω, and ϕ in

$$v(t) = V \cos(\omega t + \phi)$$

5.18   A sinusoidal current has a period of 40 μs and passes through a positive maximum of 40 mA at $t = 4$ μs. Find I, ω, and ϕ in

$$i(t) = I \cos(\omega t + \phi)$$

5.19   A sinusoidally varying charge reaches a maximum of 20 μC at $t = 8$ ms and the next negative maximum at $t = 16$ ms. Find Q, ω, and ϕ in

$$q(t) = Q \cos(\omega t + \phi)$$

5.20 A sinusoidally varying voltage has a positive maximum of 208 V at $t = 0$ and decreases to a value of 120 V at $t = 0.125$ ms. Find V, ω, and ϕ in

$$v(t) = V \cos(\omega t + \phi)$$

5.21 What are the frequency and period of a sinusoidally varying voltage that has a value of 60 V at $t = 0$ and reaches its first maximum of 120 V at $t = 2.5$ ms?

Section 5.7

5.22 Refer to Fig. P5.6, consider $T = 10$ s, and find the average and effective value of the voltage.

Figure P5.6

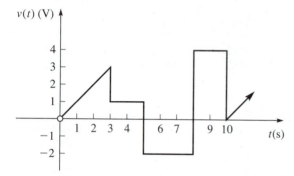

5.23 Refer to Fig. P5.7, consider $T = 10$ s, and find the average and effective value of the current.

Figure P5.7

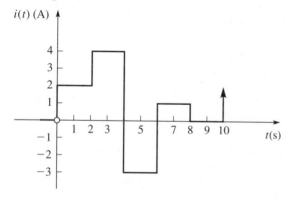

5.24 Refer to Fig. P5.8, consider $T = 10$ s, and find the average and effective value of the voltage.

Figure P5.8

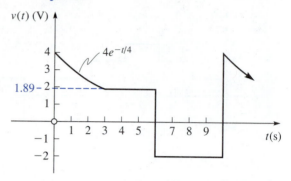

5.25 Refer to Fig. P5.9, consider $T = 10$ s, and find the average and effective value of the current.

Figure P5.9

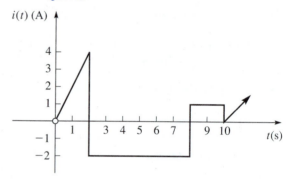

5.26 Refer to Fig. P5.10, consider $T = 40$ ms, and find the average and effective value of the voltage.

Figure P5.10

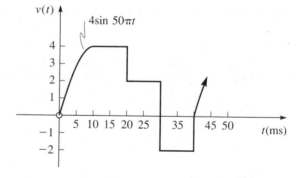

THE DIFFERENTIAL EQUATIONS
OF NETWORK THEORY

NETWORK DIFFERENTIAL EQUATIONS

6

OBJECTIVES

The objectives of this chapter are to:

- Introduce the concept of the state of a network.

- Consider the types of network response and show the existence of an exponential form of solution.

- Discuss the use of initial conditions and show how these conditions may be identified.

- Present procedures for the determination of the natural, forced, zero-input, and zero-state responses.

- Pose several examples that deal with the set-up of the differential equations for *RL* and *RC* first-order networks, and then solve these differential equations for the undriven case and a variety of forcing functions such as those presented in Chapter 5: the step, ramp, sinusoidal, and impulse functions.

INTRODUCTION

Differential equations may be classified by resorting to a consideration of their order and degree and about whether they are *ordinary* or *partial*, whether they possess *variable* or *constant coefficients*, whether they are *linear* or *nonlinear*, and whether they are *homogeneous* or *nonhomogeneous*.

It will not be necessary for the reader to keep the foregoing categorizations in mind, because all of the differential equations of network theory that will be considered in this and the next chapter have constant coefficients and are ordinary, linear, of the first degree, and either homogeneous or nonhomogeneous. For example, the ordinary second-order, homogeneous differential equation of the first degree with constant coefficients

$$\frac{d^2i}{dt^2} + 5\frac{di}{dt} + 6i = 0; \qquad t > 0$$

FIGURE 6.1   (a) An *RLC* series network with a voltage across the capacitor and (b) an identical
network except for the addition of a 12-V battery to drive the network

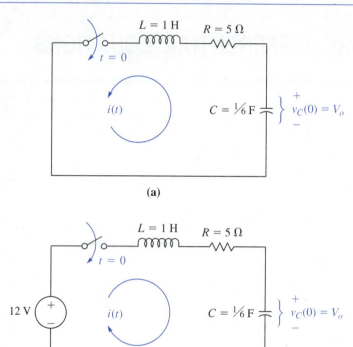

(a)

(b)

can be obtained from a consideration of the *RLC* series network shown in Fig. 6.1a.
In this equation, the *dependent variable* is the current *i*, and the *independent variable*
is the time *t*. Observe that a voltage across the capacitor terminals exists before the
switch closes at time $t = 0$. There is no forcing function, and the current that flows
when the switch is closed is due to the energy stored in the capacitor. It is a fact that
undriven networks such as the one in Fig. 6.1a and those that are energized by initial
currents through inductors always yield homogeneous differential equations. In this
undriven network, the initial voltage across the capacitor, $v_C(0) = V_o$, and the initial
current through the inductor, $i(0) = 0$, are called the *initial conditions*.

In Fig. 6.1b, a forcing function in the form of a 12-V battery is switched into
the network at $t = 0$ when the initial voltage across the capacitor is V_o. In this case,
the differential equation for the voltage across the capacitor as a function of time is

$$\frac{d^2v}{dt^2} + 5\frac{dv}{dt} + 6v = 12; \qquad t > 0$$

This differential equation is nonhomogeneous, and the nonhomogeneity is due to the
presence of the forcing function. The fact is that driven systems and driven networks
are always represented by nonhomogeneous differential equations even though the
stored energy is present in exactly the same form and magnitude as in the undriven
case.

An *RLC* series network with a voltage across the capacitor

FIGURE 6.2

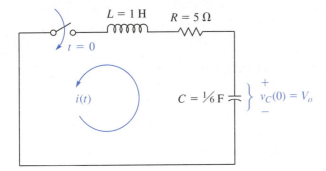

THE STATE OF A NETWORK

The *state of a network* or *system* pertains to the minimum set of quantities or variables that, because they contain enough data or information regarding the history of the network or system, enable the future performance of the network or system to be predicted. The variables themselves are called *state variables*; and in network analysis, they pertain to currents through inductors and voltages across capacitors, because the energy stored in these elements (and in the network) is a function of these variables. This selection of state variables is made because when the energy stored in the network changes, the state of the network changes.

The state variables may be assembled in a column vector called the *state vector*, even though when they are considered separately, they may appear to bear little resemblance to one another. But if at any time *t*, the variables possess definite values, the state at time *t* is considered specified. This may be the initial state of the network; and if an input is applied at $t = 0$, the state variables completely describe the behavior of the network. The state vector therefore represents the system state for all *t* if the system input or perturbation is specified and if the initial state of the network is known. For example, the initial state of the network in Fig. 6.1a, repeated here as Fig. 6.2, is $v_C(0) = V_o$ and $i(0) = 0$.

The concept of the state of a network is quite important and is to be found at the core of all of the examples that will be presented in this and in the next chapter. Analyses conducted using the concept of state are referred to as *state-space analyses* or *state-variable analyses*.

TYPES OF RESPONSE AND EXPONENTIAL SOLUTIONS

Consider the differential equation

$$a\frac{dy}{dt} + by = K; \qquad t > 0 \tag{6.1}$$

which is a linear, first-order, nonhomogeneous differential equation with constant coefficients. The variable *y* has been chosen as the dependent variable for the purpose of generality; it could just have well been chosen as *i*, *q*, or *v*. The independent value is the time *t*.

Equation (6.1) can be rewritten so that the highest-order derivative has a coefficient of unity:

$$\frac{dy}{dt} + \frac{b}{a} y = \frac{K}{a}$$

and now a little algebraic adjustment will yield

$$\frac{dy}{dt} = \frac{K}{a} - \frac{b}{a} y \tag{6.2}$$

Further rearrangement, by a technique known as *separation of the variables*, will give

$$dy = \frac{b}{a} \left(\frac{K}{b} - y \right) dt$$

and finally

$$\frac{dy}{(K/b - y)} = \frac{b}{a} dt \tag{6.3}$$

Equation (6.3) may be integrated over the time interval of interest, in this case, from $t = 0$, where $y = y(0) = y_o$, to $t = t$, where $y = y(t) = y$. Use of the dummy variables u and z for the integration, to avoid confusion between the variables of integration and the limits of integration, allows one to write

$$\int_{y_o}^{y} \frac{du}{(u - K/b)} = -\frac{b}{a} \int_{0}^{t} dz$$

and a straightforward integration gives the result

$$\ln \left[u - \frac{K}{b} \right]_{y_o}^{y} = -\frac{b}{a} z \Big|_{0}^{t}$$

With the limits substituted,

$$\ln \frac{y - K/b}{y_o - K/b} = -\frac{b}{a} t \tag{6.4}$$

The inverse natural logarithm of eq. (6.4) may be taken:

$$\frac{y - K/b}{y_o - K/b} = e^{-(b/a)t}$$

or

$$y - \frac{K}{b} = \left(y_o - \frac{K}{b} \right) e^{(-b/a)t}$$

and finally

$$y = y_o e^{-(b/a)t} + \frac{K}{b}(1 - e^{-(b/a)t}); \qquad t \geq 0 \qquad (6.5)$$

which shows the existence of an exponential in the solution.

For eq. (6.5) to be valid as a solution, its evaluation at time $t = 0$ must yield the initial condition, $y = y_o$. Substitute $t = 0$ into eq. (6.5) and obtain the inferred result:

$$y(0) = y_o + \frac{K}{b}(1 - 1) = y_o$$

Next, after a great deal of time has elapsed, the exponential will approach zero as a limit; that is

$$\lim_{t \to \infty} e^{-(b/a)t} = 0$$

so that for $t \to \infty$,

$$y(\infty) = \frac{K}{b}(1 - 0) = \frac{K}{b}$$

This shows that after a great deal of time has elapsed, the response y will be in the same form as the forcing function. This is the underlying principle behind the method of undetermined coefficients, which is described in Section 6.7.

All of the foregoing can be summarized by the following three conditions for the time response of a network that can be represented by a linear, first-order, non-homogeneous differential equation with constant coefficients:

1. It is a function of the condition or state of the network when the time interval of interest begins. This is the initial state of the network and is called the initial condition, $y(0) = y_o$.
2. It is a function of the condition or state of the system after a great deal of time has elapsed. The form of the state or dependent variable y will be the same form as the forcing function.
3. It is a function of how the variable of interest proceeds from the condition at $t = 0$ to the condition after a great deal of time has elapsed. This consideration is dominated by the descending exponential, which will be strongly influenced by the network parameters a and b.

6.3.1 Zero-Input and Zero-State Responses

A much simpler approach may now be taken. Take eq. (6.1) once again, and observe that because it represents a linear system, the principle of superposition permits writing it as

$$\frac{dy}{dt} + \frac{b}{a}y = \frac{K}{a} + 0; \qquad t > 0 \qquad (6.6)$$

The breakdown of eq. (6.6) to indicate two forcing functions, one of which is zero, suggests that the total response will be the sum of two component responses:

1. One component is a *zero-input* response, which is the response of the network with no applied input. The zero-input response depends upon the initial state of the network, which is represented by the initial condition y_o, and the characteristics of the network, a and b.

2. The second component is a *zero-state* response, which is the response of the network to an input applied at $t = 0$ subject to the condition that the system is in the zero state (all initial conditions are set equal to zero) just prior to the application of the input.

An examination of eqs. (6.5) and (6.6) indicates that the zero-input response must be the only term in eq. (6.5) that involves the initial condition or initial state, $y = y_o$:

$$y_{ZI} = y_o e^{-(b/a)t}; \qquad t > 0 \tag{6.7}$$

The zero-state response must involve the forcing function and not the initial condition or the initial state. From eq. (6.5), it is seen to be

$$y_{ZS} = \frac{K}{b}(1 - e^{-(b/a)t}); \qquad t \geq 0 \tag{6.8}$$

This can, of course, be checked by substituting eq. (6.7) and then eq. (6.8) into eq. (6.6).

6.3.2 Linearity Again

A knowledge of what is meant by the state of a network permits one to be more precise in the definition of a linear system, which was given in Section 1.5. Suppose that a system produces a response $r(t)$ when excited by an input $e(t)$. Assume that the system is in the *zero state*; that is, there are no electric or magnetic fields within the boundaries of the system. It has just been observed that if a system is initially in the zero state, its response to an arbitrary input is called the *zero-state response*. The system is considered to be *zero-state linear* if it satisfies the *homogeneity* and *superposition* conditions.

> **Homogeneity Condition:** If an arbitrary input to the system e causes a zero-state response r, then if ce is the input, the zero-state output is cr, where c is some arbitrary constant.
>
> **Superposition Condition:** If the input to the system e_1 causes a zero-state response r_1, and if an input to the system e_2 causes a zero-state response r_2, then a zero-state response $r_1 + r_2$ will occur when the input is $e_1 + e_2$.

If either the homogeneity condition or the superposition condition is not satisfied, the system is said to be *zero-state nonlinear*.

If electric and magnetic fields exist within a system when an input is applied, the system is not in the zero state. It possesses an initial state that can cause a system response in the case of a complete absence of an external input. Such a response is termed a *zero-input response*, and a system can be *zero-input linear* if it satisfies two different homogeneity and superposition conditions.

Homogeneity Condition: If an arbitrary initial state represented by the conditions $(u_1, u_2, u_3, \ldots)$, causes a zero-input response r_1, then if the initial state is represented by $(cu_1, cu_2, cu_3, \ldots)$, the zero-input response will be cr_1, where c is some arbitrary constant.

Superposition Condition: If an arbitrary initial state represented by the conditions $(u_1, u_2, u_3, \ldots)$ produces a zero-input response r_1, and if another arbitrary initial state $(z_1, z_2, z_3, \ldots)$ produces a zero-input response r_2, then a zero-input response $r_1 + r_2$ will occur when the initial state is $(u_1 + z_1, u_2 + z_2, u_3 + z_3, \ldots)$.

If a system is *both* zero-state linear and zero-input linear, it will be linear, because it satisfies what is called the *decomposition property*:

Decomposition Property: If an arbitrary input e_1 produces a zero-state response r_1, and if an arbitrary initial state $(u_1, u_2, u_3, \ldots)$ produces a zero-input response r_2, then the total response for the same arbitrary initial state and the same arbitrary input will be $r_3 = r_1 + r_2$.

6.3.3 Natural and Forced Responses

The use of the zero-input and zero-state responses can be of considerable significance. However, there is an alternative and equally fundamental approach. This involves dividing the total response, as in eq. (6.5), into terms that do and do not contain the exponential. Then, two additional types of responses can be defined:

Definition: The *natural response* is the behavior of the network that is determined by the characteristics or nature of the network and the initial conditions or the initial state of the network.

If the natural response exists at all, it will, in a stable network, eventually disappear if a dissipative element (a resistor) is present. It is intimately associated with the exponential terms, but its magnitude will depend on the magnitude of the forcing function. The time frame between $t = 0$ and the eventual disappearance of the natural response is known as the *transient* period, and because of this, some refer to the natural response as the *transient response*. This terminology, however, leads to confusion when the network forcing function contains one or more time-decaying components. These lead to time-decaying components in the forced response that cannot be lumped with the natural response. Thus, in general, it may be bad form to equate the unforced or natural response with what is often referred to as the transient response. The natural response, in a purely mathematical study, is referred to as the *complementary function*.

Definition: The *forced response* is the behavior of the network that is determined by the external forcing function.

It can be maintained indefinitely by the forcing function and it will be in the same form as the forcing function. The forced response is frequently referred to as the *steady-state response*, and in a study of differential equations, the forced response is known as the *particular integral*, although many refer to it as the particular solution.

Equation (6.5),

$$y = y_o e^{-(b/a)t} + \frac{K}{b}(1 - e^{-(b/a)t})$$ (6.5)

may be broken into two parts in accordance with the definitions of natural and forced responses. Keep in mind that the two parts of the solution must be consistent with eq. (6.6):

$$\frac{dy}{dt} + \frac{b}{a}y = \frac{K}{a} + 0$$ (6.6)

The natural response contains the initial condition y_o, and because the natural response must eventually disappear if a dissipative element is present, the natural response must contain all of the decaying exponential terms:

$$y_N = \left(y_o - \frac{K}{b}\right)e^{(-b/a)t}$$ (6.9)

The forced response is the response to K/a and is

$$y_F = \frac{K}{b}$$ (6.10)

6.3.4 The Complete or Total Response

The *complete* or *total* response is the sum of the natural and forced responses. Thus, the solution to eq. (6.6) is the sum of eqs. (6.9) and (6.10):

$$y = y_N + y_F = \left(y_o - \frac{K}{b}\right)e^{-(b/a)t} + \frac{K}{b}$$ (6.11a)

The reader may wish to verify this by substituting eq. (6.11a) into the left side of eq. (6.6).

The complete or total response is also the sum of the zero-input and zero-state responses. Thus, the solution to eq. (6.6) is also the sum of eqs. (6.7) and (6.8):

$$y = y_{ZI} + y_{ZS} = y_o e^{-(b/a)t} + \frac{K}{b}(1 - e^{-(b/a)t})$$ (6.11b)

The procedure for solving an *n*th-order linear differential equation with constant coefficients will be presented in Sections 6.5 and 6.6 after a more detailed discussion of initial conditions is provided in Section 6.4. A summary of the steps required for the determination of the zero-input and zero-state responses on the one hand and the natural (transient) and forced (steady-state) responses on the other is displayed in Fig. 6.3. This summary reveals the significant fact that the total response, which is free of all arbitrary constants, is the sum of the zero-input and zero-state responses. Alternatively, the total response is the sum of the natural and forced responses. The total response is often referred to as the *particular solution* (not the particular integral).

Summary of the steps required to obtain the zero-input and zero-state and the
natural and forced responses and the total response, which is the particular solution

FIGURE 6.3

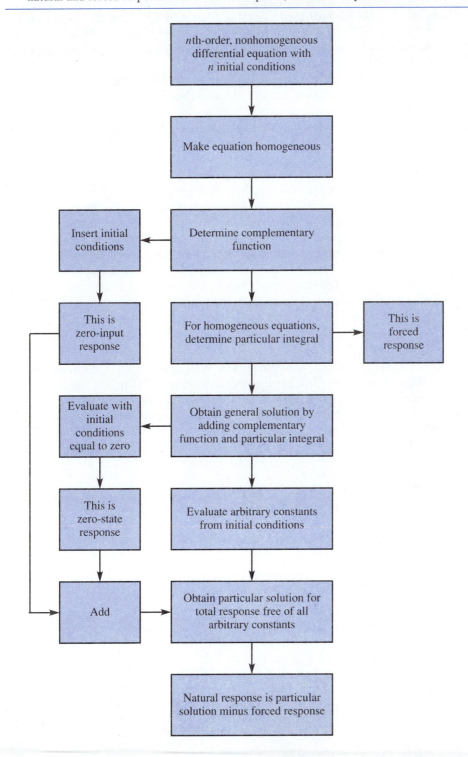

SECTION 6.4   INITIAL CONDITIONS

Because all of the conditions used to evaluate the arbitrary constants in the general solutions to the differential equations of network theory are usually taken at time $t = 0$, the equations are considered to be *initial-value problems*,[1] and the conditions at $t = 0$ are called initial conditions.

Initial conditions manifest themselves in the form of currents through inductors and the voltages across capacitors, and the initial state of the network is the value of these inductor currents and capacitor voltages at $t = 0$. The initial inductor currents and capacitor voltages evolve from flux and charge considerations, which, by themselves, may be used for initial conditions or to describe the initial or zero state.

The continuity of stored energy principle, introduced and discussed in detail in Section 1.8, prohibits jump discontinuities in inductor currents and capacitor voltages in the absence of the impulse function. This principle and KVL, KCL, and the elemental equations for the resistor, inductor, and capacitor are all that are required to evaluate the necessary initial conditions that are employed to determine the arbitrary constants in the general solution for the total response. A comprehensive example pertaining to the evaluation of initial conditions now follows. In this discussion, and throughout this book, use is made of $t = 0^-$ and $t = 0^+$, which are intended to mean the time immediately before and immediately after an event that occurs over a short interval of time. Thus, the time required for a switch to close is the very, very short interval between $t = 0^-$ and $t = 0^+$.

■ EXAMPLE 6.1

In Fig. 6.4a, the switch, which has been in position 1 for a considerable period of time, moves instantaneously to position 2. Designate the instant before the switch is closed as $t = 0^-$ and the instant after the switch is closed as $t = 0^+$. If the capacitor contains no charge at $t = 0^-$, that is, $v_C(0) = 0$, find (in any order) the following:

$$i_L(0^-) \qquad v_{R_1}(0^-) \qquad v_L(0^-)$$

$$v_C(0^-) \qquad i_L(0^+) \qquad v_{R_1}(0^+)$$

$$v_{R_2}(0^+) \qquad v_C(0^+) \qquad v_L(0^+)$$

$$\left.\frac{di_L}{dt}\right|_{t=0^+} \qquad \left.\frac{dv_C}{dt}\right|_{t=0^+}$$

Solution   The network may be considered as operating in the steady state prior to the movement of the switch. Thus, at $t = 0^-$, the situation will be as shown in Fig. 6.4b; and the current, whatever its value, is steady. This means that

$$\left.\frac{di_L}{dt}\right|_{t=0^-} = \frac{1}{L}\,v_L(0^-) = 0 \text{ A/s}$$

and

$$v_L(0^-) = 0 \text{ V}$$

---

[1] *Initial-value problem* is the terminology used, not because time $t = 0$ is involved, but because the arbitrary constants are all evaluated at the *same* value of the independent variable.

FIGURE 6.4

(a) A network in which the switch moves from position 1 to position 2 instantaneously at $t = 0$, (b) part of the network at $t = 0^-$, the instant before the switch is moved, and (c) part of the network at $t = 0^+$, the instant just after the switch is moved

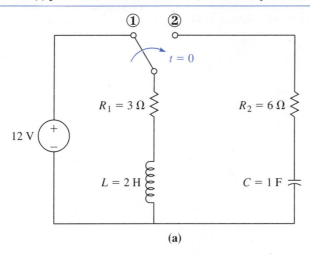

(a)

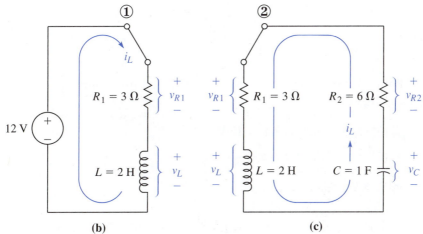

(b)   **(c)**

For the loop shown in Fig. 6.4b, KVL requires that at $t = 0^-$,

$$v_{R_1}(0^-) + v_L(0^-) - 12 = 0$$

and because $v_L(0^-) = 0$, it is observed that

$$v_{R_1}(0^-) = 12 \text{ V}$$

At $t = 0^-$, $i_L(0^-) = i_{R_1}(0^-)$, and Ohm's law can be used to show that

$$i_L(0^-) = i_{R_1}(0^-) = \frac{v_{R_1}(0^-)}{R_1} = \frac{12}{3} = 4 \text{ A}$$

The situation immediately after the switch moves to position 2 (at $t = 0^+$) is indicated in Fig. 6.4c. The switch is assumed to have moved during the instantaneous interval from $t = 0^-$ to $t = 0^+$ ($0^- \leq t \leq 0^+$), and it is seen immediately that for this series configuration,

$$i_L(0^+) = i_{R_1}(0^+) = i_{R_2}(0^+) = i_C(0^+)$$

But continuity of stored energy demands that

$$i_L(0^-) = i_L(0^+) = 4 \text{ A}$$

so that

$$i_L(0^+) = i_{R_1}(0^+) = i_{R_2}(0^+) = i_C(0^+) = 4 \text{ A}$$

Here, too, KVL must govern, so that

$$v_{R_1}(0^+) + v_L(0^+) = v_C(0^+) + v_{R_2}(0^+)$$

However, the capacitor was uncharged at $t = 0^-$, and by continuity of stored energy, it must remain uncharged at $t = 0^+$. Thus,

$$v_C(0^+) = v_C(0^-) = 0 \text{ V}$$

and

$$v_{R_1}(0^+) + v_L(0^+) = v_{R_2}(0^+)$$

Both resistor voltage drops can now be found from Ohm's law. With $i_{R_1}(0^+) = i_{R_2}(0^+) = i_L(0^+)$:

$$v_{R_1}(0^+) = R_1 i_{R_1}(0^+) = 3(4) = 12 \text{ V}$$
$$v_{R_2}(0^+) = -R_2 i_{R_2}(0^+) = 6(4) = -24 \text{ V}$$

From the KVL relationship,

$$v_L(0^+) = -v_{R_1}(0^+) + v_{R_2}(0^+) = L \frac{di_L}{dt}\bigg|_{t=0^+} = -12 - 24 = -36 \text{ V}$$

which shows that the current must begin to decay instantaneously:

$$\frac{di_L}{dt}\bigg|_{t=0^+} = \frac{v_L(0^+)}{L} = -\frac{36}{2} = -18 \text{ A/s}$$

Although there is no voltage drop across the capacitor at $t = 0^+$, the voltage across the capacitor must begin to increase, because

$$i_C(0^+) = C \frac{dv_C}{dt}\bigg|_{t=0^+}$$

or

$$\frac{dv_C}{dt}\bigg|_{t=0^+} = \frac{i_C(0^+)}{C} = \frac{4}{1} = 4 \text{ V/s}$$

Network (Exercise 6.1)

FIGURE 6.5

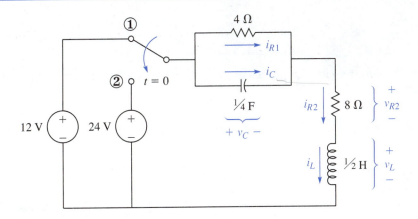

EXERCISE 6.1

In the network shown in Fig. 6.5, the switch moves instantaneously from position 1 to position 2 at $t = 0$. Find the following:

$i_{R_1}(0^-)$ $\quad$ $i_{R_2}(0^-)$ $\quad$ $i_C(0^-)$ $\quad$ $i_L(0^-)$ $\quad$ $v_C(0^-)$

$v_{R_2}(0^-)$ $\quad$ $v_L(0^-)$ $\quad$ $i_{R_1}(0^+)$ $\quad$ $i_{R_2}(0^+)$ $\quad$ $i_C(0^-)$

$i_L(0^+)$ $\quad$ $v_C(0^+)$ $\quad$ $v_L(0^+)$ $\quad$ $v_C'(0^+)$ $\quad$ $i_{R_2}'(0^+)$

Answer

$i_{R_1}(0^+) = 1$ A $\qquad$ $i_{R_2}(0^+) = 1$ A $\qquad$ $i_C(0^-) = 0$ A

$i_L(0^-) = 1$ A $\qquad$ $v_C(0^-) = 4$ V $\qquad$ $v_{R_2}(0^-) = 8$ V

$v_L(0^-) = 0$ V $\qquad$ $I_{R_1}(0^+) = 1$ A $\qquad$ $i_{R_2}(0^+) = 1$ A

$i_L(0^+) = 1$ A $\qquad$ $i_C(0^+) = 0$ A $\qquad$ $v_C(0^+) = 4$ V

$v_L(0^+) = 12$ V $\qquad$ $v_C'(0^+) = 0$ V/s $\qquad$ $i_L'(0^+) = 24$ A/s

PROCEDURE FOR SOLUTION (NATURAL AND FORCED RESPONSE)

SECTION 6.5

In this section, the terminology *complementary function* and *particular integral* will be used. The natural response derives from the complementary function, and the forced response is the particular integral. The procedure that will be delineated in this section is a general procedure that applies to an nth-order, linear differential equation with constant coefficients of the form

$$a_n \frac{d^n y}{dt^n} + a_{n-1} \frac{d^{n-1} y}{dt^{n-1}} + a_{n-2} \frac{d^{n-2} y}{dt^{n-2}} + \cdots + a_1 \frac{dy}{dt} + a_0 y = f(t) \qquad (6.12)$$

which can be decomposed into two solutions. The principles of linearity and superposition permit the left side of eq. (6.12) to be set equal to $f(t) + 0$:

$$a_n \frac{d^n y}{dt^n} + a_{n-1} \frac{d^{n-1} y}{dt^{n-1}} + a_{n-2} \frac{d^{n-2} y}{dt^{n-2}} + \cdots + a_1 \frac{dy}{dt} + a_0 y = f(t) + 0$$

so that two solutions occur. The first solution yields the complementary function, which is a solution of the homogeneous equation:

$$a_n \frac{d^n y_c}{dt^n} + a_{n-1} \frac{d^{n-1} y_c}{dt^{n-1}} + a_{n-2} \frac{d^{n-2} y_c}{dt^{n-2}} + \cdots + a_1 \frac{dy_c}{dt} + a_0 y_c = 0 \qquad (6.13)$$

The second solution is for the particular integral, which is the solution of the non-homogeneous equation:

$$a_n \frac{d^n y_{\text{pi}}}{dt^n} + a_{n-1} \frac{d^{n-1} y_{\text{pi}}}{dt^{n-1}} + a_{n-2} \frac{d^{n-2} y_{\text{pi}}}{dt^{n-2}} + \cdots + a_1 \frac{dy_{\text{pi}}}{dt} + a_0 y_{\text{pi}} = f(t) \qquad (6.14)$$

The total response is the sum of these two solutions.

$$y = y_c + y_{\text{pi}} \qquad (6.15)$$

The steps in the procedure now follow.

1. Because exponential solutions will exist, assume a solution of the form

$$y_c = Y e^{st} \qquad (6.16)$$

where Y is some real or complex, yet arbitrary, constant and s is a complex exponent. The fact that s is a complex number of the form $\sigma + j\omega$ has yet to be established and will be considered in Chapter 9.

2. Substitute eq. (6.16) into eq. (6.13), performing all differentiations and whatever algebraic operations necessary to separate the result into factors involving the arbitrary constant Y, the exponential e^{st}, and a polynomial in s that is of the same order as the differential equation. The polynomial in s is called the *characteristic polynomial*; and when it is set equal to zero, it is called the *characteristic equation*. When eq. (6.16) is substituted into eq. (6.13), the result is

$$a_n s^n Y e^{st} + a_{n-1} s^{n-1} Y e^{st} + a_{n-2} s^{n-2} Y e^{st} + \cdots + a_1 s Y e^{st} + a_0 Y e^{st} = 0$$

or

$$(a_n s^n + a_{n-1} s^{n-1} + a_{n-2} s^{n-2} + \cdots + a_1 s + a_0) Y e^{st} = 0$$

Observe that if either Y or e^{st} is equal to zero, the solution is trivial. Thus, the characteristic polynomial must equal zero, and it becomes the characteristic equation having roots that are called *eigenvalues*.

Also observe that s acts as a differential operator (the reader may encounter alternative forms such as p or D) and that multiplication by s implies a differentiation. Because y has been assumed to be of the form $y_c = Y e^{st}$,

can be written

3. Obtain the roots of the characteristic equation. These roots will be the values of the exponents in the complementary function. If the roots of the charac-

teristic equation are $s_1, s_2, s_3, \ldots, s_{n-1}$, and s_n, the complementary function for the nth-order differential equation can be written in terms of n exponential components, each with an arbitrary constant:

$$y_c = Y_1 e^{s_1 t} + Y_2 e^{s_2 t} + y_3 e^{s_3 t} + \cdots + Y_{n-1} e^{s_{n-1} t} + Y_n e^{s_n t} \qquad (6.17)$$

4. Use the *method of undetermined coefficients* or, in the case of a second-order equation, the method of variation of the parameters to evaluate the particular integral. The method of undetermined coefficients is discussed in Section 6.7 and is based on a consideration of eq. (6.14).

5. Obtain the general solution to the differential equation by adding the complementary function (step 3) to the particular integral (step 4). This solution will contain n arbitrary constants.

6. Use the initial conditions to evaluate the arbitrary constants. When all of the arbitrary constants are evaluated, the total response will be obtained. The natural- and forced-response components can then be easily identified. The natural response will contain all exponential terms that derive from the complementary function, and the forced response will be in the same form as the forcing function. The total response is sum of the natural response and the forced response.

PROCEDURE FOR SOLUTION (ZERO-INPUT AND ZERO-STATE RESPONSE)   SECTION 6.6

The procedure for finding the zero-input and zero-state responses follows the procedure outlined in the foregoing section to a point. Equation (6.17) at the end of step 3 represents the complementary function and contains n arbitrary constants; and at this point, there is a departure in the procedure:

3a. Evaluate the arbitrary constants in eq. (6.17) from the initial conditions. The result will be the zero-input response.

The next departure from the natural-response–forced-response procedure occurs at the end of step 5, where the general solution is formed by adding the complementary function found in step 3 (with n arbitrary constants) to the particular integral found in step 5:

5a. Determine the zero-state response by evaluating the arbitrary constants in the general solution with all initial conditions set equal to zero. This will yield the zero-state response.

Step 6 in the natural-response–forced-response procedure is not necessary in the zero-input–zero-state procedure. Recall that a flow chart for both procedures is shown in Fig. 6.3. However, the total response is the sum of the zero-input response and the zero-state response.

FINDING THE PARTICULAR INTEGRAL   SECTION 6.7

Because the forced response will resemble the forcing function in form, one method for obtaining the particular integral, the *method of undetermined coefficients*, requires the analyst to assume the form of the forcing function in order to obtain the particular

TABLE 6.1   The form of the assumed particular integral

$f(t)$ Form	y_{pi} Choice
a (a constant)	A
at^{α}	$At^{\alpha} + Bt^{\alpha-1} + Ct^{\alpha-2} + \cdots$
$a \cos \omega t$	$A \cos \omega t + B \sin \omega t$
$b \sin \omega t$	$A \cos \omega t + B \sin \omega t$
$ae^{\beta t}$	$Ae^{\beta t}$

Note: If any term in the forcing function $f(t)$ appears in the complementary function, it must be multiplied by the least power of t that will ensure that it is unlike any term in the complementary function.

integral. This assumed form and all of its derivatives are then substituted into the nonhomogeneous differential equation. The method has merit because only differentiations are required for its use. However, it fails dismally in an attempt to apply it to an integro-differential equation.

1. Solve the related homogeneous equation to obtain the complementary function and be sure to take note of its form.
2. Set down the variable parts of the forcing function $f(t)$ and the variable parts of any other terms obtainable by differentiating $f(t)$ in accordance with Table 6.1. This will be the assumed form of the particular integral.
3. Group the terms found in step 2 so that all terms obtainable from a single component of y_{pi} appear in only one group.
4. If, in some group, any term is identical in form to a term of the complementary function, multiply this term by the lowest possible integral power of the independent variable (in this case, the time t) that will make it differ from that term in the complementary function.
5. Substitute the assumed form of y_{pi} into the nonhomogeneous differential equation and perform the indicated differentiation or differentiations.
6. Solve for the as yet undetermined coefficients by equating the coefficients of like terms on both sides of the nonhomogeneous equation.

■ **EXAMPLE 6.2**

The differential equation

$$\frac{di}{dt} + 4i = 16t^2 + 10; \qquad t > 0$$

represents the current flow in the network of Fig. 6.6 when the switch closes at $t = 0$. Assume that the current flow at $t = 0^+$ is $i(0^+) = 2$ A, and determine the total response.

Solution   The complementary function, designated i_c, is determined from the homogeneous equation

$$\frac{di}{dt} + 4i = 0; \qquad t > 0$$

FIGURE 6.6

Network for Example 6.2

The switch closes at $t = 0$.

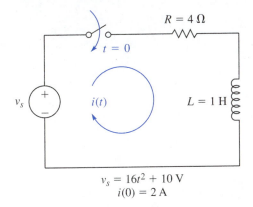

$$v_s = 16t^2 + 10 \text{ V}$$
$$i(0) = 2 \text{ A}$$

One assumes a solution of the form

$$i_c = Ie^{st}$$

so that when this and its derivative

$$\frac{di_c}{dt} = sIe^{st}$$

are substituted into the homogeneous equation, one obtains

$$sIe^{st} + 4Ie^{st} = 0$$

or

$$(s + 4)Ie^{st} = 0$$

This yields the characteristic equation

$$(s + 4) = 0$$

which has a single root, $s = -4$, making the complementary function

$$i_c = Ie^{-4t}$$

The particular integral designated as i_{pi} is determined from the non-homogeneous equation

$$\frac{di_{pi}}{dt} + 4i_{pi} = 16t^2 + 10$$

by assuming the form of the particular integral in accordance with Table 6.1. To accommodate the $16t^2$ in the forcing function, one chooses

$$At^2 + Bt + C$$

There is no need to choose a constant term to associate with the 10 in the forcing function because a constant term C has already been included in the assumed form of i_{pi}.

When i_{pi} and its derivative

$$\frac{di_{pi}}{dt} = 2At + B$$

are substituted into the nonhomogeneous differential equation, the result is

$$(2At + B) + 4(At^2 + Bt + C) = 16t^2 + 10$$

and when terms are collected, the result is

$$4At^2 + (2A + 4B)t + (B + 4C) = 16t^2 + 10$$

For the foregoing to be an identity, the coefficients of like terms on either side of the equal sign must be equal. This fact can be used to provide three simultaneous equations in the three unknown (as yet undetermined) coefficients:

$$4A \qquad\qquad = 16$$
$$2A + 4B \qquad = \ 0$$
$$B + 4C = 10$$

These simultaneous equations are not formidable and they are easily solved. The reader may verify that

$$A = 4 \qquad B = -2 \qquad C = 3$$

and the particular integral is

$$i_{pi} = 4t^2 - 2t + 3$$

The general solution is the sum of the complementary function and the particular integral:

$$i = i_c + i_{pi} = Ie^{-4t} + 4t^2 - 2t + 3$$

and the arbitrary constant may be evaluated from the initial condition,

$$i(0^+) = 2 \text{ A}$$

Thus,

$$i(0^+) = 2 = Ie^0 + 4(0) - 2(0) + 3$$
$$2 = I + 3$$

or

$$I = -1$$

This leads to the total response

$$i(t) = 4t^2 - 2t + 3 - e^{-4t} \text{ amperes}$$

$$(6.18) \quad \blacksquare$$

EXERCISE 6.2

Determine the particular solution of

$$\frac{dv}{dt} + 2v = 24t + 16; \qquad t > 0$$

when $v(t = 0^+) = 24$ V.

Answer $\quad v = 22e^{-2t} + 12t + 2$ volts.

6.7.1 Natural and Forced Responses

The natural and forced responses are easily identified from an inspection of the total response. In eq. (6.18), the natural response is

$$i_N = -e^{-4t} \text{ amperes}$$

because it is the only component of the total response associated with the exponential. Observe that the natural response eventually disappears.

The forced response is also easily identified:

$$i_F = 4t^2 - 2t + 3 \text{ amperes}$$

and the resemblance of the forced response to the forcing function may be noted.

6.7.2 Zero-Input and Zero-State Responses

The zero-input response is determined from the complementary function. In Example 6.2, where

$$i_c = Ie^{-4t}$$

use of the initial condition $i(0^+) = 2$ A gives

$$i(0^+) = 2 = Ie^0 = I$$

so that the zero-input response is

$$i_{ZI} = 2e^{-4t} \text{ amperes}$$

The zero-state response is determined from the general solution with the initial condition set equal to zero:

$$i(0^+) = 0 = Ie^0 + 4(0) + 2(0) + 3 = I + 3$$

or

$$I = -3$$

and

$$i_{ZS} = 4t^2 - 2t + 3 - 3e^{-4t} \text{ amperes}$$

The total response is the sum of the zero-input and zero-state responses:

$$i = i_{ZI} + i_{ZS} = 2e^{-4t} + 4t^2 - 2t + 3 - 3e^{-4t} \text{ amperes}$$

or

$$i(t) = 4t^2 - 2t + 3 - e^{-4t} \text{ amperes}$$

Note that the particular solution is the same in both cases and that the natural response and the zero-input response are not equal.

EXERCISE 6.3

The particular solution of

$$\frac{dv}{dt} + 2v = 12t^2 + 6e^{-t} - 2e^{-2t}; \qquad t > 0$$

when $v(t = 0^+) = 24$ V is

$$v = 15e^{-2t} + 6t^2 - 6t + 3 + 6e^{-t} - 2te^{-2t}$$

Identify the natural, forced, zero-input, and zero-state responses.

Answer

$$v_N = 15e^{-2t} \text{ volts}$$

$$v_F = 6t^2 - 6t + 3 + 6e^{-t} - 2te^{-2t} \text{ volts}$$

$$v_{ZI} = 24e^{-2t} \text{ volts}$$

$$v_{ZS} = 6t^2 - 6t + 3 + 6e^{-t} - 2te^{-2t} - 9e^{-2t} \text{ volts}$$

SECTION 6.8   *RL AND RC NETWORKS*

A first-order network is one that contains but a single energy storage element such as an inductor or capacitor. Each of these elements, either singly or in combination and in association with resistors, may be arranged in series, parallel, series-parallel, or parallel-series. Whatever the arrangement, the differential equation that governs the behavior of the network is of first order.

The procedures developed through Section 6.7 will now be utilized to predict the zero-input, zero-state, natural, and forced responses of first-order networks to various types of input or forcing functions. An example of the formulation and the solution of the governing differential equation for the undriven case (no forcing function) is presented in the next section.

THE UNDRIVEN FIRST-ORDER NETWORK SECTION 6.9

■ EXAMPLE 6.3

In the network shown in Fig. 6.7, the switch closes at $t = 0$ when the capacitor has a voltage of V_o across its terminals. Find the current as a function of time after the switch is closed.

Solution   The formulation of the differential equation begins with an application of KVL. At any time after the switch is closed,

$$v_R + v_C = 0$$

and the elemental equations for the resistor and capacitor may be substituted to give the integro-differential equation

$$Ri + \frac{1}{C} \int_{-\infty}^{t} i \, dz = 0$$

This may be differentiated to yield, after some algebra,

$$\frac{di}{dt} + \frac{1}{RC} i = 0; \qquad t > 0 \tag{6.19}$$

which is homogeneous and which, in turn, indicates that there will be no forced or zero-state responses.

Assume that

$$i = Ie^{st}$$

and substitute this and its derivative

$$\frac{di}{dt} = sIe^{st}$$

An RC series network that is excited by the energy stored in the capacitor when the switch closes instantaneously at $t = 0$ **FIGURE 6.7**

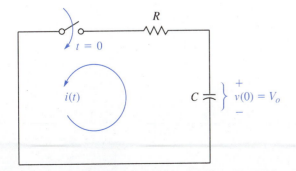

into eq. (6.19):

$$sIe^{st} + \frac{1}{RC} Ie^{st} = 0$$

so that with just a little algebra

$$\left(s + \frac{1}{RC} \right) Ie^{st} = 0$$

The characteristic equation is

$$s + \frac{1}{RC} = 0$$

and this equation has a single root,

$$s = -\frac{1}{RC}$$

so that

$$i = Ie^{-t/RC}$$

Observe that the time constant (Section 5.5) of this first-order network is $T = RC$ and that the arbitrary constant I can be determined from a consideration of the initial condition $v(0^+) = V_o$ with cognizance taken of the direction of current flow. From KVL and for all time,

$$v_R + v_C = 0$$

In particular, at time $t = 0^+$, with (note the direction of the current flow)

$$v_R(0^+) = -v_C(0^+) = -(-V_o) = V_o$$

so that

$$Ri(0^+) = V_o$$

and

$$i(0^+) = \frac{V_o}{R}$$

Thus,

$$i(0^+) = \frac{V_o}{R} = Ie^0 = I$$

and

$$i(t) = \frac{V_o}{R} e^{-t/RC} \text{ amperes}$$

This shows that the current will decay to an eventual zero value because all of the energy originally contained in the capacitor at $t = 0$ will be dissipated irreversibly by the resistor as heat. This current represents both the zero-input and natural responses. ∎

The differentiation performed to change the governing integro-differential equation to a differential equation is perfectly legal. But such differentiation can pose difficulties when a forcing function is involved and zero-state and forced responses exist. The differentiation can be rendered unnecessary by working with the capacitor voltage or the charge in the network.

To work with the capacitor voltage, write a KVL relationship,

$$v_R + v_C = 0$$

as before, but then use only the elemental equation for the resistor. With $v_C = v$,

$$Ri + v = 0$$

and with $i = C\,dv/dt$, one obtains

$$RC\frac{dv}{dt} + v = 0$$

or

$$\frac{dv}{dt} + \frac{1}{RC}v = 0; \qquad t > 0$$

By assuming

$$v = Ve^{st}$$

one can obtain

$$v = Ve^{-t/RC}$$

and with $v(0^+) = V_o$,

$$v = V_o e^{-t/RC} \text{ volts}$$

Then for this case, because of the assumed direction of current flow, as indicated in Fig. 6.7,

$$i = -C\frac{dv}{dt} = -C\frac{d}{dt}V_o e^{-t/RC} = \frac{C}{RC}V_o e^{-t/RC}$$

or

$$i(t) = \frac{V_o}{R}e^{-t/RC} \text{ amperes}$$

as before.

To work with the charge, recognize that $i = dq/dt$ and $q = \int i\, dt$, so that

$$v_R + v_C = Ri + \frac{1}{C}\int_{-\infty}^{t} i\, dz = 0$$

becomes, after some algebra,

$$\frac{dq}{dt} + \frac{1}{RC}q = 0; \qquad t > 0$$

By assuming that

$$q = Qe^{st}$$

one can obtain

$$q = Qe^{-t/RC}$$

and then with $q(0^+) = Cv(0^+) = CV_o = Q_o$,

$$q = CV_o e^{-t/RC} \text{ coulombs}$$

In this case, as indicated in Fig. 6.7, $i = -dq/dt$, and

$$i(t) = -\frac{d}{dt}\, CV_o e^{-t/RC} = \frac{CV_o}{RC}\, e^{-t/RC} = \frac{V_o}{R}\, e^{-t/RC} \text{ amperes}$$

EXERCISE 6.4

In the network of Fig. 6.8, the switch moves instantaneously from position 1 to position 2 at $t = 0$. Find $i(t)$.

Answer $i(t) = 20e^{-8t}$ amperes.

FIGURE 6.8   Network (Exercise 6.4)

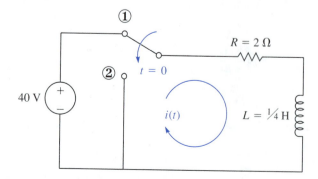

DRIVEN FIRST-ORDER NETWORKS   SECTION 6.10

The intent of this section is to show the formulation and solution of the differential equations governing the performance of *RL* and *RC* networks when subjected to a variety of forcing functions. Natural, forced, zero-input, zero-state, and total responses will be obtained in each case. In an attempt to maintain the interest of the reader, a systematic procedure will be followed; and where comments seem necessary or where questions by the reader are anticipated, remarks deemed to be pertinent will be made.

6.10.1 Excitation by a Step Input

■ EXAMPLE 6.4

In the network of Fig. 6.9, the switch has been in position 1 for a considerable period of time. At $t = 0$, the switch is instantaneously moved from position 1 to position 2. The current as a function of time is sought.

Solution

a. The differential equation for $i(t)$ is formulated via KVL:

$$v_L + v_R = v_{s2} = V_s$$

b. The elemental equations are

$$v_L = L\frac{di}{dt} \qquad v_R = Ri$$

c. The differential equation is

$$L\frac{di}{dt} + Ri = V_s; \qquad t > 0$$

d. The homogeneous differential equation is

$$L\frac{di}{dt} + Ri = 0$$

An *RL* first-order network in which the switch moves instantaneously at $t = 0$ from position 1 to position 2   **FIGURE 6.9**

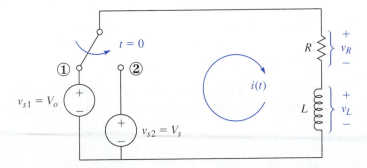

e. The characteristic equation derived from the assumption of $i_c = Ie^{st}$ is

$$Ls + R = 0$$

f. The complementary function is

$$i_c = Ie^{(-R/L)t}$$

g. The time constant is

$$T = \frac{L}{R}$$

h. The assumed form of the particular integral is

$$i_{pi} = A$$

i. The solution for the particular integral is

$$RA = V_s$$

$$i_{pi} = A = \frac{V_s}{R}$$

j. The general solution is

$$i = Ie^{(-R/L)t} + \frac{V_s}{R}$$

k. For the initial condition, with the switch in position 1 for a considerable period of time, KVL governs:

$$v_L + v_R = v_{s1} = V_o$$

and because the current is steady, $di/dt = 0$. In particular, at $t = 0^-$, the instant just before the movement of the switch, $v_L(0^-) = 0$ and

$$v_R(0^-) = Ri(0^-) = V_o$$

This makes

$$i(0^-) = \frac{V_o}{R}$$

which, by the continuity of stored energy in the inductor, is the current flow at $t = 0^+$, the instant just after the movement of the switch. Thus,

$$i(0^+) = \frac{V_o}{R}$$

l. The zero-input response from the complementary function and $i(0^+) = V_o/R$ is

$$i_{ZI} = \frac{V_o}{R} e^{-(R/L)t} \text{ amperes}$$

m. The zero-state response from the general solution with $i(0^+) = 0$ uses

$$i(0^+) = 0 = I + \frac{V_s}{R}$$

so that

$$I = -\frac{V_s}{R}$$

which gives the zero-state response:

$$i_{ZS} = \frac{V_s}{R} (1 - e^{-(R/L)t})$$

n. The natural response from the general solution, with $i(0^+) = V_o/R$, is

$$i(0^+) = \frac{V_o}{R} = Ie^0 + \frac{V_s}{R} = I + \frac{V_s}{R}$$

or

$$I = \frac{V_o - V_s}{R}$$

This value of I, when placed in the complementary function, provides the natural response

$$i(t) = \frac{V_o - V_s}{R} e^{-(R/L)t} \text{ amperes}$$

o. The forced response is the particular integral:

$$v_F = \frac{V_s}{R}$$

p. The total response is the sum of the zero-input and zero-state responses or the natural and the forced responses:

$$i(t) = \frac{V_o - V_s}{R} e^{-(R/L)t} + \frac{V_s}{R}$$

∎

EXERCISE 6.5

In Fig. 6.10, the switch closes at $t = 0$ when the voltage across the capacitor is 4 V. Determine $v(t)$.

Answer $v = 24 - 20e^{-2t}$ volts.

FIGURE 6.10   Network (Exercise 6.5)

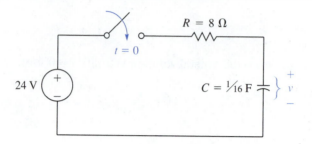

6.10.2 Excitation by a Ramp Input

■ **EXAMPLE 6.5**

The switch in the network of Fig. 6.11 closes instantaneously at $t = 0$. The source provides a current ramp of strength K $[i(t) = Kr(t)]$ to the parallel combination of resistance and capacitance. The capacitor is initially charged so that a voltage V_o appears across its terminals just prior to the closing of the switch. Find an expression for the voltage across the parallel combination as a function of time.

Solution

a. The differential equation for $i(t)$ is formulated via KCL:

$$i_C + i_R - i_s = 0$$

b. The elemental equations are

$$i_R = \frac{v}{R} \qquad i_C = C \frac{dv}{dt}$$

c. The differential equation is

$$C \frac{dv}{dt} + \frac{1}{R} v = Kr(t); \qquad t > 0$$

FIGURE 6.11   A parallel network driven by a current ramp

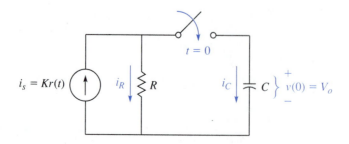

d. The homogeneous differential equation is

$$C\frac{dv}{dt} + \frac{1}{R}v = 0$$

e. The characteristic equation derived from the assumption of $v_c = Ve^{st}$ is

$$Cs + \frac{1}{R} = 0$$

f. The complementary function is

$$v_c = Ve^{-t/RC}$$

g. The time constant is

$$T = RC$$

h. The assumed form of the particular integral is

$$v_{\text{pi}} = At + B$$

i. The solution for the particular integral is

$$CA + \frac{1}{R}(At + B) = Kt$$

$$\frac{A}{R}t + \left(CA + \frac{B}{R}\right) = Kt$$

For this to be an identity, it is required that

$$\frac{A}{R} = K$$

and

$$AC + \frac{B}{R} = 0$$

Thus,

$$A = RK$$

and

$$B = -RCA = -R^2KC$$

The particular integral is therefore

$$v_{\text{pi}} = RKt - R^2KC$$

j. The general solution is

$$v = v_c + v_{\text{pi}} = Ve^{-t/RC} + RKt - R^2CK$$

k. The initial condition is given:

$$v(0^+) = V_o$$

l. The zero-input response from the complementary function and $v(0^+) = V_o$ uses

$$v(0^+) = V_o = Ve^0$$

so that the arbitrary constant V is

$$V = V_o$$

and

$$v_{ZI} = V_o e^{-t/RC} \text{ volts}$$

m. The zero-state response from the general solution, with $v(0^+) = 0$, is

$$v(0^+) = 0 = Ve^0 + RK(0) - R^2CK$$

so that

$$V = R^2CK$$

and

$$v_{ZS} = R^2CKe^{-t/RC} + RKt - R^2CK$$

n. The natural response from the general solution, with $v(0^+) = V_o$, is

$$v(0^+) = V_o = Ve^0 + RK(0) - R^2CK$$
$$V = V_o + R^2CK$$

and

$$v_N = (V_o + R^2CK)e^{-t/RC} \text{ volts}$$

o. The forced response is the particular integral:

$$v_F = RKt - R^2CK$$

p. The total response is the sum of the zero-input and zero-state responses or the natural and the forced responses:

$$v = V_o e^{-t/RC} + R^2CKe^{-t/RC} + RKt - R^2CK \text{ volts} \qquad (6.20) \quad \blacksquare$$

Remark   Observe that at $t = 0$, $v(0) = V_o$, as it should, and that after a great deal of time has passed,

$$v = RK(t - RC)$$

or because $T = RC$,

$$v = RK(t - T)$$

The voltage response of the network in Fig. 6.11 with $R = 1000\ \Omega$, $K = 2$ A/s, $C = 10\ \mu$F, and $V_o = 10$ V

FIGURE 6.12

The initial decay is due to the initial voltage across the capacitor.

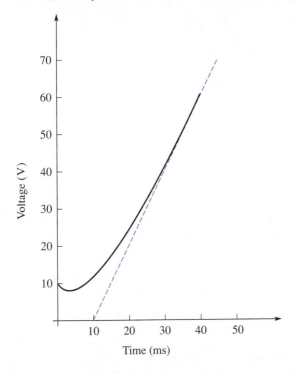

This shows that after a great deal of time has passed, the voltage across the parallel combination is a ramp. This is to be expected when a ramp forcing function is provided. Moreover, the ramp response so obtained, although modified by the value of R, is delayed from the forcing ramp by a single time constant.

Equation (6.20) is plotted in Fig. 6.12 for values of $K = 2$ A/s, $R = 1000\ \Omega$, $C = 10\ \mu$F, and $V_o = 10$ V. Observe that the ramp eventually attains a slope and a displacement, which, as shown by the dashed line, intersects the time coordinate axis at a value equal to one time constant, $T = RC = 1000(10 \times 10^{-6}) = 0.01$ s, or 10 ms.

EXERCISE 6.6

In Fig. 6.13, the switch moves from position 1 to position 2 instantaneously at $t = 0$. Find the natural, forced, zero-input, and zero-state current responses.

Answer

$$i_N = \frac{97}{16} e^{-8t} \text{ amperes}$$

$$i_F = \frac{t}{2} - \frac{1}{16} \text{ amperes}$$

$$i_{ZI} = 6e^{-8t} \text{ amperes}$$

$$i_{ZS} = \frac{1}{16} e^{-8t} + \frac{t}{2} - \frac{1}{16} \text{ amperes}$$

FIGURE 6.13   Network (Exercise 6.6)

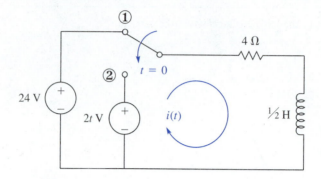

6.10.3 Excitation by a Sinusoidal Input

■ EXAMPLE 6.6

In Fig. 6.14, the switch closes at $t = 0$ when a voltage V_o exists across the capacitor terminals. The current $i(t)$ is sought.

Solution

 a. The differential equation for $i(t)$ is formulated via KVL:

$$v_R + v_C = v_s$$

 b. The elemental equations are

$$v_R = Ri \qquad v_C = \frac{1}{C} \int_{-\infty}^{t} i \, dz$$

 c. The differential equation is

$$Ri + \frac{1}{C} \int_{-\infty}^{t} i \, dz = V_m \sin \omega t; \qquad t > 0$$

FIGURE 6.14   Network excited by a sinusoidal voltage when the switch closes at $t = 0$

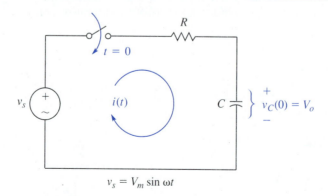

$$v_s = V_m \sin \omega t$$

The voltage across the capacitor can be treated as the dependent variable. Then, $v_C = (1/C) \int i\, dt$ and $i = C\, dv/dt$; and the differential equation becomes

$$\frac{dv}{dt} + \frac{1}{RC} v = \frac{V_m}{RC} \sin \omega t; \qquad t > 0$$

d. The homogeneous differential equation is

$$\frac{dv}{dt} + \frac{1}{RC} v = 0$$

e. The characteristic equation derived from the assumption of $v_c = Ve^{st}$ is

$$s + \frac{1}{RC} = 0$$

f. The complementary function is

$$v_c = Ve^{-t/RC}$$

g. The time constant is

$$T = RC$$

h. The assumed form of the particular integral is

$$v_{pi} = A \sin \omega t + B \cos \omega t$$

i. The solution for the particular integral uses

$$v_{pi} = A \sin \omega t + B \cos \omega t$$

and its derivative

$$v'_{pi} = \omega A \cos \omega t - \omega B \sin \omega t$$

substituted into the differential equation, giving the result

$$RC\omega(A \cos \omega t - B \sin \omega t) + A \sin \omega t + B \cos \omega t = V_m \sin \omega t$$

A regrouping of terms yields

$$(A - \omega RCB) \sin \omega t + (\omega RCA + B) \cos \omega t = V_m \sin \omega t$$

In order for the foregoing to be an identity, the coefficients of both the sine and cosine terms must match. Thus, two simultaneous algebraic equations in the unknown coefficients A and B result:

$$A - \omega RCB = V_m$$

$$\omega RCA + B = 0$$

and the reader may verify that the solutions for A and B are

$$A = \frac{1}{1 + (\omega RC)^2} V_m$$

$$B = -\frac{\omega RC}{1 + (\omega RC)^2} V_m$$

This makes

$$v_{\text{pi}} = \frac{V_m}{1 + (\omega RC)^2} \left[\sin \omega t - (\omega RC) \cos \omega t\right]$$

j. The general solution is

$$v = v_c + v_{\text{pi}} = Ve^{-t/RC} + \frac{V_m}{1 + (\omega RC)^2} \left[\sin \omega t - (\omega RC) \cos \omega t\right]$$

k. The initial condition is given:

$$v(0^+) = V_o$$

l$_1$. To find the zero-input voltage response from the complementary function and $v(0^+) = V_o$, first find the arbitrary constant V:

$$v(0^+) = V_o = Ve^0 = V$$

and then with $V = V_o$,

$$v_{\text{ZI}} = V_o e^{-t/RC}$$

l$_2$. The zero-input current response from $i = C \, dv/dt$ is

$$i_{\text{ZI}} = -\frac{V_o}{R} e^{-t/RC} \text{ amperes}$$

m$_1$. To find the zero-state voltage response from the general solution, with $v(0^+) = 0$, first find the arbitrary constant:

$$v(0^+) = 0 = Ve^0 + \frac{V_m}{1 + (\omega RC)^2} \left[\sin(0) - (\omega RC)(1)\right]$$

so that

$$V = \frac{(\omega RC)V_m}{1 + (\omega RC)^2}$$

and

$$v_{\text{zs}} = \frac{V_m}{1 + (\omega RC)^2} \left[(\omega RC)e^{-t/RC} + \sin \omega t - (\omega RC) \cos \omega t\right] \text{ volts}$$

m₂. The zero-state current response from $i = C\,dv/dt$ is (which the reader may wish to verify)

$$i_{zs} = \frac{\omega C V_m}{1 + (\omega RC)^2}\left[(\omega RC)\sin \omega t + \cos \omega t\right] - \frac{\omega C V_m}{1 + (\omega RC)^2}\,e^{-t/RC}$$

n₁. To find the natural voltage response from the general solution, with $v(0^+) = V_o$, first obtain the arbitrary constant V:

$$v(0^+) = V_o = Ve^0 + \frac{V_m}{1 + (\omega RC)^2}\left[\sin(0) - (\omega RC)\cos(0)\right]$$

so that

$$V = V_o + \frac{\omega RC V_m}{1 + (\omega RC)^2}$$

and

$$v_N = \left[V_o + \frac{\omega RC V_m}{1 + (\omega RC)^2}\right]e^{-t/RC}\ \text{volts}$$

n₂. The natural current response from $i = C\,dv/dt$ is (which the reader may wish to verify)

$$i_N = -\left[\frac{V_o}{R} + \frac{\omega C V_m}{1 + (\omega RC)^2}\right]e^{-t/RC}\ \text{amperes}$$

o₁. The forced voltage response is the particular integral:

$$v_F = v_{pi} = \frac{V_m}{1 + (\omega RC)^2}\left[\sin \omega t - (\omega RC)\cos \omega t\right]$$

o₂. The forced current response from $i = C\,dv/dt$ is (which the reader may wish to verify)

$$i_F = \frac{(\omega C)V_m}{1 + (\omega RC)^2}\left[(\omega RC)\sin \omega t + \cos \omega t\right]\ \text{amperes}$$

p. The total response is the sum of the zero-input and zero-state responses or the natural and forced responses:

$$i(t) = -\frac{V_o}{R}e^{-t/RC} + \frac{\omega C V_m}{1 + (\omega RC)^2}\left[(\omega RC)\sin \omega t + \cos \omega t\right]$$

$$-\frac{\omega C V_m}{1 + (\omega RC)^2}\,e^{-t/RC}$$

∎

EXERCISE 6.7

In Fig. 6.15, the switch closes instantaneously at $t = 0$ when the voltage across the capacitor terminals is $v(0) = 12$ V. Find the natural, forced, zero-input, and zero-state voltage responses.

FIGURE 6.15   Network (Exercise 6.7)

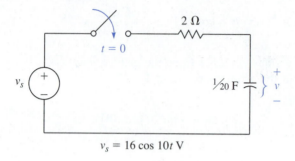

$$v_s = 16 \cos 10t \text{ V}$$

Answer

$$v_N = 4e^{-10t} \text{ volts}$$

$$v_F = 8 \cos 10t + 8 \sin 10t \text{ volts}$$

$$v_{ZI} = 12e^{-10t} \text{ volts}$$

$$v_{ZS} = -8e^{-10t} + 8 \cos 10t + 8 \sin 10t \text{ volts}$$

6.10.4 Excitation by an Impulse Input

■ **EXAMPLE 6.7**

Figure 6.16 shows a series network with an inductor and a resistor that is driven by a unit impulse of voltage that is applied at $t = 0$. No current flows in the network prior to the application of the impulse. Find the current as a function of time.

Solution

a. The differential equation for $i(t)$ is formulated via KVL:

$$v_L + v_R = v_s$$

FIGURE 6.16   An RL series network excited by a unit impulse of voltage applied at $t = 0$

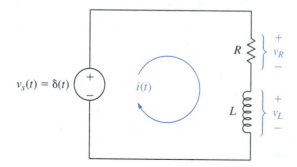

b. The elemental equations are

$$v_L = L\frac{di}{dt} \qquad v_R = Ri$$

c. The differential equation is

$$L\frac{di}{dt} + Ri = \delta(t)$$

which will hold only for the duration of the impulse between $0^- \le t \le 0^+$. For all time after $t = 0^+$, the governing differential equation will be

$$L\frac{di}{dt} + Ri = 0; \qquad t > 0$$

d. The differential equation is homogeneous:

$$L\frac{di}{dt} + Ri = 0; \qquad t > 0$$

e. The characteristic equation derived from the assumption of $i_c = Ie^{st}$ is

$$Ls + R = 0$$

f. The complementary function is

$$i_c = Ie^{-(R/L)t}$$

g. The time constant is

$$T = \frac{L}{R}$$

h. There is no assumed form of the particular integral. The governing differential equation shows that there will be no particular integral, because the equation is homogeneous. The total response will be the complementary function, which represents both the natural response and the zero-input response.

i. The solution for the particular integral is not required.

j. The general solution is

$$i = Ie^{-(R/L)t}$$

k. To determine the initial condition, one must consider the impulse acting between $0^- \le t \le 0^+$. During this fleeting instant, the continuity of stored energy principle, in its demand that the current through the inductor may not change, forces consideration of the inductor as an open circuit that prohibits the passage of current. If no current can flow, there can be no voltage drop across the resistor. Hence, by KVL, the applied voltage must be equal to the voltage across the inductor.

The current flow at $t = 0^+$ can now be obtained:

$$i(0^+) = \frac{1}{L} \int_{0^-}^{0^+} v_L \, dt = \frac{1}{L} \int_{0^-}^{0^+} \delta(t) \, dt$$

Here, the second integral is precisely the definition of the unit impulse as given by eq. (5.8b), repeated here for convenience:

$$\int_{0^-}^{0^+} \delta(t) \, dt = 1 \qquad\qquad\qquad (5.8b)$$

Hence,

$$i(0^+) = \frac{1}{L}$$

which shows that the unit impulse, in passing through, has left a current in its wake.

l. The zero-input response from the complementary function and $i(0^+) = 1/L$ is

$$i_{ZI} = \frac{1}{L} e^{-(R/L)t}$$

m. There will be no zero-state response.

n. The natural response from the general solution, with $i(0^+) = 1/L$, is

$$i_N = \frac{1}{L} e^{-(R/L)t} \text{ amperes}$$

Notice that this is identical to the zero-input response.

o. There will be no forced response.

p. In this case, the total response is equal to the zero-input and natural responses:

$$i = \frac{1}{L} e^{-(R/L)t}$$

■

Remark   The relationship between the step and impulse responses is further proof of the impact of linearity. In a linear network, if a response is produced by a certain stimulus, then for the same linear network, the derivative of the response is produced by the derivative of the stimulus.

Once again, consider Fig. 6.16, but this time, imagine that the input is not the unit impulse $\delta(t)$ but a unit step $u(t)$ applied at $t = 0$ when the initial current $i(0) = 0$. In a development that is similar to the development in Example 6.4, it can be shown that

$$i = \frac{1}{R}(1 - e^{(-R/L)t})u(t) \text{ amperes}$$

FIGURE 6.17

Network (Exercise 6.8)

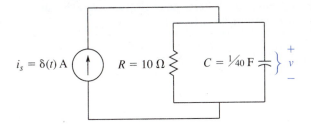

$i_s = \delta(t)$ A $R = 10\,\Omega$ $C = \frac{1}{40}$ F $+\, v\, -$

Observe that the derivative is

$$\frac{d}{dt}\left[\frac{1}{R}(1 - e^{-(R/L)t})u(t)\right] = \frac{1}{L}e^{-(R/L)t}u(t) + \frac{1}{R}(1 - e^{-(R/L)t})\delta(t)$$

which is precisely the response to the unit impulse. This indicates that the response to the unit impulse is the derivative of the response to the unit step. In general, responses to the derivatives of stimuli are the derivatives of the responses to stimuli. This fact will be put to use in Chapter 17.

EXERCISE 6.8

In Fig. 6.17, the current impulse strikes the network at $t = 0$ when the voltage across the parallel combination of the resistor and capacitor is $v(0) = 0$ V. Find the natural, forced, zero-input, and zero-state voltage responses.

Answer

$$v_N = v_{ZI} = 40e^{-4t} \text{ volts}$$

$$v_F = v_{ZS} = 0 \text{ V}$$

MORE ADVANCED CASES

As long as a network contains a single inductor or a single capacitor (or single equivalent inductors or capacitors), it does not matter how many resistors are present. Consider, for example, the network shown in Fig. 6.18a, where the voltage across the inductor is sought. The network is termed an STC network (STC for *single time constant*), and the single state variable that will completely describe the performance of the entire network is the current through the inductor.

If the inductor is considered as the load, it may be removed and a Thévenin equivalent can be found to represent the rest of the network. This is shown in Fig. 6.18b, and it is seen that the eventual voltage across the inductor when it is reconnected to the Thévenin equivalent will be partially governed by the time constant $T = L/R_T$. The Thévenin equivalent voltage is found from the network shown in Fig. 6.18c. The resistances R_1 and R_2 are

$$R_1 = 4 + 3 + 5 = 12\,\Omega$$

$$R_2 = \frac{12(12)}{12 + 12} = 6\,\Omega$$

FIGURE 6.18

(a) A more advanced network containing many resistors and one inductor, (b) the form of its Thévenin equivalent, (c) some aids to the establishment of the Thévenin equivalent voltage, (d) the evaluation of the Thévenin equivalent resistance, and (e) the final Thévenin equivalent with the load L.

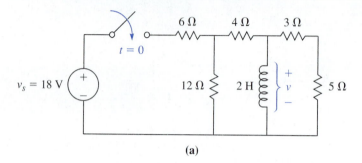

(a)

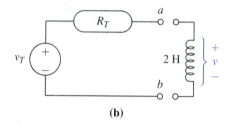

(b)

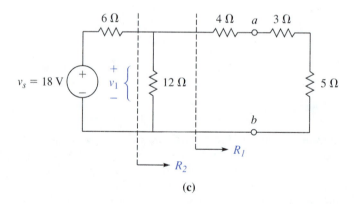

(c)

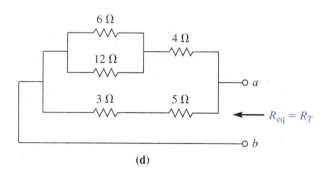

(d)

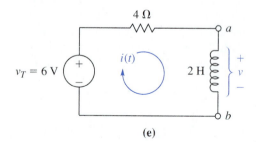

(e)

Then, two applications of voltage division provide

$$v_1 = \left(\frac{R_2}{6 + R_2}\right)v_s = \left(\frac{6}{6 + 6}\right)18 = \frac{1}{2}(18) = 9 \text{ V}$$

and

$$v_T = v_{oc} = \left(\frac{3 + 5}{4 + 3 + 5}\right)v_1 = \frac{8}{12}(9) = 6 \text{ V}$$

The Thévenin equivalent resistance is found by looking back into terminals a–b with the 18-V source replaced by a short circuit, as indicated in Fig. 6.18d. With the parallel combination of the 6- and 12-Ω resistors equivalent to $(6)(12)/(6 + 12) = 4 \, \Omega$,

$$R_{eq} = R_T = \frac{(4 + 4)(3 + 5)}{(4 + 4) + (3 + 5)} = \frac{8(8)}{16} = 4 \, \Omega$$

The final equivalent with the inductor reconnected between a and b is indicated in Fig. 6.12e. Now, the current through the inductor can be found, and the voltage across the inductor can be determined via the simple elemental equation $v_L = L \, di/dt$.

The differential equation for the current derives from an application of KVL and is

$$2\frac{di}{dt} + 4i = 6; \qquad t > 0$$

or

$$\frac{di}{dt} + 2i = 3; \qquad t > 0$$

and with a time constant of $T = L/R_T = 2/4 = 1/2$ s, the complementary function and the particular integral are, respectively,

$$i_c = Ie^{-2t} \text{ amperes}$$

and

$$i_{pi} = \frac{3}{2} \text{ A}$$

The general solution is

$$i = Ie^{-2t} + \frac{3}{2} \text{ amperes}$$

and with $i(0^+) = 0$ A, the total response becomes

$$i = \frac{3}{2}(1 - e^{-2t}) \text{ amperes} \qquad\qquad (6.21)$$

which shows that the natural and forced responses are

$$i_N = -\frac{3}{2} e^{-2t} \text{ amperes}$$

$$i_F = \frac{3}{2} \text{ A}$$

respectively.

Because the initial condition is $i(0) = 0$, there is no zero-input response; and eq. (6.21), in its entirety, represents the zero-state current response.

The voltage across the inductor eventually must decay to zero, because the steady forced-response current, $i_F = \frac{3}{2}$ A, leads, via $v = L\, di/dt$, to the forced-response voltage, $v_F = 0$ V. Thus, the natural-response voltage will be

$$v = L\frac{di}{dt} = L\frac{d}{dt}\left[\frac{3}{2}(1 - e^{-2t})\right] = 6e^{-2t} \text{ volts}$$

The Thévenin equivalent network can also be employed in cases where the first-order network contains controlled sources. Example 6.8 provides an illustration.

■ **EXAMPLE 6.8**

For the network of Fig. 6.19a, find the voltage across the capacitor (initially uncharged) for the time period that begins immediately after the switch closes instantaneously at $t = 0$.

Solution   The Thévenin equivalent of the resistive portion of the network with the capacitor as the load is shown in Fig. 6.19b. One way of finding the Thévenin equivalent voltage is by a simple mesh analysis, using the network of Fig. 6.19c. The mesh equations, with $i_a = i_2$, are

$$9i_1 - 3i_2 = 36$$
$$-3i_1 + 9i_2 = -4i_2$$

or

$$9i_1 - 3i_2 = 36$$
$$-3i_1 + 13i_2 = 0$$

A Cramer's rule solution for i_2,

$$i_2 = \frac{\begin{vmatrix} 9 & 36 \\ -3 & 0 \end{vmatrix}}{\begin{vmatrix} 9 & -3 \\ -3 & 13 \end{vmatrix}} = \frac{108}{108} = 1 \text{ A}$$

leads to the Thévenin equivalent voltage,

$$v_T = v_{oc} = 2i_2 + 4i_2 = 6i_2 = 6(1) = 6 \text{ V}$$

(a) A first-order network with a single controlled source, (b) a proposed Thévenin equivalent, (c) the network used to find the Thévenin equivalent voltage, (d) the network used to find the short circuit or *Norton* current and (e) the final equivalent

FIGURE 6.19

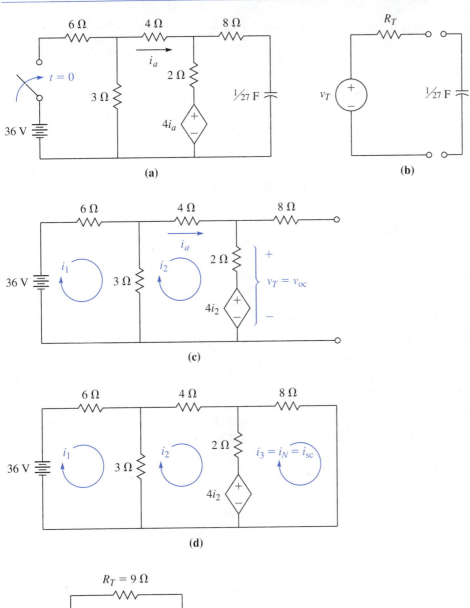

The Thévenin equivalent resistance can be determined by finding the Norton equivalent current, $i_N = i_{sc}$ in Fig. 6.19d. Another set of mesh equations,

$$9i_1 - 3i_2 \qquad\quad = \quad 36$$
$$-3i_1 + 9i_2 - \ 2i_3 = -4i_2$$
$$-2i_2 + 10i_3 = \quad 4i_2$$

or

$$9i_1 - \ 3i_2 \qquad\quad = 36$$
$$-3i_1 + 13i_2 - \ 2i_3 = \ 0$$
$$- \ 6i_2 + 10i_3 = \ 0$$

leads to a Cramer's rule solution for $i_3 = i_N$:

$$i_3 = \frac{\begin{vmatrix} 9 & -3 & 36 \\ -3 & 13 & 0 \\ 0 & -6 & 0 \end{vmatrix}}{\begin{vmatrix} 9 & -3 & 0 \\ -3 & 13 & -2 \\ 0 & -6 & 10 \end{vmatrix}} = \frac{648}{972} = \frac{2}{3}\, \text{A}$$

By eq. (4.5),

$$R_T = \frac{v_T}{i_N} = \frac{6}{\frac{2}{3}} = 9\,\Omega$$

The problem now involves the network of Fig. 6.19e, where the differential equation for the capacitor voltage after the switch closes is

$$R_T C \frac{dv}{dt} + v = v_T$$

or with the time constant $T = R_T C = 9(\frac{1}{27}) = \frac{1}{3}$ s,

$$\frac{dv}{dt} + 3v = 18; \qquad t > 0$$

Here, the complementary function and the particular integral are, respectively,

$$v_c = Ve^{-3t}\ \text{volts}$$

and

$$v_{pi} = 6\ \text{V}$$

The general solution is

$$v = Ve^{-3t} + 6\ \text{volts}$$

and with $v(0^+) = 0$ V, the total response becomes

$$v = 6(1 - e^{-3t}) \text{ volts} \qquad (6.22)$$

which shows that the natural and forced responses are

$$v_N = -6e^{-3t} \text{ volts}$$

and

$$v_F = 6 \text{ V}$$

respectively.

Because the initial condition is $v(0) = 0$, there is no zero-input response; and eq. (6.22), in its entirety, represents the zero-state voltage response. ■

SPICE EXAMPLES   SECTION 6.12

Reading: In addition to reading Appendix Sections C.1 through C.5, which were recommended for the SPICE examples in Chapters 2 and 3, one should read and understand Section C.6 before proceeding to the SPICE examples that follow.

EXAMPLE S6.1

In Fig. S6.1, the initial voltage across the capacitor is 12 V. The current $i(t)$ is to be found after the switch closes at $t = 0$. The network made ready for analysis with PSPICE is shown in Fig. S6.2, the input file is reproduced in Fig. S6.3, and pertinent extracts from the output file are presented in Fig. S6.4.

Figure S6.1   Undriven network for a first-order transient analysis

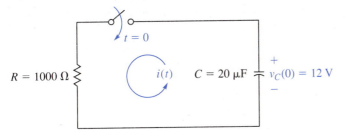

Figure S6.2   The network of Fig. S6.1 ready for PSPICE analysis

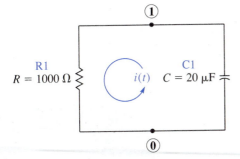

(continues)

Example S6.1 (continued)

Figure S6.3   Input PSPICE file for analysis of the network of Fig. S6.1

```
SPICE EXAMPLE - CHAPTER 6 - NUMBER 1 - RC CIRCUIT - NATURAL RESPONSE
R1        0        1          1000
C1        1        0          20U        IC=12V
*THE LINES BETWEEN THE ASTERISKS ARE THE CONTROL STATEMENTS
*************************************************************
.TRAN     750U     50M        UIC
.PLOT     TRAN     V(1)
.END
*************************************************************
```

Figure S6.4   Pertinent output for the PSPICE analysis of the network in Fig. S6.1

```
SPICE EXAMPLE - CHAPTER 6 - NUMBER 1 - RC CIRCUIT - NATURAL RESPONSE


****        TRANSIENT ANALYSIS                    TEMPERATURE =    27.000 DEG C


*****************************************************************************


TIME           V(1)
(*)---------    0.0000E+00    5.0000E+00    1.0000E+01    1.5000E+01    2.0000E+01

0.000E+00    1.200E+01  . - - - - - - - - - - - - - * - - - - - - - - - - -
7.500E-04    1.156E+01  .                    .                 .  *      .
1.500E-03    1.114E+01  .                    .                 . *       .
2.250E-03    1.072E+01  .                    .                 .*        .
3.000E-03    1.033E+01  .                    .               .*          .
3.750E-03    9.951E+00  .                    .               *           .
4.500E-03    9.584E+00  .                    .             *.            .
5.250E-03    9.229E+00  .                    .            *              .
6.000E-03    8.891E+00  .                    .          *.               .
6.750E-03    8.565E+00  .                    .        *                  .
7.500E-03    8.248E+00  .                    .      *                    .
8.250E-03    7.943E+00  .                    .      *                    .
9.000E-03    7.653E+00  .                    .    *                      .
9.750E-03    7.371E+00  .                    .  *                        .
1.050E-02    7.099E+00  .                    . *                         .
1.125E-02    6.837E+00  .                    .*                          .
1.200E-02    6.586E+00  .                   .*                           .
1.275E-02    6.344E+00  .                 . *                            .
1.350E-02    6.110E+00  .                 . *                            .
1.425E-02    5.884E+00  .               . *                              .
1.500E-02    5.669E+00  .               . *                              .
1.575E-02    5.460E+00  .              .*                                .
1.650E-02    5.259E+00  .              .*                                .
1.725E-02    5.064E+00  .             *                                  .
1.800E-02    4.879E+00  .             *                                  .
1.875E-02    4.700E+00  .          *.                                    .
1.950E-02    4.526E+00  .          *.                                    .
2.025E-02    4.359E+00  .         * .                                    .
2.100E-02    4.199E+00  .         * .                                    .
2.175E-02    4.045E+00  .        * .                                     .
2.250E-02    3.896E+00  .       *                                        .
2.325E-02    3.751E+00  .       *                                        .
2.400E-02    3.614E+00  .     *   .                                      .
2.475E-02    3.481E+00  .     *   .                                      .
2.550E-02    3.353E+00  .     *   .                                      .
2.625E-02    3.229E+00  .    *    .                                      .
2.700E-02    3.111E+00  .    *    .                                      .
2.775E-02    2.996E+00  .    *    .                                      .
2.850E-02    2.886E+00  .    *    .                                      .
2.925E-02    2.779E+00  .  *      .                                      .
3.000E-02    2.677E+00  .  *      .                                      .
3.075E-02    2.579E+00  .  *      .                                      .
3.150E-02    2.484E+00  . *       .                                      .
3.225E-02    2.392E+00  . *       .                                      .
```

Example S6.1, Figure S6.4 (continued)

```
3.300E-02  2.304E+00 .        *      .         .            .         .
3.375E-02  2.219E+00 .        *      .         .            .         .
3.450E-02  2.138E+00 .        *      .         .            .         .
3.525E-02  2.058E+00 .       *       .         .            .         .
3.600E-02  1.983E+00 .       *       .         .            .         .
3.675E-02  1.910E+00 .       *       .         .            .         .
3.750E-02  1.840E+00 .       *       .         .            .         .
3.825E-02  1.772E+00 .       *    /  .         .            .         .
3.900E-02  1.707E+00 .      *       .         .            .         .
3.975E-02  1.644E+00 .      *       .         .            .         .
4.050E-02  1.583E+00 .      *       .         .            .         .
4.125E-02  1.525E+00 .      *       .         .            .         .
4.200E-02  1.469E+00 .      *       .         .            .         .
4.275E-02  1.415E+00 .      *       .         .            .         .
4.350E-02  1.363E+00 .      *       .         .            .         .
4.425E-02  1.312E+00 .     *        .         .            .         .
4.500E-02  1.264E+00 .     *        .         .            .         .
4.575E-02  1.218E+00 .     *        .         .            .         .
4.650E-02  1.173E+00 .     *        .         .            .         .
4.725E-02  1.130E+00 .     *        .         .            .         .
4.800E-02  1.088E+00 .     *        .         .            .         .
4.875E-02  1.048E+00 .     *        .         .            .         .
4.950E-02  1.009E+00 .     *        .         .            .         .
5.000E-02  9.844E-01 .     *        .         .            .         .
                     - - - - - - - - - - - - - - - - - - - - - - - - -
```

                    JOB CONCLUDED

EXAMPLE S6.2

In Fig. S6.5, the initial voltage across the capacitor is 4 V. The voltage across the capacitor $v(t)$ is to be found after the switch closes at $t = 0$. The network made ready for analysis with PSPICE is shown in Fig. S6.6, the input file is reproduced in Fig. S6.7, and pertinent extracts from the output file are presented in Fig. S6.8.

Figure S6.5   Network driven by a 24-V battery for a   **Figure S6.6**   The network of Fig. S6.5 ready for
first-order transient analysis   PSPICE analysis

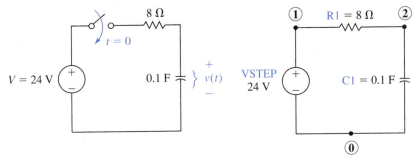

Figure S6.7   Input PSPICE file for analysis of the network of Fig. S6.5

```
SPICE EXAMPLE - CHAPTER 6 - NUMBER 2 - RC CIRCUIT - STEP EXCITATION
VSTEP    1      0          PWL(0 0 5M 0 10M 24)
R1       1      2          8
C1       2      0          .1         IC=4
*THE LINES BETWEEN THE ASTERISKS ARE THE CONTROL STATEMENTS
*****************************************************************
.TRAN    5M     150M    UIC
.PLOT    TRAN   V(1)    V(2)
.END
*****************************************************************
```

                                                                              (continues)

Example S6.2 (continued)

Figure S6.8 Pertinent output for the PSPICE analysis of the network in Fig. S6.5

```
SPICE EXAMPLE - CHAPTER 6 - NUMBER 2 - RC CIRCUIT - STEP EXCITATION


****        TRANSIENT ANALYSIS                      TEMPERATURE =    27.000 DEG C


*********************************************************************************


LEGEND:

*: V(1)
+: V(2)

 TIME            V(1)
(*)---------    0.0000E+00   1.0000E+01   2.0000E+01   3.0000E+01   4.0000E+01
(+)---------    2.0000E+00   4.0000E+00   6.0000E+00   8.0000E+00   1.0000E+01

0.000E+00   0.000E+00 *  - - - - -+ - - - - - - .- - - - - - .- - - - - - -
5.000E-03   1.440E-08 *           +            .            .            .
1.000E-02   2.400E+01 .           +            .        *   .            .
1.500E-02   2.400E+01 .          .+            .        *   .            .
2.000E-02   2.400E+01 .          . +           .        *   .            .
2.500E-02   2.400E+01 .          .  +          .        *   .            .
3.000E-02   2.400E+01 .          .  +          .        *   .            .
3.500E-02   2.400E+01 .          .    +        .        *   .            .
4.000E-02   2.400E+01 .          .     +       .        *   .            .
4.500E-02   2.400E+01 .          .      +      .        *   .            .
5.000E-02   2.400E+01 .          .       +     .        *   .            .
5.500E-02   2.400E+01 .          .       +     .        *   .            .
6.000E-02   2.400E+01 .          .        +    .        *   .            .
6.500E-02   2.400E+01 .          .         +   .        *   .            .
7.000E-02   2.400E+01 .          .          +  .        *   .            .
7.500E-02   2.400E+01 .          .          +  .        *   .            .
8.000E-02   2.400E+01 .          .           + .        *   .            .
8.500E-02   2.400E+01 .          .           +.        *   .            .
9.000E-02   2.400E+01 .          .           + .        *   .            .
9.500E-02   2.400E+01 .          .           + .        *   .            .
1.000E-01   2.400E+01 .          .           .+        *   .            .
1.050E-01   2.400E+01 .          .           . +       *   .            .
1.100E-01   2.400E+01 .          .           . +       *   .            .
1.150E-01   2.400E+01 .          .           .  +     * .            .
1.200E-01   2.400E+01 .          .           .       +*   .            .
1.250E-01   2.400E+01 .          .           .        X   .            .
1.300E-01   2.400E+01 .          .           .        X   .            .
1.350E-01   2.400E+01 .          .           .        *+  .            .
1.400E-01   2.400E+01 .          .           .        * + .            .
1.450E-01   2.400E+01 .          .           .        * + .            .
1.500E-01   2.400E+01 .          .           .        *  + .           .
                       - - - - - - - - - - - .- - - - - - -+ - - - - - - .

        JOB CONCLUDED
```

CHAPTER 6

SUMMARY

- The response of a network may be composed of a zero-input and a zero-state response.

   —The zero-input response is the response of the network with no applied input. The zero-input response depends upon the initial state of the network, which is represented by the initial condition and the characteristics (characteristic frequencies) of the network.

—The zero-state response is the response of the network to an input applied at $t = 0$ subject to the condition that the system is in the zero state (all initial conditions are set equal to zero) just prior to the application of the input.

- The response of a network may also be composed of a natural and a forced response.

  —The natural response is the behavior of the network that is determined by the characteristics of the network and the initial conditions or the initial state of the network. If the natural response exists at all, it will eventually disappear, provided resistors are present. It is intimately associated with time-decaying terms. The time frame between $t = 0$ and the eventual disappearance of the natural response is known as the transient period, and the natural response is frequently referred to as the complementary function.

  —The forced response is the behavior of the network that is determined by the external forcing function. It can be maintained indefinitely by the forcing function, and it will be in the same form as the forcing function. The forced response is frequently referred to as the steady-state response and is often called the particular integral.

- The evaluation of the arbitrary constants that appear in the general solution of a differential equation requires the use of initial conditions.

- The procedure for finding the natural and forced responses is given in Section 6.5.

- The procedure for finding the zero-input and zero-state responses is given in Section 6.6.

- The procedure for finding the particular integral by the method of undetermined coefficients is given in Section 6.7. Guidance about the form to assume for the particular integral may be obtained by referring to Table 6.1.

- The procedure for the determination of the complete or total response of a first-order (RL or RC) network is as follows:

  —Write the governing differential or integro-differential equation from an application of KVL or KCL.

  —Reduce the governing differential or integro-differential equation to a homogeneous equation.

  —Assume an exponential solution containing an undetermined exponent and an arbitrary constant.

  —Use the assumed exponential solution to determine the complementary function.

  —Use the method of undetermined coefficients to establish the particular integral. Remember that the form of the forcing function dictates the form of the assumed value of the particular integral.

  —Add the complementary function and the particular integral to form the general solution for the total response.

  —If the initial condition is not given in the "right" form (it must be specified somehow), establish the initial condition to be used for the evaluation of the arbitrary constant in the general solution.

  —Use the initial condition in the general solution to obtain the total response.

■ The natural response is the response that contains an exponential term that does not resemble any exponential term that appears in the forcing function.

■ The forced response is that part of the total response that does not contain the natural response and is in the form of the forcing function.

■ The zero-input response is determined from the complementary function, using the actual initial condition.

■ The zero-state response is determined from the general solution for the total response, with the initial condition set equal to zero.

■ The sum of the natural and forced responses must equal the sum of the zero-input and zero-state responses.

■ Thévenin's theorem (or Norton's theorem) can be used to advantage when the first-order network contains a single inductor or capacitor and several resistors and/or controlled sources. The Thévenin or Norton equivalents are also useful when several inductors can be combined to form an equivalent single inductance or when several capacitors can be combined to form an equivalent single capacitance.

Additional Readings

Blackwell, W.A., and L.L. Grigsby. *Introductory Network Theory*. Boston: PWS Engineering, 1985, pp. 217–231, 264–286.

Bobrow, L.S. *Elementary Linear Circuit Analysis*. 2d ed. New York: Holt, Rinehart and Winston, 1987, pp. 201–207, 218–266.

Del Toro, V. *Engineering Circuits*. Englewood Cliffs, N.J.: Prentice-Hall, 1987, pp. 257–276, 290–314.

Dorf, R.C. *Introduction to Electric Circuits*. New York: Wiley, 1989, pp. 208–230, 240–262.

Hayt, W.H, Jr., and J.E. Kemmerly. *Engineering Circuit Analysis*. 4th ed. New York: McGraw-Hill, 1986, pp. 145–160, 173–185.

Irwin, J.D. *Basic Engineering Circuit Analysis*. 3d ed. New York: Macmillan, 1989, pp. 255–306.

Johnson, D.E., J.L. Hilburn, and J.R. Johnson. *Basic Electric Circuit Analysis*. 4th ed. Englewood Cliffs, N.J.: Prentice-Hall, 1989, pp. 205–246.

Karni, S. *Applied Circuit Analysis*. New York: Wiley, 1988, pp. 176–201.

Madhu, S. *Linear Circuit Analysis*. Englewood Cliffs, N.J.: Prentice-Hall, 1988, pp. 270–297.

Nilsson, J.W. *Electric Circuits*. 3d ed. Reading, Mass.: Addison-Wesley, 1990, pp. 202–219, 225–244.

Paul, C.R. *Analysis of Linear Circuits*. New York: McGraw-Hill, 1989, pp. 407–433, 442–472.

PROBLEMS   CHAPTER 6

Section 6.4

6.1   In Fig. P6.1, the switch has been in position a for a long period of time. At $t = 0^-$, it moves instantaneously from position a to position b at $t = 0^+$. Find the following:

$i_R(0^-)$	$i_L(0^-)$	$i_C(0^-)$	$v_C(0^-)$
$v_L(0^-)$	$i_R(0^+)$	$i_L(0^+)$	$i_C(0^+)$
$v_C(0^+)$	$v_L(0^+)$	$\ddot{i}_L(0^+)$	

Figure P6.1

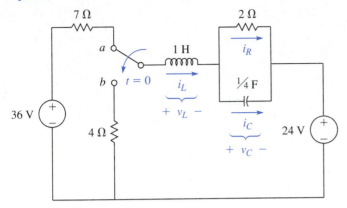

6.2   In Fig. P6.2, the switch has been closed for a long
period of time. At $t = 0$, it opens instantaneously.
Consider $t = 0^-$ and $t = 0^+$ as the instants just
before and just after the switch opens. Find the
following:

$i_{R_1}(0^-)$	$i_L(0^-)$	$i_C(0^-)$	$i_{R_2}(0^-)$
$v_L(0^-)$	$v_{R_1}(0^-)$	$v_C(0^-)$	$v_{R_2}(0^-)$
$i_{R_1}(0^+)$	$i_L(0^+)$	$i_C(0^+)$	$i_{R_2}(0^+)$
$v_L(0^+)$	$v_{R_1}(0^+)$	$v_C(0^+)$	$v'_C(0^+)$

Figure P6.2

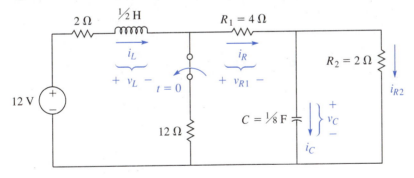

6.3   In Fig. P6.3, the switch has been open for a long
period of time, and the voltage across the $\frac{1}{20}$-F
capacitor is 40 V. At $t = 0$, the switch closes
instantaneously. Consider $t = 0^-$ and $t = 0^+$ as
the instants just before and just after the switch
closes. Find the following:

$i_{R_1}(0^+)$	$i_L(0^+)$	$i_C(0^+)$	$i_{R_2}(0^+)$
$v_L(0^+)$	$v_{R_1}(0^+)$	$v_C(0^+)$	$i'_L(0^+)$

Figure P6.3

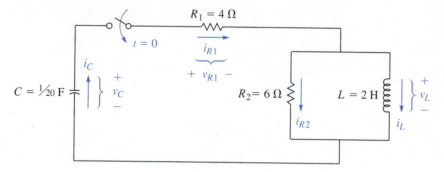

6.4 In Fig. P6.4, the switch has been open for a long period of time, and the voltage across the 1/100-F capacitor is 80 V. At $t = 0$, it closes instantaneously. Consider $t = 0^-$ and $t = 0^+$ as the instants just before and just after the switch closes. Find the following:

$i_{R_1}(0^+)$ $v_{R_1}(0^+)$ $v'_C(0^+)$

$i_{R_3}(0^+)$ $i_C(0^+)$ $i_{R_2}(0^+)$

$v_{R_3}(0^+)$ $v_{R_2}(0^+)$ $v_C(0^+)$

Figure P6.4

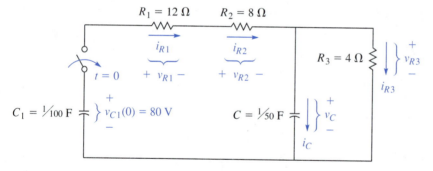

6.5 In Fig. P6.5, the switch has been in position 1 for a long period of time. At $t = 0^-$, it moves instantaneously from position 1 to position 2 at $t = 0^+$. Find the following:

$i_L(0^-)$ $i_C(0^+)$ $v_R(0^+)$

$i_R(0^-)$ $i_L(0^+)$ $i'_L(0^+)$

$v_C(0^-)$ $i_R(0^+)$ $v'_C(0^-)$

$v_L(0^-)$ $v_C(0^+)$ $i_C(0^-)$

$v_R(0^-)$ $v_L(0^+)$ $v'_C(0^+)$

Figure P6.5

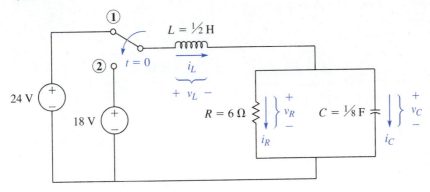

Section 6.7

The following differential equations with the indicated initial conditions have been written for $t > 0$ to describe the behavior of specific networks subjected to a variety of forcing functions. In all cases, determine the natural, forced, zero-input, and zero-state responses.

6.6 $\quad \dfrac{di}{dt} + 3i = 18; \; i(0^+) = 2 \text{ A}$

6.7 $\quad \dfrac{dv}{dt} + 4v = 6 \cos 16t; \; v(0^+) = 12 \text{ V}$

6.8 $\quad \dfrac{di}{dt} + 2i = 4t + 6; \; i(0^+) = 4 \text{ A}$

6.9 $\quad \dfrac{di}{dt} + 8i = t^2 + 4t + 4; \; i(0^+) = 2 \text{ A}$

6.10 $\quad \dfrac{dv}{dt} + 40v = 120 \cos 16t + 60; \; v(0^+) = 80 \text{ V}$

6.11 $\quad \dfrac{dv}{dt} + 50v = 24e^{-40t} + 36; \; v(0^+) = 16 \text{ V}$

6.12 $\quad \dfrac{di}{dt} + i = 8 \sin 4t + 4e^{-t}; \; i(0^+) = 1 \text{ A}$

6.13 $\quad \dfrac{dv}{dt} + 24v = 18t + 40 \sin 10t; \; v(0^+) = 20 \text{ V}$

6.14 $\quad \dfrac{dq}{dt} + 8q = 0; \; q(0^+) = 8 \; \mu\text{C}$

6.15 $\quad \dfrac{di}{dt} + 12i = 6t + 4 \cos 8t; \; i(0^+) = 0 \text{ A}$

6.16 $\quad \dfrac{dv}{dt} + 100v = 8t + 16e^{-2t}; \; v(0^+) = 100 \text{ V}$

6.17 $\quad \dfrac{dv}{dt} + 200v = 80 + 20e^{-200t}; \; v(0^+) = 0 \text{ V}$

6.18 $\quad \dfrac{dv}{dt} + 64v = 16 \cos 200t + 80e^{-64t}; \; t > 0 \; v(0^+) = 0 \text{ V}$

Section 6.9

6.19 In the network of Fig. P6.6, the switch opens instantaneously at $t = 0$. Find the current response for all $t \geq 0$.

Figure P6.6

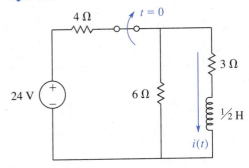

6.20 In the network of Fig. P6.7, the switch opens instantaneously at $t = 0$. Find the voltage response for all $t \geq 0$.

Figure P6.7

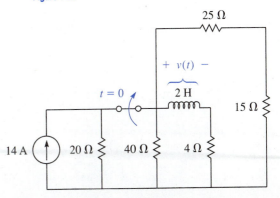

6.21 In the network of Fig. P6.8, the switch opens instantaneously at $t = 0$. Find the voltage response for all $t \geq 0$.

Figure P6.8

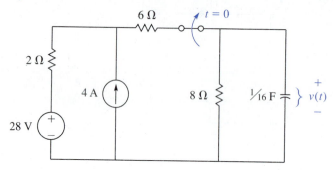

6.22 In the network of Fig. P6.9, the switch opens instantaneously at $t = 0$. Find the voltage response for all $t \geq 0$.

Figure P6.9

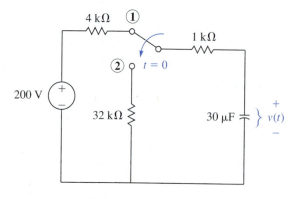

6.23 In the network of Fig. P6.10, the switch moves instantaneously from position 1 to position 2 at $t = 0$. Find the voltage response for all $t \geq 0$.

Figure P6.10

Section 6.10

6.24 In the network of Fig. P6.11, the switch moves instantaneously from position 1 to position 2 at $t = 0$. Find the current response for all $t \geq 0$.

Figure P6.11

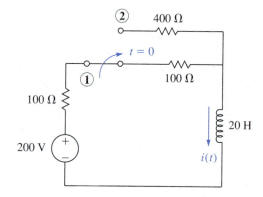

Problems 6.25 to 6.29 are based upon Fig. P6.12, where the current source designated by i_s is applied at $t = 0$. In all cases, $i(t)$ is to be found for the specified value of the current source i_s.

Figure P6.12

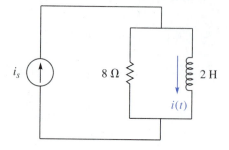

6.25 $i_s = 4\delta(t) + 8$ amperes

6.26 $i_s = 8t^2 + 4t + 12$ amperes

6.27 $i_s = 10t + 16 \sin 2t$ amperes

6.28 $i_s = 12t + 24 + 8e^{-4t}$ amperes

6.29 $i_s = 24t^3 + 18e^{-t}$ amperes

Problems 6.30 to 6.34 are based upon Fig. P6.13, where the switch closes instantaneously at $t = 0$. In all cases, $v(t)$ is to be found for the specified forcing function v_s.

Figure P6.13

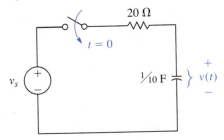

6.30 $v_s = 18t + 12 + 24 \cos t$ volts

6.31 $v_s = 16t + 4e^{-t}$ volts

6.32 $v_s = 8$ V

6.33 $v_s = 8 + 12e^{-2t} + 4 \sin 2t$ volts

6.34 $v_s = 12t^2 + 64$ volts

6.35 In the network of Fig. P6.14, the switch closes instantaneously at $t = 0$. Find the natural, forced, zero-input, and zero-state current responses for both $i_1(t)$ and $i_2(t)$ for all $t \geq 0$.

Figure P6.14

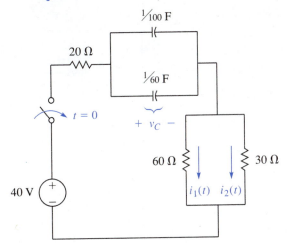

6.36 In the network of Fig. P6.15, the switch closes instantaneously at $t = 0$. Find the natural, forced, zero-input, and zero-state voltage responses for both $v_1(t)$ and $v_2(t)$ for all $t \geq 0$.

Figure P6.15

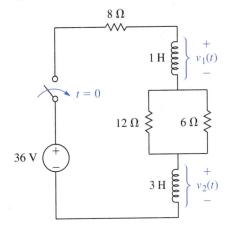

6.37 Figure P6.16 shows a network that can be used to control the current through the 4 H inductor to values between 1.1 and 1.3 A. Switch 1 (S_1) closes at $t = 0$ and switch 2 (S_2) opens at $t = t_1$ when $i(t) = 1.3$ A. Switch 2 remains open until $t = t_2$ when $i(t) = 1.1$ A, at which time it closes and remains closed until $t = t_3$, when, once again, $i(t) = 1.3$ A. Find the times t_1, t_2, and t_3 and the steady-operation ratio of switch 2 *open time* to *closed time*.

Figure P6.16

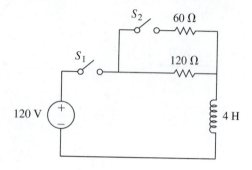

6.38 The purpose of the network in Fig. P6.17 is to use switch 1 and R_1 to permit the capacitor to charge during a specified time interval and to use switch 2 and R_2 to hold the charge at the specified level. Select a value for R_1 to make the capacitor charge to 16 V in 20 ms, and then select the value of R_2 to hold the capacitor voltage to 16 V.

Figure P6.17

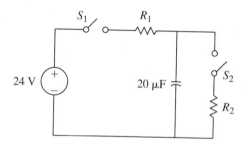

6.39 In the network of Fig. P6.18, the switch closes instantaneously at $t = 0$. Find the natural, forced, zero-input, and zero-state voltage response for $v(t)$ for all $t \geq 0$.

Figure P6.18

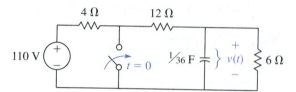

6.40 In the network of Fig. P6.19, the switch closes instantaneously at $t = 0$. Find the natural, forced, zero-input, and zero-state current responses for all $t \geq 0$.

Figure P6.19

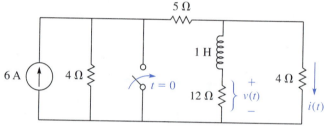

Section 6.11

6.41 In the network of Fig. P6.20, both switches close instantaneously at $t = 0$. Find the total voltage response for all $t \geq 0$.

Figure P6.20

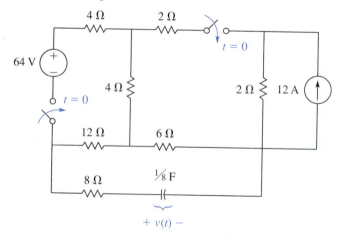

6.42 In the network of Fig. P6.21, the switch moves instantaneously from position 1 to position 2 at $t = 0$. Find the total voltage response for all $t \geq 0$.

Figure P6.20

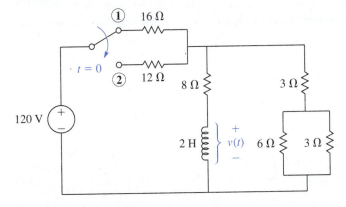

6.43 In the network of Fig. P6.22, the switch closes instantaneously at $t = 0$. Find the total current response for all $t \geq 0$.

Figure P6.22

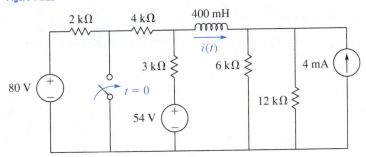

6.44 In the network of Fig. P6.23, the switch opens instantaneously at $t = 0$. Find the total voltage response for all $t \geq 0$.

Figure P6.23

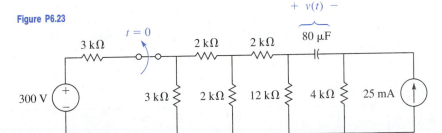

7

RLC AND HIGHER-ORDER NETWORKS

OBJECTIVES

The objectives of this chapter are to:

- Show how to formulate and solve the differential equations that govern the response of networks containing more than one energy storage element such as *RLC* networks.

- Consider the forms of response and obtain an appreciation for the underdamped, overdamped, critically damped, and nondamped forms of network response.

- Pose several examples that deal with the setup of the differential equations for *RLC* second-order networks and then solve these differential equations for the undriven case and a variety of forcing functions such as those presented in Chapter 5: the step, ramp, sinusoidal, and impulse functions.

- Treat the case of an *RLC* network with a pulsed input.

- Consider the formulation and solution of the differential equations obtained from node and mesh analysis of more advanced networks.

SECTION 7.1   **INTRODUCTION**

The study of network differential equations that began in the previous chapter continues here. More complex networks consisting of combinations of resistors, inductors, and capacitors are considered in this chapter and, as in Chapter 6, the responses to various forms of a forcing function are observed. Here, however, it will be noted that the form of the response will be dictated, in part, by component (R, L, or C) values.

SECTION 7.2   **FORMS OF RESPONSE**

The procedure developed in Section 6.5 can be used to determine the solution to a second-order differential equation. Consider

$$a\frac{d^2y}{dt^2} + b\frac{dy}{dt} + cy = f(t) \tag{7.1}$$

and look first at the complementary function, which will be obtained from

$$a\frac{d^2y_c}{dt^2} + b\frac{dy_c}{dt} + cy_c = 0 \qquad (7.2)$$

If a solution of the form

$$y_c = Ye^{st}$$

is assumed, it may be substituted into eq. (7.2), and the result will be

$$as^2 Ye^{st} + bs Ye^{st} + c Ye^{st} = 0$$

or

$$(as^2 + bs + c)Ye^{st} = 0$$

A nontrivial solution will occur only if

$$as^2 + bs + c = 0 \qquad (7.3)$$

and this is the *characteristic equation*, which may be written as

$$s^2 + 2\alpha s + \omega_n^2 = 0 \qquad (7.4)$$

where α (frequently referred to as the *attenuation*) is defined by

$$\alpha \equiv \frac{b}{2a} \qquad (7.5)$$

and where ω_n is the *natural angular frequency*, defined by

$$\omega_n^2 \equiv \frac{c}{a} \qquad (7.6)$$

There is a method for the solution of a quadratic equation known as *completing the square*. If this method is applied to eq. (7.4), the procedure is as follows:

$$s^2 + 2\alpha s + \alpha^2 + \omega_n^2 - \alpha^2 = 0$$

$$(s + \alpha)^2 = \alpha^2 - \omega_n^2$$

$$s + \alpha = \pm\sqrt{\alpha^2 - \omega_n^2}$$

and the roots are seen to be

$$s_1, s_2 = -\alpha \pm \sqrt{\alpha^2 - \omega_n^2} \qquad (7.7)$$

The radicand in eq. (7.7) requires considerable attention. It is called the *discriminant* and can be positive, zero, or negative. Three different cases are therefore apparent.

7.2.1 The Overdamped Case

The overdamped case occurs when $\alpha^2 > \omega_n^2$. In this case, the roots of eq. (7.4) are both real, negative, and unequal. Under these circumstances, the complementary function will be

$$y_c \equiv Y_1 e^{s_1 t} + Y_2 e^{s_2 t} \tag{7.8}$$

where the constants Y_1 and Y_2 are real and where

$$s_1 = -\alpha + \sqrt{\alpha^2 - \omega_n^2}$$
$$s_2 = -\alpha - \sqrt{\alpha^2 - \omega_n^2}$$

7.2.2 The Critically Damped Case

The critically damped case occurs when $\alpha = \omega_n$. Here, $\alpha^2 = \omega_n^2$ and the roots of eq. (7.4) are real, equal, and negative,

$$s_1 = s_2 = s = -\alpha$$

An attempt to write the complementary function as

$$y_c = Y_1 e^{-\alpha t} + Y_2 e^{-\alpha t}$$

is doomed to failure, because

$$y_c = (Y_1 + Y_2) e^{-\alpha t} = Y e^{-\alpha t}$$

which indicates that the complementary function for this second-order equation contains only one solution component. This clearly is inconsistent with the theory of differential equations.

One component of the complementary function is certainly

$$Y_{c1} = Y_1 e^{-\alpha t}$$

The other component can be found by taking eq. (7.4) and considering s_1 and s_2 as its roots regardless of their form. In this event, it is certainly true that

$$y_c = Y_1 e^{s_1 t} + Y_2 e^{s_2 t}$$

for all values of the constants Y_1 and Y_2, because these constants are perfectly arbitrary. Thus, it is possible to let $Y_1 = -Y_2$ and assume that

$$Y_1 = -Y_2 = \frac{1}{s_1 - s_2}$$

Then,

$$y_c = Y_1 e^{s_1 t} + Y_2 e^{s_2 t} = \frac{e^{s_1 t} - e^{s_2 t}}{s_1 - s_2} \tag{7.9}$$

In the limiting case where $s_1 \to s_2$, eq. (7.9) becomes indeterminate (0/0). However, invoking *L'Hôpital's rule* and differentiating both numerator and denominator with respect to s_2 yields

$$y_c = \frac{-te^{s_2 t}}{-1} = te^{s_2 t}$$

and when $s_1 \to s_2$,

$$y_c = te^{s_1 t}$$

is the other solution component of y_c. Thus, both $Y_1 e^{s_1 t}$ and $Y_2 te^{s_1 t}$ are components of the complementary function; and for the critically damped case, where $s_1 = s_2 = s = -\alpha$ and Y_1 and Y_2 are real,

$$y_c = Y_1 e^{-\alpha t} + Y_2 te^{-\alpha t} \tag{7.10}$$

7.2.3 The Underdamped Case

The underdamped case occurs when $\alpha < \omega_n$. This makes $\alpha^2 - \omega_n^2$ negative, and the roots of eq. (7.4) will be complex conjugates[1]. Define a *damped angular frequency*

$$\omega_d \equiv \sqrt{\omega_n^2 - \alpha^2} \tag{7.11}$$

and note that the roots of eq. (7.4) as defined by eq. (7.7) will be

$$s_1, s_2 = -\alpha \pm \sqrt{-\omega_d^2} = -\alpha \pm j\omega_d \tag{7.12}$$

where $j = \sqrt{-1}$, and for this underdamped case,

$$y_c = Y_1 e^{(-\alpha + j\omega_d)t} + Y_2 e^{(-\alpha - j\omega_d)t} \tag{7.13}$$

where Y_1 and Y_2 are complex conjugates.

The form of eq. (7.13) is not very useful. A better form can be obtained by employing the Euler relationships (see Appendix A),

$$e^{j\omega t} = \cos \omega t + j \sin \omega t$$

and

$$e^{-j\omega t} = \cos \omega t - j \sin \omega t$$

Equation (7.13) may be written as

$$y_c = e^{-\alpha t}(Y_1 e^{j\omega t} + Y_2 e^{-j\omega t})$$

and the Euler relationships may be used to provide

$$y_c = e^{-\alpha t}[Y_1(\cos \omega t + j \sin \omega t) + Y_2(\cos \omega t - j \sin \omega t)]$$

---

[1] Appendix A provides a review of complex numbers and complex number algebra.

A little algebra will then show that

$$y_c = e^{-\alpha t}[(Y_1 + Y_2) \cos \omega_d t + j(Y_1 - Y_2) \sin \omega_d t]$$

But Y_1 and Y_2 are perfectly arbitrary, so that their sum, $Y_a = Y_1 + Y_2$, and their difference, $Y_b = Y_1 - Y_2$, are also perfectly arbitrary. Thus, the complementary function may be written as

$$y_c = e^{-\alpha t}(Y_a \cos \omega_d t + Y_b \sin \omega_d t) \tag{7.14}$$

which is a much neater form than that of eq. (7.13).

The combination of sine and cosine terms displayed in eq. (7.14) can be adjusted to the amplitude–phase angle form,

$$y_c = Ye^{-\alpha t} \sin(\omega_d t + \theta)$$

where, in the terminology of eq. (7.14),

$$Y = \sqrt{Y_a^2 + Y_b^2} \tag{7.15}$$

$$\theta = \arctan \frac{Y_a}{Y_b} \tag{7.16}$$

If it is desired to use the form

$$y_c = Ye^{-\alpha t} \cos(\omega_d t + \phi)$$

then Y is given by eq. (7.15) and

$$\phi = \arctan \frac{-Y_b}{Y_a} \tag{7.17}$$

EXERCISE 7.1

If

$$y(t) = e^{-2t}(4 \cos 3t - 3 \sin 3t)$$

what are Y, ω_d, θ, and ϕ in

$$y_c = Ye^{-2t} \sin(\omega_d t + \theta)$$

and

$$y_c = Ye^{-2t} \cos(\omega_d t + \phi)$$

Answer $Y = 5$, $\omega_d = 3$ rad/s, $\theta = 126.87°$, and $\phi = 36.87°$.

7.2.4 The Damping Factor

The alert reader has undoubtedly recognized that the trick is in the recognition of the form of the natural response and to act accordingly. For assistance in this recognition process, eq. (7.4) is written in the form

$$s^2 + 2\zeta\omega_n s + \omega_n^2 = 0$$

and a damping factor is defined as

$$\zeta \equiv \frac{\alpha}{\omega_n} = \frac{b/2a}{\sqrt{c/a}} = \frac{b}{2a\sqrt{c/a}} = \frac{b}{2\sqrt{ca}} \qquad (7.18)$$

Here, the values of a, b, and c come from the form of the characteristic equation in eq. (7.3). Observe that the type of response can be pinpointed by merely examining the damping factor. For the overdamped case, $\alpha > \omega_n$ so that $\zeta > 1$. For the critically damped case, $\alpha = \omega_n$ so that $\zeta = 1$. For the underdamped case, $\alpha < \omega_n$ and $\zeta < 1$.

EXERCISE 7.2

The undriven voltage response of a second-order network is governed by the differential equation

$$\frac{d^2v}{dt^2} + 250\frac{dv}{dt} + 40{,}000v = 0$$

Determine α, ω_n, the damping factor, and the form of response.

Answer $\alpha = 125$, $\omega_n = 200$ rad/s, $\zeta = 0.625$, and the response is under-damped.

THE UNDRIVEN SECOND-ORDER NETWORK   SECTION 7.3

■ **EXAMPLE 7.1**

In the RLC parallel network shown in Fig. 7.1, the switch closes instantaneously at $t = 0$ when a voltage V_o exists across the capacitor terminals and no current is flowing through the inductor. Find the voltage across the parallel combination as a function of time.

Solution   The governing differential equation is derived from a consideration of KCL at node A in Fig. 7.1:

$$i_C + i_R + i_L = 0$$

With $i_C = C\,dv/dt$, $i_R = v/R$, and $i_L = (1/L)\int_{-\infty}^{t} v\,dz$, the system integro-differential equation,

$$C\frac{dv}{dt} + \frac{1}{R}v + \frac{1}{L}\int_{-\infty}^{t} v\,dz = 0$$

FIGURE 7.1 RLC parallel network in which the switch closes instantaneously at $t = 0$

Here, $C = \frac{1}{3}$ F, $R = \frac{3}{4} \Omega$, $L = 1$ H, and $V_o = 2$ V.

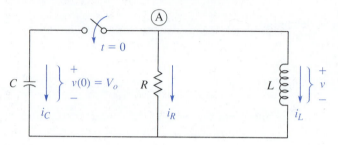

is easily formulated. This may be differentiated once to obtain

$$C\frac{d^2v}{dt^2} + \frac{1}{R}\frac{dv}{dt} + \frac{1}{L}v = 0; \qquad t > 0 \tag{7.19}$$

which is a homogeneous, second-order differential equation that can have only a natural response or a zero-input response.

Therefore, assume that

$$v_c = Ve^{st}$$

and substitute this into eq. (7.19):

$$Cs^2Ve^{st} + \frac{1}{R}sVe^{st} + \frac{1}{L}Ve^{st} = 0$$

Algebraic adjustment provides

$$\left(Cs^2 + \frac{1}{R}s + \frac{1}{L}\right)Ve^{st} = 0$$

and to avoid trivial solutions, one writes

$$Cs^2 + \frac{1}{R}s + \frac{1}{L} = 0$$

which is the characteristic equation. With $C = \frac{1}{3}$ F, $R = \frac{3}{4} \Omega$, and $L = 1$ H, the characteristic equation becomes

$$s^2 + 4s + 3 = 0 \tag{7.20}$$

Note here that $\alpha = \frac{4}{2} = 2$, $\omega_n^2 = 3$, $\omega_n = \sqrt{3}$, and $\zeta = \alpha/\omega_n = 2/\sqrt{3} > 1$. Hence, eq. (7.20) will possess unequal negative roots. It can be factored to

$$(s + 1)(s + 3) = 0$$

which gives two real and negative roots, $s_1 = -1$ and $s_2 = -3$; so that the two components of the natural response and zero-input response are

$$v_N = v_{ZI} = V_1 e^{-t} + V_2 e^{-3t} \qquad (7.21)$$

This is the general solution, and it remains to evaluate the arbitrary constants V_1 and V_2 from the initial conditions. One of these, $v(0^+) = V_o = 2$ V, is available. Another must be found from Kirchhoff law considerations and the $v - i$ relations of the elements.

At the instant the switch is closed, current begins to flow out of the capacitor. The continuity of stored energy principle requires that this current must flow through the resistor because a sudden or instantaneous change of current through the inductor is prohibited. Because $i_L(0^-) = i_L(0^+) = 0$, the KCL relationship with $i_L(0^+) = 0$

$$i_R(0^+) + i_C(0^+) = 0$$

shows that $i_C(0^+) = -i_R(0^+)$, or

$$C \frac{dv}{dt}\Big|_{t=0^+} = -\frac{v(0^+)}{R}$$

This says that

$$\frac{dv}{dt}\Big|_{t=0^+} = -\frac{v(0^+)}{RC} = -\frac{2}{(\frac{3}{4})(\frac{1}{3})} = -8 \text{ V/s}$$

is the required second initial condition.

When eq. (7.21) and its derivative

$$\frac{dv}{dt} = -V_1 e^{-t} - 3V_2 e^{-3t}$$

are evaluated at $t = 0$, a pair of simultaneous algebraic equations in V_1 and V_2 are obtained:

$$V_1 + V_2 = 2$$
$$-V_1 - 3V_2 = -8$$

which when solved give

$$V_1 = -1 \qquad V_2 = 3$$

The total response, which contains only natural and zero-input response components, is therefore

$$v = 3e^{-3t} - e^{-t} \text{ volts} \qquad (7.22)$$

Equation (7.22) is plotted in Fig. 7.2, where the two voltage components and their sum can be noted. Observe that at $t = 0$, $v(0) = 3 - 1 = 2$ V, as it should. ∎

FIGURE 7.2   The voltage across the parallel combination shown in Fig. 7.1 after the switch closes instantaneously at $t = 0$

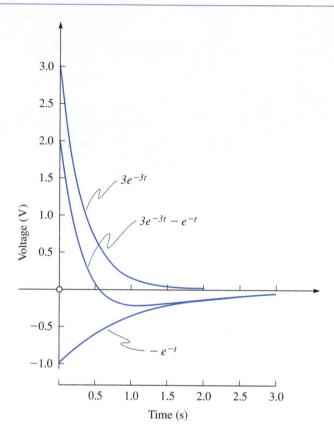

EXERCISE 7.3

In Fig. 7.3, the switch closes instantaneously at $t = 0$ with 10 V across the capacitor terminals and no current flowing through the inductor. Find $i(t)$.

Answer $i(t) = 8te^{-40t}$ amperes.

FIGURE 7.3   A second-order undriven network (Exercise 7.3)

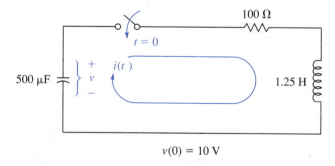

DRIVEN SECOND-ORDER NETWORKS

The intent of this section is to show the formulation and solution of the differential equations governing the performance of *RLC* networks when subjected to a variety of forcing functions. The procedure adopted in Chapter 6 will be followed; and natural, forced, zero-input, zero-state, and total responses will be obtained in each case. In an attempt to maintain the interest of the reader, a systematic procedure will be followed, and where comments seem necessary or where questions by the reader are anticipated, remarks deemed to be pertinent will be made.

7.4.1 Excitation by a Step Input

■ **EXAMPLE 7.2**

The switch in the network shown in Fig. 7.4 closes instantaneously at $t = 0$. Just before the switch is closed, the capacitor has a voltage of $v_C(0^-) = 2$ V across its terminals, and no current flows in the inductor. The current as a function of time is to be investigated.

Solution

a. The differential equation for $i(t)$ is formulated via KVL:

$$v_L + v_R + v_C - v_s = 0$$

b. The elemental equations are

$$v_L = L\frac{di}{dt} \qquad v_R = Ri \qquad v_C = \frac{1}{C}\int_{-\infty}^{t} i\,dz$$

c. With $R = 4\,\Omega$, $L = 1$ H, $C = \frac{1}{4}$ F, and $v_s = 6$ V, use of the elemental equations yields the integro-differential equation,

$$\frac{di}{dt} + 4i + 4\int_{-\infty}^{t} i\,dz = 6; \qquad t > 0$$

An *RLC* series network driven by a constant voltage after the switch closes instantaneously at $t = 0$

FIGURE 7.4

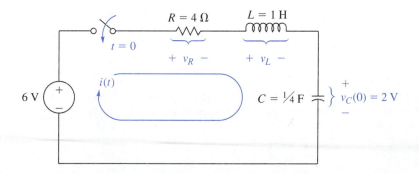

If the method of undetermined coefficients is to be used for the evaluation of the particular integral, the integral representing the voltage across the capacitor can lead to solution complications. It may be removed by a term-by-term differentiation that is mathematically correct but apparently makes the constant term, which represents the forcing function, vanish. Two approaches may be taken to resolve this difficulty.

The first involves writing the integro-differential equation as

$$\frac{di}{dt} + 4i + 4 \int_{-\infty}^{t} i \, dz = 6u(t)$$

and then differentiating to obtain

$$\frac{d^2i}{dt^2} + 4\frac{di}{dt} + 4i = 6\delta(t)$$

The solution to this equation will be the derivative of the solution to the integro-differential equation, because the unit impulse is the derivative of the unit step. When the solution is determined, the solution to the integro-differential equation can be obtained via an integration.

The second approach is the one taken in Chapter 6. The differential equation may be written in terms of the capacitor voltage or in terms of the charge in the system. In this case, the troublesome integral is removed without changing the forcing function. In this approach, it may be necessary to transform the initial conditions and the sought-after current history must be obtained by a differentiation.

Here, one may develop the solution in the charge domain by recognizing that $i = dq/dt$ and $q = \int_{-\infty}^{t} i \, dz$. Then the integro-differential equation can be adjusted to a differential equation.

d. The homogeneous differential equation is

$$\frac{d^2q}{dt^2} + 4\frac{dq}{dt} + 4q = 6; \qquad t > 0$$

e. The characteristic equation is derived from the assumption of $q_c = Qe^{st}$:

$$s^2 + 4s + 4 = 0$$

with $\alpha = 2$, $\omega_n^2 = 4$, $\omega_n = 2$, and $\zeta = \alpha/\omega_n = 2/2 = 1$. The response of this second-order network will be critically damped.

f. The complementary function is derived from the characteristic equation, which can be factored as

$$s^2 + 4s + 4 = (s + 2)^2 = 0$$

which indicates two real, negative, and equal roots, $s_1 = s_2 = s = -2$. The complementary function is represented by

$$q_c = Q_1 e^{-2t} + Q_2 t e^{-2t}$$

g. The assumed form of the particular integral is

$$q_{pi} = A$$

h. The solution for the particular integral is

$$q_{pi} = \frac{3}{2}$$

i. The general solution is

$$q = q_c + q_{pi} = Q_1 e^{-2t} + Q_2 t e^{-2t} + \frac{3}{2}$$

j. The arbitrary constants are determined from a consideration of the initial conditions,

$$q(0^+) = C v_c(0^+) = \frac{1}{4}(2) = \frac{1}{2} \text{ C}$$

and by the continuity of stored energy principle,

$$i(0^+) = \left. \frac{dq}{dt} \right|_{t=0^+} = 0$$

because the current cannot change instantaneously in the inductor.

k. The zero-input response from the complementary function, $q(0^+) = \frac{1}{2}$ C and $i(0^+) = q'(0^+) = 0$, is

$$q(0^+) = \frac{1}{2} = Q_1$$

$$i(0^+) = q'(0^+) = 0 = -2Q_1 + Q_2$$

This makes $Q_2 = 2Q_1 = 2(\frac{1}{2}) = 1$ so that

$$q_{ZI} = \frac{1}{2} e^{-2t} + t e^{-2t}$$

Then, via $i = dq/dt$,

$$i_{ZI} = -2t e^{-2t} \text{ amperes}$$

l. The zero-state response from the general solution, with $q(0^+) = 0$ and $i(0^+) = q'(0^+) = 0$, is

$$q(0^+) = Q_1 + \frac{3}{2} = 0$$

$$i(0^+) = q'(0^+) = -2Q_1 + Q_2 = 0$$

A simultaneous solution of these provides $Q_1 = -\frac{3}{2}$ and $Q_2 = -3$, so that

$$q_{zs} = -\frac{3}{2}e^{-2t} - 3te^{-2t} + \frac{3}{2}$$

And again, via $i = dq/dt$,

$$i_{zs} = 6te^{-2t} \text{ amperes}$$

m. The natural response from the general solution, with $q(0^+) = \frac{1}{2}$ and $i(0^+) = q'(0^+) = 0$, begins with

$$Q_1 + \frac{3}{2} = \frac{1}{2}$$

$$-2Q_1 + Q_2 = 0$$

From these, it is seen that $Q_1 = -1$ and $Q_2 = -2$, so that the particular solution for the charge flow will be

$$q(t) = -e^{-2t} - 2te^{-2t} + \frac{3}{2}$$

and the natural response involves the exponential terms

$$q_N(t) = -e^{-2t} - 2te^{-2t}$$

A differentiation then provides the natural-response current:

$$i_N(t) = 4te^{-2t} \text{ amperes}$$

n. The forced response is the particular integral,

$$q_F(t) = \frac{3}{2}$$

A differentiation then shows that there is no forced-response current:

$$i_F(t) = 0 \text{ A}$$

o. The total response is the sum of the zero-input and zero-state responses or the natural and forced responses. For the charge,

$$q(t) = -e^{-2t} - 2te^{-2t} + \frac{3}{2} \text{ coulombs}$$

and for the current,

$$i(t) = 4te^{-2t} \text{ amperes}$$

The current response for the network in Fig. 7.4

FIGURE 7.5

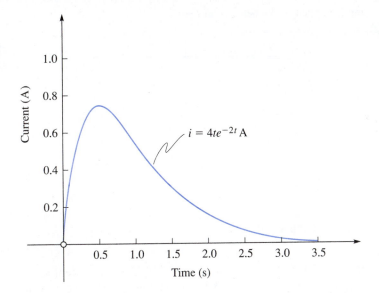

The solution for the total response current is plotted in Fig. 7.5, where it can be observed that both $i(0) = 0$ A and $i(\infty) = 0$ A. This confirms the values that are dictated by continuity of stored energy at $t = 0$, and it also confirms that the forced-response current must be $i(\infty) = 0$, because the capacitor blocks direct current.

EXERCISE 7.4

In the network shown in Fig. 7.6, the switch closes instantaneously when the voltage across the capacitor terminals is $v(0^+) = 8$ V and there is no current flow in the inductor. What is the total response for $v(t)$?

Answer $v(t) = 12 - \dfrac{1}{5}\, e^{-8t}(20 \cos 20t + 8 \sin 20t)$ volts.

A second-order network excited by a voltage step (Exercise 7.4)

FIGURE 7.6

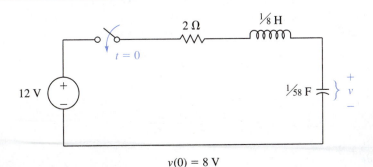

$v(0) = 8$ V

FIGURE 7.7   Parallel *RLC* network

The current ramp is applied to the network at $t = 0$.

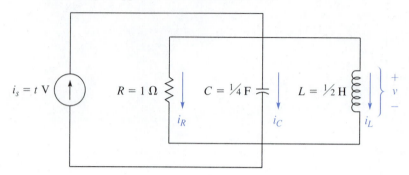

7.4.2 Excitation by a Ramp Input

■ **EXAMPLE 7.3**

Suppose that the parallel *RLC* network shown in Fig. 7.7 has a current ramp $i = Kt$ ($K = 1$ A/s) applied to it at time $t = 0$. At this instant, there is no charge on the capacitor and no current through the inductor. What is the voltage across the parallel combination?

Solution

a. The differential equation for $i(t)$ is formulated via *KCL*:

$$i_C + i_R + i_L - i_s = 0$$

b. The elemental equations are

$$v_L = L\frac{di}{dt} \qquad v_R = Ri \qquad v_C = \frac{1}{C}\int_{-\infty}^{t} i\,dz$$

c. The differential equation, with $R = 1\,\Omega$, $C = \frac{1}{4}$ F, $L = \frac{1}{2}$ H, and $i_s = t$, is

$$\frac{1}{4}\frac{dv}{dt} + \frac{v}{1} + 2\int_{-\infty}^{t} v\,dz = t$$

which is an integro-differential equation.

Alternatively, one may keep the inductor current (call it i) and eliminate the ungainly integral, noting that $v = L\,di/dt$ and $dv/dt = L\,d^2i/dt^2$. With these and $L = \frac{1}{2}$ H, the integro-differential equation can be adjusted to

$$\frac{1}{8}\frac{d^2i}{dt^2} + \frac{1}{2}\frac{di}{dt} + i = t$$

or

$$\frac{d^2i}{dt^2} + 4\frac{di}{dt} + 8i = 8t; \qquad t > 0$$

d. The homogeneous differential equation is

$$\frac{d^2i}{dt^2} + 4\frac{di}{dt} + 8i = 0$$

e. The characteristic equation, derived from the assumption $i_c = Ie^{st}$, is

$$s^2 + 4s + 8 = 0$$

f. For the complementary function in this case, $\alpha = 2$, $\omega_n^2 = 8$, $\omega_n = \sqrt{8}$, and $\zeta = \alpha/\omega_n = 2/\sqrt{8} < 1$. The response is underdamped, and the roots of the characteristic equation are complex conjugates. The damped natural frequency will be

$$\omega_d = \sqrt{\omega_n^2 - \alpha^2} = \sqrt{8 - (2)^2} = \sqrt{4} = 2 \text{ rad/s}$$

and

$$s_1, s_2 = -\alpha \pm j\omega_d = -2 \pm j2$$

The complementary function is therefore

$$i_c = e^{-2t}(I_1 \cos 2t + I_2 \sin 2t) \text{ amperes}$$

g. The assumed form of the particular integral (see Table 6.1) is

$$i_{pi} = At + B$$

h. The solution for the particular integral uses the assumed form of the particular integral along with any of its possible derivatives, which may be substituted into the differential equation to yield

$$0 + 4A + 8(At + B) = 8t$$

When the coefficients of like terms are equated, two equations in the two unknowns, A and B, evolve:

$$8A \qquad = 8$$
$$4A + 8B = 0$$

and a simultaneous solution gives $A = 1$ and $B = -\frac{1}{2}$. The particular integral is therefore

$$i_{pi} = t - \frac{1}{2}$$

i. The general solution is

$$i = i_c + i_{pi} = e^{-2t}(I_1 \cos 2t + I_2 \sin 2t) + t - \frac{1}{2}$$

j. The continuity of stored energy dictates both initial conditions. The current through the inductor and the voltage across the capacitor may not change instantaneously. Thus, $i_L(0^+) = i(0^-) = 0$ and $v(0^+) = 0$; and because $v(0^+) = Li'(0^+) = 0$, the two initial conditions are

$$i(0^+) = 0 \qquad i'(0^+) = 0$$

k. The zero-input response, from the complementary function, with the actual initial conditions, $i(0^+) = 0$, and $i'(0^+) = 0$, is

$$i_c(0^+) = 0 = e^0[I_1(1) + I_2(0)]$$

and hence,

$$I_1 = 0$$

This reduces the complementary function to

$$i_c = I_2 e^{-2t} \sin 2t$$

The derivative of this equation is

$$i_c' = 2I_2 e^{-2t} \cos 2t - 2I_2 e^{-2t} \sin 2t$$

and at $t = 0$,

$$i_c'(0^+) = 0 = 2I_2(1)(1) - 2I_2(1)(0)$$

or

$$I_2 = 0$$

This indicates that there is no zero-input current response and no zero-input voltage response.

l. The zero-state response is derived from the general solution with the initial conditions set equal to zero, which, by coincidence, is the actual case here. The general solution is

$$i = e^{-2t}(I_1 \cos 2t + I_2 \sin 2t) + t - \frac{1}{2}$$

With $i(0^+) = 0$,

$$i(0^+) = 0 = (1)[I_1(1) + I_2(0)] + 1(0) - \frac{1}{2}$$

or

$$I_1 = \frac{1}{2}$$

The derivative of the general solution is

$$i'(t) = 2e^{-2t}(I_2 \cos 2t - I_1 \sin 2t) - 2e^{-2t}(I_1 \cos 2t + I_2 \sin 2t) + 1$$

and when this is set equal to zero,

$$i'(0^+) = 0 = 2(1)[I_2(1) - I_1(0)] - 2(1)[I_1(1) + I_2(0)] + 1$$

With $I_1 = \frac{1}{2}$, it is seen that

$$I_2 = 0$$

The zero-state current response is

$$i_{zs} = \frac{1}{2} e^{-2t} \cos 2t + t - \frac{1}{2} \text{ amperes}$$

and the zero-state voltage response is then determined from $v = L \, di/dt$:

$$v_{zs} = L \frac{di}{dt} = \frac{1}{2} \frac{d}{dt} \left(\frac{1}{2} e^{-2t} \cos 2t + t - \frac{1}{2} \right)$$

$$= \frac{1}{2} (-e^{-2t} \cos 2t - e^{-2t} \sin 2t + 1)$$

or

$$v_{zs} = \frac{1}{2} - \frac{1}{2} e^{-2t}(\cos 2t + \sin 2t) \text{ volts}$$

m. The natural response is derived from the general solution with $i(0^+) = 0$ and $i'(0^+) = 0$. The development in item 1 shows that

$$i_N = \frac{1}{2} e^{-2t} \cos 2t \text{ amperes}$$

is the natural current response, and the natural voltage response is then determined from $v = L \, di/dt$:

$$v_N = L \frac{di}{dt} = \frac{1}{2} \frac{d}{dt} \left(\frac{1}{2} e^{-2t} \cos 2t \right) = \frac{1}{2} (-e^{-2t} \cos 2t - e^{-2t} \sin 2t)$$

or

$$v_N = -\frac{1}{2} e^{-2t}(\cos 2t + \sin 2t) \text{ volts}$$

n. The forced current response is the particular integral,

$$i_{pi} = t - \frac{1}{2} \text{ amperes}$$

and the forced voltage response is then determined from $v = L\, di/dt$:

$$v_{\text{pi}} = L\frac{di}{dt} = \frac{1}{2}\frac{d}{dt}\left(t - \frac{1}{2}\right) = \frac{1}{2}\ \text{V}$$

o. The total response is the sum of the zero-input and zero-state responses or the natural and forced responses. For the current,

$$i = \frac{1}{2}e^{-2t}\cos 2t + t - \frac{1}{2}\ \text{amperes}$$

and for the voltage,

$$v = \frac{1}{2} - \frac{1}{2}e^{-2t}(\cos 2t + \sin 2t)\ \text{volts}$$

■

Remark   The natural voltage response can be put into the amplitude–phase angle form by the procedure provided in Section 7.2.3. Here, $Y_a = 1$ and $Y_b = 1$, so that

$$Y = \sqrt{Y_a^2 + Y_b^2} = \sqrt{1 + 1} = \sqrt{2}$$

$$\theta = \arctan\frac{Y_a}{Y_b} = \arctan 1 = 45°$$

$$\phi = \arctan\frac{-Y_b}{Y_a} = \arctan(-1) = -45°$$

Thus,

$$v_N = -\frac{\sqrt{2}}{2}e^{-2t}\sin(2t + 45°)\ \text{amperes}$$

or

$$v_N = -\frac{\sqrt{2}}{2}e^{-2t}\cos(2t - 45°)\ \text{amperes}$$

are also valid representations for the natural voltage response. Either of these forms could have been obtained directly, as Example 7.4 clearly shows.

■ **EXAMPLE 7.4**

Consider the differential equation governing the current flow in Fig. 7.7 after the switch closes instantaneously at $t = 0$:

$$\frac{d^2i}{dt^2} + 4\frac{di}{dt} + 8i = 8t; \qquad t > 0$$

with $i(0^+) = i'(0^+) = 0$. Show that the particular solution (the total response) for $i(t)$, using the amplitude–phase angle form

$$i_N = Ie^{-\alpha t}\cos(\omega_d t + \phi)\ \text{amperes}$$

is as determined in Example 7.3, namely,

$$i = \frac{1}{2} e^{-2t} \cos 2t + t - \frac{1}{2} \text{ amperes}$$

Solution In the auxiliary equation obtained from the assumption that $i_c = Ie^{st}$,

$$s^2 + 4s + 8 = 0$$

$\alpha = -2$, $\omega_d = 2$, and $\zeta = \frac{1}{2}$. The response is underdamped, and

$$i_C = Ie^{-2t} \cos(2t + \phi) \text{ amperes}$$

The particular integral is

$$i_{pi} = t - \frac{1}{2} \text{ amperes}$$

and the general solution may be written as

$$i(t) = Ie^{-2t} \cos(2t + \phi) + t - \frac{1}{2}$$

This has a derivative

$$i'(t) = -2Ie^{-2t} \cos(2t + \phi) - 2Ie^{-2t} \sin(2t + \phi) + 1$$

Use of the initial condition, $i(0^+) = i'(0^+) = 0$, provides the pair of equations

$$0 = I \cos \phi - \frac{1}{2}$$

$$0 = -2I \cos \phi - 2I \sin \phi + 1$$

With the first equation showing that

$$I \cos \phi = \frac{1}{2}$$

the second equation may be written as

$$0 = -2\left(\frac{1}{2}\right) - 2I \sin \phi + 1$$

or

$$I \sin \phi = 0$$

From these, it is noted that $\phi = 0$ and $I = \frac{1}{2}$ and that the total solution is

$$i(t) = \frac{1}{2} e^{-2t} \cos 2t + t - \frac{1}{2} \text{ amperes}$$

FIGURE 7.8   A second-order network excited by a voltage ramp (Exercise 7.5)

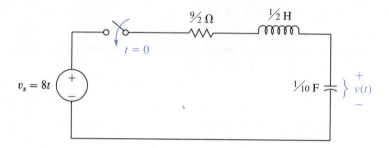

EXERCISE 7.5

In the network shown in Fig. 7.8, the switch closes instantaneously when the voltage across the capacitor terminals is $v(0) = 4$ V. What is the total response for $v(t)$?

Answer $v(t) = \dfrac{1}{5}(40t - 18 + 150e^{-4t} - 112e^{-5t})$ volts.

7.4.3 Excitation by a Sinusoidal Input

■ **EXAMPLE 7.5**

Suppose that a sinusoidal current $i_s = 2 \sin 4t$ is applied to the network of Fig. 7.7 at time $t = 0$. If the initial conditions of no current through the inductor and no voltage across the capacitor are maintained, as in the case of excitation by a ramp as presented in Example 7.4, what is the voltage response?

Solution   The procedure is similar to that used in Example 7.4.

a. The differential equation for $i(t)$ is formulated via KCL:

$$i_C + i_R + i_L - i_s = 0$$

b. The elemental equations are

$$v_L = L\frac{di}{dt} \qquad v_R = Ri \qquad v_C = \frac{1}{C}\int_{-\infty}^{t} i\,dz$$

c. The differential equation is derived by using the element values of Example 7.4 and following the same reasoning for the use of the current through the inductor as the dependent variable with the sinusoidal forcing function:

$$\frac{d^2i}{dt^2} + 4\frac{di}{dt} + 8i = 16 \sin 4t; \qquad t > 0$$

d. The homogeneous differential equation is

$$\frac{d^2i}{dt^2} + 4\frac{di}{dt} + 8i = 0$$

e. The characteristic equation, derived from the assumption $i_c = Ie^{st}$, is

$$s^2 + 4s + 8 = 0$$

f. As in Example 7.4, the response is underdamped, the roots of the characteristic equation are complex conjugates, and the damped natural frequency will be

$$\omega_d = \sqrt{\omega_n^2 - \alpha^2} = \sqrt{8 - (2)^2} = \sqrt{4} = 2 \text{ rad/s}$$

The complementary function is therefore

$$i_c = e^{-2t}(I_1 \cos 2t + I_2 \sin 2t) \text{ amperes}$$

g. The assumed form of the particular integral is

$$i_{\text{pi}} = A \cos 4t + B \sin 4t$$

which, in accordance with Table 6.1, contains the form of the forcing function and any of the possible derivatives of the forcing function.

h. The particular integral and its first two derivatives,

$$i'_{\text{pi}} = 4(B \cos 4t - A \sin 4t)$$
$$i''_{\text{pi}} = -16(A \cos 4t + B \sin 4t)$$

are put into the differential equation to obtain

$$-16(A \cos 4t + B \sin 4t) + 4(4)(B \cos 4t - A \sin 4t)$$
$$+ 8(A \cos 4t + B \sin 4t) = 16 \sin 4t$$

A rearrangement gives

$$(-8A + 16B) \cos 4t - (16A + 8B) \sin 4t = 16 \sin 4t$$

and for this to be an equality, the coefficients of the sine and cosine terms on both sides of the equal sign must match. Two simultaneous equations in the two undetermined coefficients result:

$$-8A + 16B = 0$$
$$-16A - 8B = 16$$

A solution of these gives

$$A = -\frac{4}{5} \qquad B = -\frac{2}{5}$$

so that

$$i_{\text{pi}} = -\frac{1}{5}(4 \cos 4t + 2 \sin 4t)$$

i. The general solution is

$$i = i_c + i_{pi} = e^{-2t}(I_1 \cos 2t + I_2 \sin 2t) - \frac{1}{5}(4 \cos 4t + 2 \sin 4t) \text{ amperes}$$

j. The initial conditions are the same as those for Example 7.4:

$$i(0^+) = 0 \qquad i'(0^+) = 0$$

k. The zero-input response is derived from the complementary function with the actual initial conditions, $i(0^+) = 0$, and $i'(0^+) = 0$, which leads to

$$i(0^+) = 0 = 1[I_1(1) + I_2(0)]$$

or $I_1 = 0$. This means that i_c reduces to

$$i_c = I_2 e^{-2t} \sin 2t$$

and with

$$i_c' = 2I_2 e^{-2t}(\cos 2t - \sin 2t)$$

then,

$$i_c'(0^+) = 0 = 2I_2(1 - 0)$$

or

$$I_2 = 0$$

With $I_1 = I_2 = 0$ deriving from $i(0^+) = i'(0^+) = 0$, there can be no zero-input inductor current or voltage responses.

l. The zero-state response is derived from the general solution with the initial conditions set equal to zero, and by coincidence, this is the actual case here:

$$i(0^+) = 0 = 1[I_1(1) + I_2(0)] - \frac{1}{5}[4(1) + 2(0)]$$

or

$$I_1 = \frac{4}{5}$$

Then with

$$i'(t) = 2e^{-2t}(-I_1 \sin 2t + I_2 \cos 2t) - 2e^{-2t}(I_1 \cos 2t + I_2 \sin 2t)$$

$$- \frac{4}{5}(2 \cos 4t - 4 \sin 4t)$$

one obtains, at $t = 0^+$,

$$i'(0^+) = 0 = 2(1)[-I_1(0) + I_2(1)] - 2(1)[I_1(1) + I_2(0)] - \frac{4}{5}[2(1) - 4(0)]$$

or

$$i'(0^+) = 0 = 2I_2 - 2I_1 - \frac{8}{5}$$

and

$$I_2 = I_1 + \frac{4}{5} = \frac{4}{5} + \frac{4}{5} = \frac{8}{5}$$

With I_1 and I_2 so determined, the zero-state current response is

$$i_{ZS} = \frac{1}{5} \left[e^{-2t}(4 \cos 2t + 8 \sin 2t) - 4 \cos 4t - 2 \sin 4t \right] \text{ amperes}$$

and the zero-state voltage response is obtained from $v = L \, di/dt$:

$$v_{ZS} = \frac{1}{10} \left[e^{-2t}(8 \cos 2t - 24 \sin 2t) + 16 \sin 4t - 8 \cos 4t \right] \text{ volts}$$

m. The natural response is derived from the general solution with $i(0^+) = 0$ and $i'(0^+) = 0$. The development in item 1 shows that the natural current response is

$$i_N = \frac{1}{5} \left[e^{-2t}(4 \cos 2t + 8 \sin 2t) \right] \text{ amperes}$$

and the natural voltage response is obtained from $v = L \, di/dt$:

$$v_N = \frac{1}{10} \left[e^{-2t}(8 \cos 2t - 24 \sin 2t) \right] \text{ volts}$$

n. The forced current response is the particular integral:

$$i_F = -\frac{1}{5}(4 \cos 4t + 2 \sin 4t) \text{ amperes}$$

The forced voltage response is obtained from $v = L \, di/dt$:

$$v_F = \frac{1}{10}(16 \sin 4t - 8 \cos 4t) \text{ volts}$$

o. The total response is the sum of the zero-input and zero-state responses or the natural and forced responses:

$$i = \frac{1}{5} \left[e^{-2t}(4 \cos 2t + 8 \sin 2t) - 4 \cos 4t - 2 \sin 4t \right] \text{ amperes}$$

and

$$v_N = \frac{1}{10} \left[e^{-2t}(8 \cos 2t - 24 \sin 2t) + 16 \sin 4t - 8 \cos 4t \right] \text{ volts}$$

FIGURE 7.9   A second-order network excited by a voltage sinusoid (Exercise 7.6)

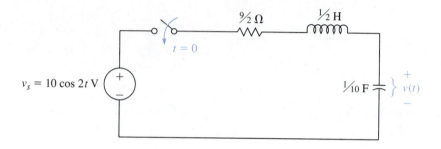

EXERCISE 7.6

In the network shown in Fig. 7.9, the switch closes instantaneously when the voltage across the capacitor terminals is $v(0) = 4$ V. What is the total response for $v(t)$?

Answer $v(t) = \dfrac{1}{29}\,(536e^{-5t} - 580e^{-4t} + 160\cos 2t + 180\sin 2t)$ volts.

7.4.4 Excitation by an Impulse

■ **EXAMPLE 7.6**

Consider the network shown in Fig. 7.10. At $t = 0$, an impulse of voltage $v_{\text{in}} = 10\delta(t)$ is impressed upon the network. The capacitor is initially uncharged, and no current is flowing in the network at $t = 0$. What is the current as a function of time?

Solution

a. The differential equation for $i(t)$ is formulated via KVL. The voltage impulse strikes the network during the period $0^- \le t \le 0^+$. For all time greater than $t = 0^+$,

$$v_L + v_R + v_C = 0$$

FIGURE 7.10   *RLC* series network subjected to a voltage impulse at $t = 0$

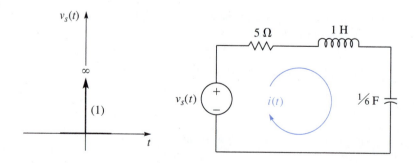

b. The elemental equations, with $R = 5\,\Omega$, $L = 1$ H, and $C = \frac{1}{6}$ F, are

$$v_L = \frac{di}{dt} \qquad v_R = 5i \qquad v_C = C \int_{-\infty}^{t} i\,dz$$

c. The differential equation is

$$\frac{di}{dt} + 5i + 6 \int_{-\infty}^{t} i\,dz = 0$$

Because this integro-differential equation is homogeneous, a single differentiation will yield a differential equation.

d. The homogeneous differential equation is

$$\frac{d^2i}{dt^2} + 5\frac{di}{dt} + 6i = 0; \qquad t > 0$$

e. The characteristic equation, derived from the assumption $i_c = Ie^{st}$, is

$$s^2 + 5s + 6 = 0$$

f. For the complementary function, with $\alpha = \frac{5}{2}$, $\omega_n^2 = 6$, $\omega_n = \sqrt{6}$, and $\zeta = \alpha/\omega_n = 5/(2\sqrt{6}) > 1$, an overdamped condition is indicated; and with $s_1 = -2$ and $s_2 = -3$,

$$i_c = I_1 e^{-2t} + I_2 e^{-3t}$$

g. The assumed form of the particular integral is not required.
h. The solution for the particular integral is not required.
i. The general solution is the complementary function,

$$i_c = I_1 e^{-2t} + I_2 e^{-3t}$$

j. For the initial conditions, at $t = 0^+$, $i(0^+)$ has a finite value, because the impulse that strikes at $t = 0^-$ leaves a current in its wake after it disappears at $t = 0^+$. During the period $0^- \le t \le 0^+$, the inductor behaves as an open circuit, and all of the impulse voltage appears across the inductor. This means that

$$i(0^+) = \frac{1}{L} \int_{0^-}^{0^+} 10\delta(t)\,dt = \frac{10}{L} \int_{0^-}^{0^+} \delta(t)\,dt$$

and because the second definite integral in the foregoing, as shown in eq. (5.8b), is the definition of the unit impulse,

$$i(0^+) = \frac{10}{L} = \frac{10}{1} = 10 \text{ A}$$

At $t = 0^+$, no charge has had a chance to accumulate on the capacitor. Thus, $v_C(0^+) = 0$; and KVL then points out that

$$v_C(0^+) + v_R(0^+) + v_L(0^+) = v_L(0^+) + v_R(0^+) = 0$$

and

$$v_L(0^+) = L\left.\frac{di}{dt}\right|_{t=0^+} = -Ri(0^+)$$

or

$$\left.\frac{di}{dt}\right|_{t=0^+} = -\frac{R}{L}i(0^+) = -\left(\frac{5}{1}\right)(10) = -50 \text{ A/s}$$

k. The zero-input response is obtained from the complementary function and its derivative,

$$\frac{di}{dt} = -2I_1 e^{-2t} - 3I_2 e^{-3t}$$

and $i(0^+) = 10$ A and $i'(0^+) = -50$ A/s. Use of the initial conditions provides

$$I_1 + I_2 = 10$$

$$-2I_1 - 3I_2 = -50$$

A simultaneous solution of these yields $I_1 = -20$ and $I_2 = 30$, so that the zero-input response is

$$i(t) = 30e^{-3t} - 20e^{-2t} \text{ amperes}$$

l. The zero-state response is derived from the general solution with $i(0^+) = 0$ and $i'(0^+) = 0$. Because there is no forcing function, there is no zero-state response.

m. The natural response is derived from the general solution with $i(0^+) = 10$ A and $i'(0^+) = -50$ A/s. The natural response is equal to the zero-input response:

$$i(t) = 30e^{-3t} - 20e^{-2t} \text{ ampere}$$

n. The forced response is the particular integral, which in this case does not exist.

o. The total response is the sum of the zero-input and zero-state responses or the natural and forced responses. Because the forced and zero-state responses do not exist,

$$i(t) = 30e^{-3t} - 20e^{-2t} \text{ amperes}$$ ∎

EXERCISE 7.7

Recall that the unit impulse is the derivative of the unit step. Show that the total response obtained in Example 7.6 for the case of an input $10\delta(t)$ at $t = 0$ can be obtained by letting the network in Fig. 7.10 be subjected to the voltage input $v_s(t) = 10u(t)$ volts at $t = 0$.

Answer   If $v_s(t) = 10u(t)$ V, $i(t) = 10(e^{-2t} - e^{-3t})$ amperes.

FIGURE 7.11

An *RLC* series network excited by a voltage impulse (Exercise 7.8)

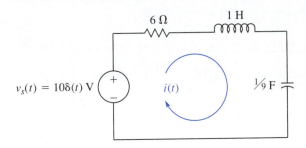

EXERCISE 7.8

In the network shown in Fig. 7.11, the voltage impulse strikes at $t = 0$.
Find $i(t)$.

Answer $i(t) = 10e^{-3t} - 30te^{-3t}$ amperes.

7.4.5 Excitation by a Pulse

The case of excitation by a pulse affords an opportunity to illustrate how an analysis of a network can be decomposed into solutions over more than one time interval. In the example that follows, the total response during two time intervals is considered.

■ **EXAMPLE 7.7**

At $t = 0$, the *RLC* combination with $v(0^+) = 0$ and $i_L(0^+) = 0$ in Fig. 7.12a is subjected to the single current pulse shown in Fig. 7.12b. The pulse has an amplitude of 10 mA and a duration of 10 ms. Find the voltage across the parallel combination for all $t \geq 0$.

FIGURE 7.12

(a) An *RLC* parallel network subjected to the single current pulse shown in (b)

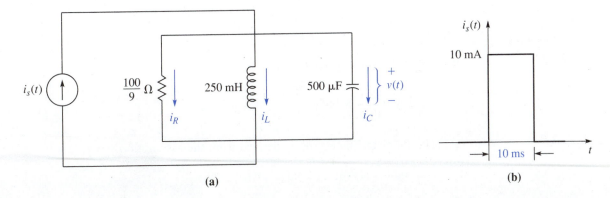

(a)   (b)

Solution

a. Here the strategy will be to find the inductor current, $i_2(t) = i(t)$. The differential equation for $i(t)$ is formulated via KCL:

$$i_C + i_R + i_L = i_s$$

b. To find the elemental equations, let $i_L = i$ so that $v = L\,di/dt$ and $dv/dt = L\,d^2i/dt^2$. Then,

$$i_C = C\frac{dv}{dt} = LC\frac{d^2i}{dt^2}$$

and

$$i_R = \frac{1}{R}v = \frac{L}{R}\frac{di}{dt}$$

c. The differential equation is

$$LC\frac{d^2i}{dt^2} + \frac{L}{R}\frac{di}{dt} + i = i_s$$

With $R = 100/9\ \Omega$, $L = 250\ \text{mH}$, $C = 500\ \mu\text{F}$, and

$$i_s = 0.010[u(t) - u(t - 0.01)]\ \text{amperes}$$

the differential equation to be solved is

$$\frac{d^2i}{dt^2} + 180\frac{di}{dt} + 8000i = 80[u(t) - u(t - 0.01)]; \qquad t > 0$$

Remark   The solution for $i(t)$ consists of two parts. The first part is for the time period $0 \le t \le 0.01$ s. The second part is for $t \ge 0.01$ s. The part of the solution, for the time period $0 \le t \le 0.01$ s, is considered first.

c_1. The differential equation is

$$\frac{d^2i}{dt^2} + 180\frac{di}{dt} + 8000i = 80; \qquad t > 0$$

d_1. The homogeneous differential equation is

$$\frac{d^2i}{dt^2} + 180\frac{di}{dt} + 8000i = 0$$

e_1. The characteristic equation, derived from the assumption $i_c = Ie^{st}$, is

$$s^2 + 180s + 8000 = (s + 80)(s + 100) = 0$$

f_1. The complementary function is

$$i_c = I_1 e + I_2 e^{-100t} \text{ amperes}$$

g_1. The assumed particular integral is a constant, $i_{pi} = A$ (Table 6.1).

h_1. The solution for the particular integral is found by putting the assumed form and its derivatives into the differential equation. Because a constant has no derivatives,

$$8000 \, A = 80$$

or

$$i_{pi} = 0.010 \text{ A}$$

i_1. The general solution is

$$i(t) = I_1 e^{-80t} + I_2 e^{-100t} + 0.010 \text{ amperes}$$

j_1. The given initial conditions are $i(0^+) = 0$ and because $v(0^+) = 0$, $i'(0^+) = 0$.

k_1. The total response is derived from the general solution, the complementary function, and $i(0^+) = i'(0^+) = 0$. With

$$i(t) = I_1 e^{-80t} + I_2 e^{-100t} + 0.010$$

and

$$i'(t) = -80 I_1 e^{-80t} - 100 I_2 e^{-100t}$$

and with $i(0^+) = i'(0^+) = 0$, the pair of equations

$$I_1 + \quad I_2 = -0.010$$
$$-80 I_1 - 100 I_2 = 0$$

leads to $I_1 = -0.05$ and $I_2 = 0.04$. So for $0 \le t \le 0.01$ s,

$$i(t) = 0.04 e^{-100t} - 0.05 e^{-80t} + 0.010 \text{ amperes}$$

$$\frac{di}{dt} = 4 e^{-80t} - 4 e^{-100t} \text{ amperes per second}$$

and

$$v(t) = 0.25 \frac{di}{dt} = 0.25(4 e^{-80t} - 4 e^{-100t}) = e^{-80t} - e^{-100t} \text{ volts}$$

Remark At $t = 0.01$ s,

$$i(0.01) = 0.04 e^{-1} - 0.05 e^{-0.8} + 0.010 = 2.25 \text{ mA}$$

and

$$i'(0.01) = 4 e^{-0.8} - 4 e^{-1} = 0.3258 \text{ A/s}$$

These are the initial conditions for the second part of the problem, which pertains to the time frame $t \geq 0.01$ s. Next, for $t \geq 0.01$ s, with $\tau = t - 0.01$ as the independent variable, the solution is:

c_2. The differential equation is

$$\frac{d^2i}{d\tau^2} + 180\frac{di}{d\tau} + 8000i = 0; \qquad t \geq 0.01 \text{ s}$$

d_2. The homogeneous differential equation is the same as the differential equation.

e_2. The characteristic equation, derived from the assumption $i_c = Ie^{st}$, is

$$s^2 + 180s + 8000 = (s + 80)(s + 100) = 0$$

f_2. The complementary function is

$$i_c = I_1e^{-80\tau} + I_2e^{-100\tau}$$

g_2. Because the differential equation is homogeneous, the particular integral is not required.

h_2. The solution for the particular integral is not required.

i_2. The general solution is

$$i(t) = I_1e^{-80\tau} + I_2e^{-100\tau}$$

j_2. initial conditions are

$$i(\tau = 0^+) = 2.25 \text{ mA} \qquad i'(\tau = 0^+) = 0.3258 \text{ A/s}$$

k_2. The total response is derived from the general solution, the complementary function, $i(\tau = 0^+) = 2.25$ mA, and $i'(\tau = 0^+) = 0.3258$ A/s. Use of these initial conditions leads to the pair of simultaneous equations

$$I_1 + \quad I_2 = 2.25 \times 10^{-3}$$
$$-80I_1 - 100I_2 = 0.3258$$

which makes $I_1 = 0.02754$ and $I_2 = -0.02059$. Thus, for $t \geq 0.01$ s ($\tau \geq 0$ s),

$$i(t) = 0.02754e^{-80\tau} - 0.02529e^{-100\tau} \text{ amperes}$$

And because $v = L \, di/d\tau$,

$$v(t) = 0.63225e^{-100\tau} - 0.55080e^{-80\tau} \text{ volts}$$

The inductor current and the voltage across the parallel combination are plotted in Fig. 7.13.

FIGURE 7.13

Inductor current and voltage for the network in Fig. 7.12a

Notice how the inductor current continues to rise after the current pulse is removed, due to the energy stored in the capacitor.

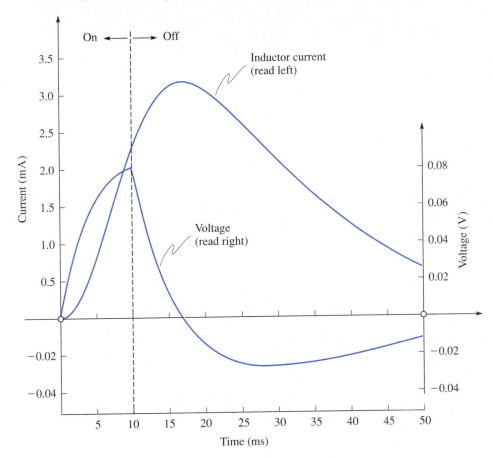

THE SECOND-ORDER NETWORK WITH NO DAMPING

The term *damping* associated with the damping factor and used so extensively in consideration of second-order networks refers to the dissipating elements in the network. In the context of this book, where only ideal network elements are employed, inductors and capacitors are only capable of storing energy; and the only network element that can dissipate energy is the resistor. It is apparent, therefore, that a second-order network that has no damping must consist only of inductors and capacitors. Such a network is shown in Fig. 7.14. Note that the switch closes at $t = 0$ when there is no current through the inductor but when the capacitor has a voltage V_o across its terminals.

The governing differential equation stems from a consideration of KVL:

$$v_C + v_L = 0$$

FIGURE 7.14

An *LC* network that is excited when the switch closes at $t = 0$, by the charge stored in the capacitor

The counterclockwise loop represents the current that tends to flow after $t = 0$.

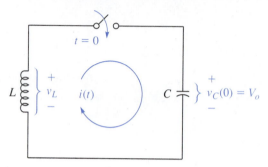

With $v_L = L\,di/dt$ and $v_C = (1/C)\int_{-\infty}^{t} i\,dz$, the network differential equation may be written as

$$L\frac{di}{dt} + \frac{1}{C}\int_{-\infty}^{t} i\,dz = 0 \qquad (7.23)$$

which is a homogeneous yet integro-differential equation that can have no zero-state or forced response. A differentiation and a division by L provides

$$\frac{d^2i}{dt^2} + \frac{1}{LC}i = 0 \qquad (7.24)$$

and the complementary function, which will yield both the zero-input and the natural response, can be obtained by assuming that

$$i_c = Ie^{st}$$

This form of the complementary function, when substituted into eq. (7.24), leads to the characteristic equation

$$s^2 + \frac{1}{LC} = 0$$

Here $\alpha = 0$, $\omega_n^2 = 1/LC$, $\omega_n = \sqrt{1/LC}$, and $\zeta = \alpha/\omega_n = 0$. This indicates that there is no damping and that the roots of the characteristic equation are

$$s_1, s_2 = \pm\sqrt{-\omega_n} = \pm j\omega_n = \pm j\frac{1}{\sqrt{LC}}$$

The complementary function in *amplitude–phase angle* form is given by

$$i_c = I\sin(\omega_n t + \theta)\ \text{amperes} \qquad (7.25)$$

Observe that because $\alpha = 0$, there is no exponential term in the complementary function.

At $t = 0^+$, $i(0^+) = 0$, so that by using this in eq. (7.25), one may obtain

$$I \sin \theta = 0$$

If $I = 0$, the trivial solution occurs. This possibility must therefore be excluded, and the alternative is that $\sin \theta = 0$, which means that $\theta = 0$. As a result, the complementary function can be written as

$$i_c = I \sin \omega_n t$$

and its derivative will be

$$i_c' = \omega_n I \cos \omega_n t$$

At $t = 0^+$, KVL requires that $v_L(0^+) + v_C(0^+) = 0$. With $v_C(0^+) = -V_o$, because of the assumed direction of current flow in Fig. 7.14,

$$v_L(0^+) = L \left.\frac{di}{dt}\right|_{t=0^+} = -v_C(0^+) = -(-V_o) = V_o$$

which means that

$$\left.\frac{di}{dt}\right|_{t=0^+} = \frac{V_o}{L} = \omega_n I$$

Thus,

$$I = \frac{V_o}{L\omega_n}$$

and

$$i = \frac{V_o}{L\omega_n} \sin \omega_n t$$

or with $\omega_n = 1/\sqrt{LC}$ radians per second,

$$i = V_o \sqrt{\frac{C}{L}} \sin \omega_n t \text{ amperes} \tag{7.26}$$

Note that at $t = 0$, $i = 0$, as it should.

MORE ADVANCED CASES

SECTION 7.6

In Fig. 7.15, the two capacitor voltages, v_1 and v_2 and in Fig. 7.16, both of the currents, i_1 and i_2, are of interest. The inevitable result in the analysis of such networks is a system of simultaneous linear differential equations. The general solution for each of

FIGURE 7.15   A more advanced network containing two capacitors in which the switch closes
at $t = \phi$

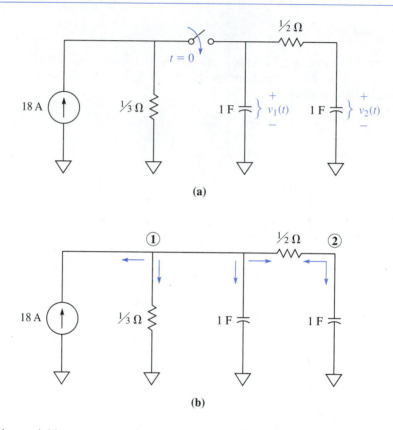

(a)

(b)

the variables may contain as many natural-response components, with arbitrary
constants, as the total number of L's and C's in the network. The arbitrary constants
are related, and an additional procedure is required to determine the relationship
between them. All of these considerations will be treated in the examples that follow.

FIGURE 7.16   A more advanced *RLC* network in which the switch closes instantaneously at
$t = 0$

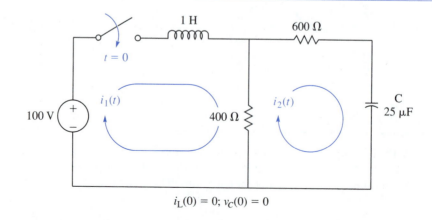

$i_L(0) = 0;\ v_C(0) = 0$

■ EXAMPLE 7.8

In the network shown in Fig. 7.15a, the switch closes when $v_1(0^+) = v_2(0^+) = 0$. The two capacitor voltages are to be determined.

Solution The capacitor voltages are the node voltages shown in Fig. 7.15b. An application of KCL using the current convention shown in Fig. 7.15b yields two node equations

$$-18 + \frac{v_1}{\frac{1}{3}} + 1\frac{dv_1}{dt} + \frac{v_1 - v_2}{\frac{1}{2}} = 0$$

$$\frac{v_2 - v_1}{\frac{1}{2}} + 1\frac{dv_2}{dt} = 0$$

or

$$\frac{dv_1}{dt} + 5v_1 \qquad - 2v_2 = 18 \qquad\qquad (7.27a)$$

$$-2v_1 + \frac{dv_2}{dt} + 2v_2 = 0 \qquad\qquad (7.27b)$$

Assume that $v_{1c} = V_1 e^{st}$ and $v_{2c} = V_2 e^{st}$ and substitute these and their derivatives into eqs. (7.27)

$$(s + 5)v_{1c} \qquad - 2v_{2c} = 18$$

$$-2v_{1c} + (s + 2)v_{2c} = 0$$

A Cramer's rule solution for both v_{1c} and v_{2c}

$$v_{1c} = \frac{\begin{vmatrix} 18 & -2 \\ 0 & (s + 2) \end{vmatrix}}{\begin{vmatrix} (s + 5) & -2 \\ -2 & (s + 2) \end{vmatrix}}$$

$$v_{2c} = \frac{\begin{vmatrix} (s + 5) & 18 \\ -2 & 0 \end{vmatrix}}{\begin{vmatrix} (s + 5) & -2 \\ -2 & 0 \end{vmatrix}}$$

shows that there is a characteristic determinant

$$\begin{vmatrix} (s + 5) & -2 \\ -2 & (s + 2) \end{vmatrix} = s^2 + 7s + 10 - 4 = s^2 + 7s + 6$$

The characteristic equation

$$s^2 + 7s + 6 = 0$$

has roots $s_1 = -1$ and $s_2 = -6$ and the complementary functions for both v_1 and v_2 are

$$v_{1c} = C_{11}e^{-t} + C_{12}e^{-6t} \tag{7.28a}$$

$$v_{2c} = C_{21}e^{-t} + C_{22}e^{-6t} \tag{7.28b}$$

The theory of differential equations points out that the arbitrary constants are related. The relationship between C_{11} and C_{21} and between C_{12} and C_{22} can be obtained by substituting eqs. (7.28) into eq. (7.27b). With

$$\frac{dv_{2c}}{dt} = -C_{21}e^{-t} - 6C_{22}e^{-6t}$$

the substitution yields

$$-2(C_{11}e^{-t} + C_{12}e^{-6t}) - (C_{21}e^{-t} + 6C_{22}e^{-6t}) + 2(C_{21}e^{-t} + C_{22}e^{-6t}) = 0$$

or

$$(C_{21} - 2C_{11})e^{-t} - (4C_{22} + 2C_{12})e^{-6t} = 0$$

This requires that

$$C_{21} - 2C_{11} = 0$$

$$4C_{22} + 2C_{12} = 0$$

or that

$$C_{21} = 2C_{11}$$

$$C_{22} = -\frac{1}{2}C_{12}$$

This establishes the relationship between the arbitrary constants and the two complementary functions can be written in terms of a single pair of arbitrary constants. Let $C_{11} = C_1$ and $C_{12} = C_2$ so that

$$v_{1c} = C_1e^{-t} + C_2e^{-6t} \tag{7.29}$$

$$v_{2c} = 2C_1e^{-t} - \frac{1}{2}C_2e^{-6t} \tag{7.30}$$

It must be assumed that both v_1 and v_2 will possess a particular integral. Therefore, assume, in accordance with Table 6.1, that $v_{1,\,pi} = K_1$ and $v_{2,\,pi} = K_2$ and put these into eqs. (7.27). The result is

$$5K_1 - 2K_2 = 18$$

$$-2K_1 + 2K_2 = 0$$

These show that $K_1 = K_2 = 6$ and the general solutions for the total responses will be

$$v_1 = C_1 e^{-t} + C_2 e^{-6t} + 6$$

$$v_2 = 2C_1 e^{-t} - \frac{1}{2} C_2 e^{-6t} + 6$$

The arbitrary constants can be evaluated from the initial conditions $v_1(0^+) = v_2(0^+) = 0$

$$v_1(0^+) = 0 = C_1 + C_2 + 6$$

$$v_2(0^+) = 0 = 2C_1 - \frac{1}{2} C_2 + 6$$

These form a pair of simultaneous equations

$$C_1 + \quad C_2 = -6$$

$$2C_1 - \frac{1}{2} C_2 = -6$$

and the reader may verify that the solution is $C_1 = -\frac{18}{5}$ and $C_2 = -\frac{12}{5}$. The solution for $v_1(t)$ and $v_2(t)$ is therefore

$$v_1(t) = 6 - \frac{18}{5} e^{-t} - \frac{12}{5} e^{-6t} \text{ volts} \qquad (7.31)$$

and

$$v_2(t) = 6 - \frac{36}{5} e^{-t} + \frac{6}{5} e^{-6t} \text{ volts} \qquad (7.32)$$

■ **EXAMPLE 7.9**

In the network shown in Fig. 7.16, the switch closes at $t = 0$ when there is no current in the inductor and no voltage across the capacitor. The currents i_1 and i_2 are to be determined.

Solution　The currents i_1 and i_2 are the mesh currents in Fig. 7.16. Application of KVL to each mesh results in two mesh equations:

$$\frac{di_1}{dt} + 400i_1 - 400i_2 = 100$$

$$-400i_1 + 1000i_2 + 40{,}000 \int_{-\infty}^{t} i_2 \, dz = 0$$

If $q_2 = \int_{-\infty}^{t} i_2\, dz$, then $i_2 = dq_2/dt$ and two simultaneous, linear differential equations result in i_1 and q_2:

$$\frac{di_1}{dt} + 400i_1 - 400\frac{dq_2}{dt} = 100 \qquad (7.33a)$$

$$-400i_1 + 1000\frac{dq_2}{dt} + 40{,}000q_2 = 0 \qquad (7.33b)$$

Assume that $i_{1c} = Ie^{st}$ and $q_{2c} = Qe^{st}$. Then, by substitution into eqs. (7.33), the pair of equations

$$(s + 400)i_{1c} - 400sq_{2c} = 100$$
$$-400i_{1c} + (1000s + 40{,}000)q_{2c} = 0$$

is obtained. A Cramer's rule solution for both i_{1c} and q_{2c} indicates that

$$i_{1c} = \frac{\begin{vmatrix} 100 & -400s \\ 0 & (100s + 40{,}000) \end{vmatrix}}{\begin{vmatrix} (s + 400) & -400s \\ -400 & (1000s + 40{,}000) \end{vmatrix}}$$

$$q_{1c} = \frac{\begin{vmatrix} (s + 400) & 100 \\ -400 & 0 \end{vmatrix}}{\begin{vmatrix} (s + 400) & -400s \\ -400 & (1000s + 40{,}000) \end{vmatrix}}$$

and shows that the characteristic determinant will be

$$\begin{vmatrix} (s + 400) & -400s \\ -400 & (1000s + 40{,}000) \end{vmatrix}$$

Evaluation of this determinant yields the *characteristic equation*:

$$1000s^2 + 440{,}000s + 16 \times 10^6 - 160{,}000s = 0$$

or

$$s^2 + 280s + 16{,}000 = 0$$

The complementary functions for both i_1 and q_2 derive from the roots of the characteristic equation, $s_1 = -80$ and $s_2 = -200$. Thus, the complementary functions are

$$i_{1c} = C_{11}e^{-80t} + C_{12}e^{-200t} \qquad (7.34a)$$

$$q_{2c} = C_{21}e^{-80t} + C_{22}e^{-200t} \qquad (7.34b)$$

The relationship between C_{11} and C_{21} and between C_{12} and C_{22} can be established by substituting eqs. (7.34) into eq. (7.33b). With

$$\frac{dq_{2c}}{dt} = -80C_{21}e^{-80t} - 200C_{22}e^{-200t}$$

the result is

$$-400(C_{11}e^{-80t} + C_{12}e^{-200t}) - 1000(80C_{21}e^{-80t} + 200C_{22}e^{-200t})$$
$$+ 40{,}000(C_{21}e^{-80t} + C_{22}e^{-200t}) = 0$$

or

$$-(400C_{11} + 40{,}000C_{21})e^{-80t} - (400C_{12} + 160{,}000C_{22})e^{-200t} = 0$$

This requires that

$$400C_{11} + 40{,}000C_{21} = 0$$
$$400C_{12} + 160{,}000C_{22} = 0$$

or that

$$C_{11} = -100C_{21}$$
$$C_{12} = -400C_{22}$$

This establishes the relationships between the arbitrary contants and permits the writing of the two complementary functions in terms of a single pair of arbitrary constants ($C_1 = C_{21}$ and $C_2 = C_{22}$):

$$i_{1c} = -100C_1e^{-80t} - 400C_2e^{-200t} \qquad (7.35a)$$
$$q_{2c} = C_1e^{-80t} + C_2e^{-200t} \qquad (7.35b)$$

Now, in accordance with Table 6.1, assume that the form of the particular integrals will be $i_{1,\,pi} = K_1$ and, because there is no reason to assume that $q_{2,\,pi} = 0$ merely because there is no forcing function in eq. (7.33b), $q_{2,\,pi} = K_2$. When these are put into eqs. (7.33), the result is

$$400K_1 = 100$$

from which $K_1 = 1/4$, and

$$-400K_1 + 40{,}000K_2 = 0$$

or

$$40{,}000K_2 = 400(1/4) = 100$$

and $K_2 = 1/400$.

These give the general solutions for the total responses i_1 and q_2 as

$$i_1 = -100C_1e^{-80t} - 400C_2e^{-200t} + \frac{1}{4} \qquad (7.36a)$$

$$q_2 = C_1e^{-80t} + C_2e^{-200t} + \frac{1}{400} \qquad (7.36b)$$

The arbitrary constants can be determined by evaluating these equations at $t = 0^+$, where $i_1(0^+) = q_2(0^+) = 0$. This results in a pair of simultaneous algebraic equations in C_1 and C_2:

$$100C_1 + 400C_2 = \frac{1}{4}$$

$$C_1 + \quad C_2 = -\frac{1}{400}$$

When these are solved, the result is $C_1 = -\frac{1}{240}$ and $C_2 = \frac{1}{600}$, so that

$$i_1 = \frac{5}{12}e^{-80t} - \frac{2}{3}e^{-200t} + \frac{1}{4} \text{ amperes} \qquad (7.37)$$

$$q_2 = -\frac{1}{240}e^{-80t} + \frac{1}{600}e^{-200t} + \frac{1}{400} \text{ coulombs} \qquad (7.38)$$

And a single differentiation provides the sought-after i_2:

$$i_2 = \frac{1}{3}e^{-80t} - \frac{1}{3}e^{-200t} \text{ amperes} \qquad (7.39)$$

■

SECTION 7.7   **TWO IMPORTANT OBSERVATIONS**

It is well to stop for a moment to assess the extent of just what has been accomplished in Chapters 6 and 7. Analytical techniques have been presented for the solution of simple *RL* or *RC* networks (Chapter 6) and simple *RLC* series or parallel networks (Chapter 7). Examples for these simple networks have been provided to show how to obtain the natural, forced, zero-input, and zero-state responses for six different forms of excitations:

- Excitation by initial energy storage
- Excitation by a step input
- Excitation by a ramp input
- Excitation by a sinusoidal input
- Excitation by an impulse input
- Excitation by a pulsed input

The total response of more advanced networks requiring node and mesh analysis has also been presented. The examples employed in Chapter 7 demonstrate the solution procedure and is interesting for two reasons:

1. They show how much work is involved to obtain a solution. The extent of this labor is the reason for advocating the use of the Laplace transformation, which will be presented in detail in Chapters 15 and 16. The labor saving will be particularly apparent when the manipulation of the arbitrary constants of integration is eliminated by the Laplace transform method.

2. The second example shows how the total responses for the two currents, $i_1(t)$ and $i_2(t)$, are obtained from a mesh analysis. Although this is straightforward, an integration is required to yield the capacitor voltage. A method, called the *state variable method*, does exist in which capacitor voltages and inductor currents are defined as state variables. Problems may then be formulated in terms of these state variables and after capacitor voltages are obtained, currents through capacitors, if of interest at all, may be obtained by a simple differentiation.

SPICE EXAMPLES

SECTION 7.8

Reading: For the PSPICE examples in this chapter, no additional reading is required.

EXAMPLE S7.1

In Fig. S7.1, the initial voltage across the capacitor is 12 V. The voltage across the capacitor is to be found after the switch closes at $t = 0$. The network made ready for analysis with PSPICE is shown in Fig. S7.2, the input file is reproduced in Fig. S7.3, and pertinent extracts from the output file are presented in Fig. S7.4.

Figure S7.1 Undriven network for second-order transient analysis

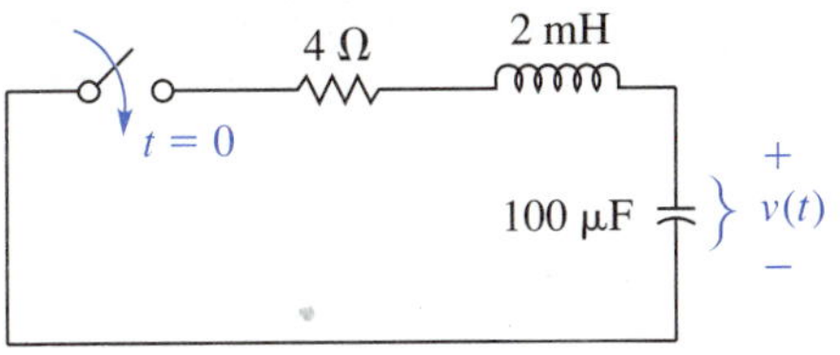

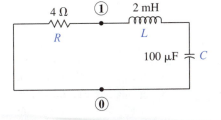

Figure S7.2 The network of Fig. S7.1 ready for PSPICE analysis

(continues)

Example S7.1 (continued)

Figure S7.3 Input PSPICE file for analysis of the network of Fig. S7.1

```
SPICE PROBLEM - CHAPTER 7 - NUMBER 1 - UNDRIVEN SECOND ORDER NETWORK
R       1       0       4
L       1       2       2E-3    IC=0
C       2       0       100E-6  IC=12
*THE LINES BETWEEN THE ASTERISKS ARE THE CONTROL STATEMENTS
********************************************************************
.TRAN   20E-5   5E-3    UIC
.PLOT   TRAN V(2)
.END
********************************************************************
```

Figure S7.4 Pertinent output for the PSPICE analysis of the network in Fig. S7.1

```
SPICE PROBLEM - CHAPTER 7 - NUMBER 1 - UNDRIVEN SECOND ORDER NETWORK


****        TRANSIENT ANALYSIS                  TEMPERATURE =    27.000 DEG C


*********************************************************************************


    TIME        V(2)
    (*)---------      -5.0000E+00     0.0000E+00     5.0000E+00     1.0000E+01     1.5000E+01

    0.000E+00    1.200E+01 . - - - - - - - - - - - - - - - - - - - . - - - * - - - .
    2.000E-04    1.094E+01 .              .              .          . * .
    4.000E-04    8.506E+00 .              .              .     *     .
    6.000E-04    5.496E+00 .              .         .*        .
    8.000E-04    2.582E+00 .              .     *          .
    1.000E-03    2.026E-01 .              .  .*            .
    1.200E-03   -1.433E+00 .         *    .              .
    1.400E-03   -2.303E+00 .       *      .              .
    1.600E-03   -2.518E+00 .      *       .              .
    1.800E-03   -2.256E+00 .       *      .              .
    2.000E-03   -1.716E+00 .         *    .              .
    2.200E-03   -1.075E+00 .          *   .              .
    2.400E-03   -4.698E-01 .             *.              .
    2.600E-03    1.365E-02 .              *              .
    2.800E-03    3.371E-01 .              .*             .
    3.000E-03    5.006E-01 .              .*             .
    3.200E-03    5.300E-01 .              .*             .
    3.400E-03    4.642E-01 .              .*             .
    3.600E-03    3.450E-01 .              .*             .
    3.800E-03    2.090E-01 .              .*             .
    4.000E-03    8.372E-02 .              *              .
    4.200E-03   -1.420E-02 .              *              .
    4.400E-03   -7.787E-02 .              *              .
    4.600E-03   -1.082E-01 .              *              .
    4.800E-03   -1.112E-01 .              *              .
    5.000E-03   -9.545E-02 .              *              .
                             - - - - - - - - - - - - - - - - - - - - - - - - .


        JOB  CONCLUDED
```

EXAMPLE S7.2

In Fig. S7.5a, the initial voltage across the capacitor is 0 V, and there is no current flowing in the inductor. The voltage across the capacitor, *v(t)*, is to be found when the network is subjected to the current pulse shown in Fig. S7.5b. The network made ready for analysis with PSPICE is shown in Fig. S7.6, the input file is reproduced in Fig. S7.7, and pertinent extracts from the output file are presented in Fig. S7.8.

Figure S7.5 Network driven by a voltage pulse for second-order transient analysis

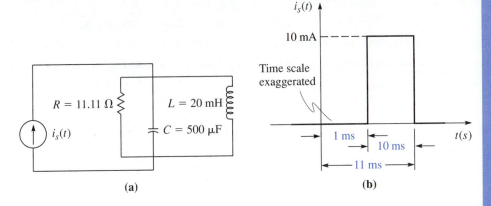

(a)   (b)

Figure S7.6 The network of Fig. S7.5 ready for PSPICE analysis.

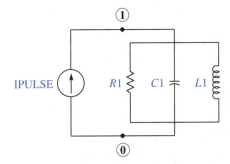

Figure S7.7 Input PSPICE file for analysis of the network of Fig. S7.5

```
SPICE EXAMPLE - CHAPTER 7 - NUMBER 2 - PULSED INPUT, 2ND ORDER
R1      1        0           11.11
C1      1        0           500U      IC=0
L1      1        0           20M       IC=0
****************************************************************
*HERE IS THE PULSED INPUT
IPULSE 0         1           PULSE(0 10M 1M 1U 1U 10M 33.33M)
****************************************************************
*THE LINES BETWEEN THE ASTERISKS ARE THE CONTROL STATEMENTS
****************************************************************
.TRAN   .5M      20M         UIC
.PLOT   TRAN V(1)
.END
****************************************************************
```

(continues)

Example S7.2 (continued)

Figure S7.8   Pertinent output for the PSPICE analysis of the network in Fig. S7.5

```
SPICE EXAMPLE - CHAPTER 7 - NUMBER 2 - PULSED INPUT, 2ND ORDER

****      TRANSIENT ANALYSIS                    TEMPERATURE =    27.000 DEG C

****************************************************************************

    TIME        V(1)
 (*)----------    -1.0000E-01   -5.0000E-02    0.0000E+00    5.0000E-02    1.0000E-01

  0.000E+00   0.000E+00  - - - - - - - - - - - - - * - - - - - - - - - - - -
  5.000E-04   0.000E+00  .              .              *              .              .
  1.000E-03   1.031E-10  .              .              *              .              .
  1.500E-03   9.447E-03  .              .              . *            .              .
  2.000E-03   1.789E-02  .              .              .   *          .              .
  2.500E-03   2.524E-02  .              .              .      *       .              .
  3.000E-03   3.137E-02  .              .              .        *     .              .
  3.500E-03   3.612E-02  .              .              .         *    .              .
  4.000E-03   3.966E-02  .              .              .          *   .              .
  4.500E-03   4.199E-02  .              .              .           * .             .
  5.000E-03   4.313E-02  .              .              .           * .             .
  5.500E-03   4.306E-02  .              .              .           * .             .
  6.000E-03   4.201E-02  .              .              .           * .             .
  6.500E-03   4.008E-02  .              .              .          *  .             .
  7.000E-03   3.737E-02  .              .              .          *  .             .
  7.500E-03   3.395E-02  .              .              .         *   .             .
  8.000E-03   3.003E-02  .              .              .        *    .             .
  8.500E-03   2.575E-02  .              .              .       *     .             .
  9.000E-03   2.123E-02  .              .              .     *       .             .
  9.500E-03   1.657E-02  .              .              .   *         .             .
  1.000E-02   1.193E-02  .              .              .  *          .             .
  1.050E-02   7.393E-03  .              .              . *           .             .
  1.100E-02   3.059E-03  .              .              .*            .             .
  1.150E-02  -1.039E-02  .              .         *    .             .             .
  1.200E-02  -2.254E-02  .              .       *      .             .             .
  1.250E-02  -3.310E-02  .              .    *         .             .             .
  1.300E-02  -4.188E-02  .              . *            .             .             .
  1.350E-02  -4.897E-02  .              *              .             .             .
  1.400E-02  -5.437E-02  .            *.               .             .             .
  1.450E-02  -5.796E-02  .           *.                .             .             .
  1.500E-02  -5.978E-02  .          * .                .             .             .
  1.550E-02  -6.008E-02  .          * .                .             .             .
  1.600E-02  -5.898E-02  .          * .                .             .             .
  1.650E-02  -5.654E-02  .          * .                .             .             .
  1.700E-02  -5.291E-02  .           *.                .             .             .
  1.750E-02  -4.837E-02  .             *               .             .             .
  1.800E-02  -4.309E-02  .             . *             .             .             .
  1.850E-02  -3.722E-02  .             .  *            .             .             .
  1.900E-02  -3.095E-02  .             .      *        .             .             .
  1.950E-02  -2.450E-02  .             .         *     .             .             .
  2.000E-02  -1.801E-02  .  - - - - - - - - - *- - - - - - - - - - - - - - - -

           JOB CONCLUDED
```

CHAPTER 7

SUMMARY

▪ Networks described by a differential equation of the form

$$a\frac{d^2y}{dt^2} + b\frac{dy}{dt} + cy = f(t)$$

will possess an auxiliary equation of the form

$$s^2 + 2\alpha s + \omega_n^2 = 0$$

where the *attenuation* is $\alpha = b/2a$ and the *natural frequency* is $\omega_n = \sqrt{c/a}$.

- A damping factor is defined as

$$\zeta = \frac{\alpha}{\omega_n} = \frac{b}{2\sqrt{ca}}$$

- The form of response depends on the value of the *discriminant* $\alpha^2 - \omega_n^2$.

  —If $\alpha > \omega_n$, the response is *overdamped*, and the auxiliary equation has two real, unequal, and negative solutions,

  $$s_1, s_2 = -\alpha \pm \sqrt{\alpha^2 - \omega_n^2}$$

  and

  $$y_c = Y_1 e^{s_1 t} + Y_2 e^{s_2 t}$$

  —If $\alpha = \omega_n$, the response is *critically damped*, and the auxiliary equation has two real, equal, and negative roots,

  $$s_1 = s_2 = s = -\alpha$$

  and

  $$y_c = Y_1 e^{-\alpha t} + Y_2 t e^{-\alpha t}$$

  —If $\alpha < \omega_n$, then the *damped natural frequency* is

  $$\omega_d = \sqrt{\omega_n^2 - \alpha^2}$$

  The solutions are

  $$s_1, s_2 = -\alpha \pm j\omega_d$$

  and the response is *underdamped*. The response can take any of four forms:

  $$y_c = Y_1 e^{(-\alpha + j\omega_d)t} + Y_2 e^{(-\alpha - j\omega_d)t}$$
  $$y_c = e^{-\alpha t}(Y_a \cos \omega_d t + Y_b \sin \omega_d t)$$
  $$y_c = Y e^{-\alpha t} \sin(\omega_d t + \theta)$$
  $$y_c = Y e^{-\alpha t} \cos(\omega_d t + \phi)$$

- The procedure for the determination of the response to a second-order network is the same as that used for a first-order network.

  —Write the governing differential or integro-differential equation from an application of KVL or KCL.

—Reduce the governing differential or integro-differential equation to a homogeneous equation.

—Assume an exponential solution containing an undetermined exponent and an arbitrary constant.

—Use the assumed exponential solution to determine the complementary function.

—Use the method of undetermined coefficients to establish the particular integral. Remember that the form of the forcing function dictates the form of the assumed value of the particular integral.

—Add the complementary function and the particular integral to form the general solution for the total response.

—If the initial conditions are not given, establish the initial conditions to be used for the evaluation of the arbitrary constants in the general solution.

—Use the initial conditions in the general solution for the total response to obtain the particular solution for the total response.

▪ Just as in the first-order network, the second-order network has the following responses:

—The zero-input response is determined from the complementary function, using the actual initial condition.

—The zero-state response is determined from the general solution for the total response, with the initial condition set equal to zero.

—The natural response is the response that contains an exponential term that does not resemble any exponential term that appears in the forcing function.

—The forced response is that part of the total response that does not contain the natural response.

—The sum of the natural and forced responses must equal the sum of the zero-input and zero-state responses.

Additional Readings

Blackwell, W.A., and L.L. Grigsby. *Introductory Network Theory*, Boston: PWS Engineering, 1985, pp. 293–318.

Bobrow, L.S. *Elementary Linear Circuit Analysis*. 2d ed. New York: Holt, Rinehart and Winston, 1987, pp. 273–308.

Del Toro, V. *Engineering Circuits*. Englewood Cliffs, N.J.: Prentice-Hall, 1987, pp. 276–282, 315–335.

Dorf, R.C. *Introduction to Electric Circuits*. New York: Wiley, 1989, pp. 274–299.

Hayt, W.H., Jr., and J.E. Kemmerly. *Engineering Circuit Analysis*. 4th ed. New York: McGraw-Hill, 1986, pp. 195–220.

Irwin, J.D. *Basic Engineering Circuit Analysis*. 3d ed. New York: Macmillan, 1989, pp. 318–347.

Johnson, D.E., J.L. Hilburn, and J.R. Johnson. *Basic Electric Circuit Analysis*. 4th ed. Englewood Cliffs, N.J.: Prentice-Hall, 1989, pp. 256–293.

Karni, S. *Applied Circuit Analysis*. New York: Wiley, 1988, pp. 211–238.

Madhu, S. *Linear Circuit Analysis*. Englewood Cliffs, N.J.: Prentice-Hall, 1988, pp. 297–307.

Nilsson, J.W. *Electric Circuits*. 3d ed. Reading, Mass.: Addison-Wesley, 1990, pp. 176–192, 266–297.

Paul, C.R. *Analysis of Linear Circuits*. New York: McGraw-Hill, 1989, pp. 498–552, 562–575.

Section 7.2

7.1 For an RLC series network, eq. (7.2) becomes

$$L\frac{d^2i}{dt^2} + R\frac{di}{dt} + \frac{1}{C}i = 0$$

If $R = 4\,\Omega$ and $C = 1/40$ F, what value of L is required to make the response critically damped?

7.2 For an RLC series network, eq. (7.2) becomes

$$L\frac{d^2i}{dt^2} + R\frac{di}{dt} + \frac{1}{C}i = 0$$

If $R = 8\,\Omega$ and $C = 1/100$ F, what value of L is required to make the response underdamped with a damped natural frequency of 6 rad/s? What are the natural frequency and the damping factor?

7.3 For an RLC parallel network, eq. (7.2) becomes

$$C\frac{d^2v}{dt^2} + \frac{1}{R}\frac{dv}{dt} + \frac{1}{L}v = 0$$

If $R = 4\,\Omega$ and $L = 1/8$ H, what value of C is required to give an overdamped response with $\zeta = 2.0$?

7.4 For an RLC parallel network, eq. (7.2) becomes

$$C\frac{d^2v}{dt^2} + \frac{1}{R}\frac{dv}{dt} + \frac{1}{L}v = 0$$

If $R = 4\,\Omega$ and $L = 70.82$ mH, what value of C is required to yield a damped natural frequency of 36 rad/s? What is the damping factor?

7.5 In the network of Fig. P7.1 where eq. (7.3) takes the form

$$C\frac{d^2v}{dt^2} + \frac{1}{R_{eq}}\frac{dv}{dt} + \frac{1}{L}v = 0$$

Figure P7.1

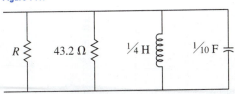

what is the value of R to make the response overdamped with a damping factor of $\zeta = 2.50$?

7.6 For the network of Fig. P7.2, eq. (7.2) becomes

$$L\frac{d^2i}{dt^2} + R\frac{di}{dt} + \frac{1}{C_{eq}}i = 0$$

What value of C is required to make the response critically damped?

Figure P7.2

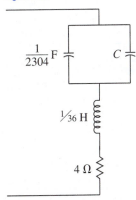

Section 7.3

7.7 In the network of Fig. P7.3, the switch opens instantaneously at $t = 0$. Find the current response for $t \geq 0$.

Figure P7.3

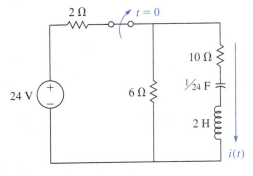

7.8   In the network of Fig. P7.4, the switch opens

instantaneously at $t = 0$. Find the voltage response for $t \geq 0$.

Figure P7.4

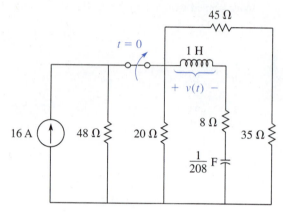

7.9   In the network of Fig. P7.5, the switch moves

instantaneously from position 1 to position 2 at
$t = 0$. Find the voltage response for $t \geq 0$.

Figure P7.5

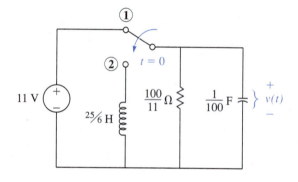

7.10   In the network of Fig. P7.6, the switch opens
instantaneously at $t = 0$. Find the current response for $t \geq 0$.

Figure P7.6

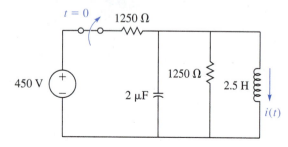

7.11   In the network of Fig. P7.7, the switch moves

instantaneously from position 1 to position 2 at
$t = 0$ when the capacitor is uncharged. Find the
voltage response for $t \geq 0$.

Figure P7.7

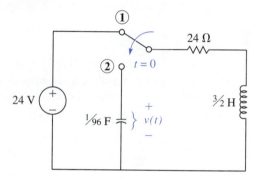

7.12   In the network of Fig. P7.8, the switch moves
instantaneously from position 1 to position 2 at
$t = 0$ when the capacitor is uncharged. Find the
voltage response for $t \geq 0$.

Figure P7.8

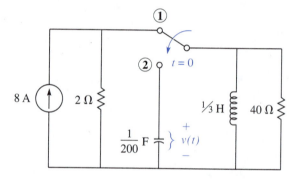

Section 7.4

Problems 7.13 to 7.17 are based upon Fig. P7.9,
where the current i_s is applied at $t = 0$ with no
current flowing through the inductor. In all cases,
$v(t)$ is to be found for the specified forcing function i_s.

Figure P7.9

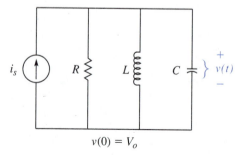

$v(0) = V_o$

7.13 $R = \frac{4}{5}\,\Omega$, $L = \frac{1}{3}\,$H, $C = \frac{1}{8}\,$F, $V_o = 16\,$V, and $i_s =$ $6t$ amperes.

7.14 $R = \frac{5}{16}\,\Omega$, $L = \frac{1}{5}\,$H, $C = \frac{1}{5}\,$F, $V_o = 12\,$V, and $i_s =$ $12\cos 2t$ amperes.

7.15 $R = 3\,\Omega$, $L = \frac{4}{3}\,$H, $C = \frac{1}{36}\,$F, $V_o = 10\,$V, and $i_s =$ $6e^{-4t} - 4$ amperes.

7.16 $R = 1\,\Omega$, $L = 1\,$H, $C = \frac{1}{4}\,$F, $V_o = 8\,$V, and $i_s =$ $12t^2$ amperes.

7.17 $R = 2\,\Omega$, $L = \frac{3}{2}\,$H, $C = \frac{1}{12}\,$F, $V_o = 6\,$V, and $i_s =$ $4e^{-2t}$ amperes.

Problems 7.18 to 7.22 are based upon Fig. P7.10, where the switch closes instantaneously at $t = 0$ with no current flowing through the inductor. In all cases, $i(t)$ is to be found for the specified forcing function v_s.

Figure P7.10

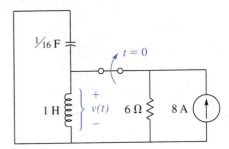

$$v(0) = V_o$$

7.18 $R = 2\,\Omega$, $L = 1\,$H, $C = \frac{1}{5}\,$F, $V_o = 2\,$V, and $v_s =$ $16 - 12\cos t$ volts.

7.19 $R = 2\,\Omega$, $L = \frac{1}{2}\,$H, $C = \frac{1}{2}\,$F, $V_o = 4\,$V, and $v_s =$ $12t + 4e^{-t}$ volts.

7.20 $R = 7\,\Omega$, $L = 1\,$H, $C = \frac{1}{12}\,$F, $V_o = 6\,$V, and $v_s =$ $6e^{-2t} + 8e^{-4t}$ volts.

7.21 $R = 4\,\Omega$, $L = 1\,$H, $C = \frac{1}{13}\,$F, $V_o = 8\,$V, and $v_s =$ $16t^2$ volts.

7.22 $R = \frac{5}{2}\,\Omega$, $L = \frac{1}{2}\,$H, $C = \frac{1}{3}\,$F, $V_o = 10\,$V, and $v_s =$ $6e^{-2t} - 8t + 12$ volts.

Section 7.5

7.23 In the network of Fig. P7.11, the switch moves from position 1 to position 2 instantaneously at $t = 0$. Find $v(t)$ for $t \geq 0$.

Figure P7.11

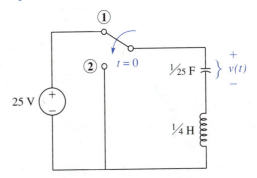

7.24 In the network of Fig. P7.12, the switch opens instantaneously at $t = 0$. Find $v(t)$ for $t \geq 0$.

Figure P7.12

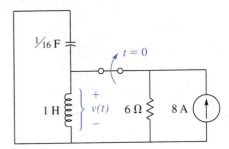

ANALOGS, DUALITY, AND OPERATIONAL AMPLIFIERS

OBJECTIVES

The objectives of this chapter are to:

- Consider the subject of duality and dual quantities, and show how to construct a dual network.

- Provide a detailed discussion of the operational amplifier, its principle of operation, and the way it may be utilized as a network element.

SECTION 8.1   **INTRODUCTION**

At present, the computational world is a digital one; and the use of the digital computer, which has forms ranging from the abacus through the hand-held calculator to the large-scale mainframe computer, as a means for the solution of a variety of real-world problems need not be discussed here. However, it is a fact that at one time, the analog computer played a significant role in the world of engineering analysis.

Although analog computers, which contain many operational amplifiers as fundamental elements, are rarely used, the usefulness of analog devices in such diverse areas as control systems, active filters, and even digital devices cannot be overemphasized. In the analog computer, the problem need not be translated into electrical terms, even though the analog computer is an electric device. The analyst, with a knowledge of the analogies among physical systems, merely programs or patches his problem into the machine and observes the result or output on a pen recorder. There are many more examples of the use of the operational amplifier and several of these are considered in this chapter. Of considerable interest in a network analysis framework is the use of the operational amplifier in active filters; this will be discussed in Chapter 20.

Thus, the highly reliable and versatile operational amplifier—which, in the present semiconductor age, can be purchased very cheaply—can be treated as a network element because of its use in a variety of applications. This is the reason the operational amplifier is considered here.

DUALITY

The principal of duality asserts that for any theorem in electric network analysis, there is a dual theorem in which one replaces quantities in the original theorem with dual quantities. Note, for example, that current and voltage, impedance and admittance, and meshes and nodes are examples of dual quantities; other dual quantities are displayed in Table 8.1. Thus, two networks can be said to be duals of one another if the mesh equations of one have the same form as the node equations of the other. The network of Fig. 8.1a, which has a pair of mesh equations

$$(R_1 + R_2 + R_3)i_1 \quad - R_3 i_2 \quad = v_s$$
$$- R_3 i_1 \quad + (R_3 + R_4)i_2 = 0$$

is the dual of the network shown in Fig. 8.1b, which possesses a pair of node equations

$$(G_1 + G_2 + G_3)e_1 \quad - G_3 e_2 \quad = i_s$$
$$- G_3 e_1 \quad + (G_3 + G_4)e_2 = 0$$

The similarity and one-for-one correspondence is immediately observed, and a duality between resistance and conductance and voltage sources and current sources as well as mesh currents and node voltages can be noted.

The principle of duality extends to network elements, network configurations, and network theorems. Some dual quantities are listed in Table 8.1 in which network 2 is the dual of network 1. Because the principle of duality provides a one-to-one transformation between the two networks, it is seen that network 1 is also the dual of network 2; the transformation from one to another may be effected in either direction.

Some dual elements and concepts

TABLE 8.1

Network 1	Network 2
Resistance	Conductance
Inductance	Capacitance
Voltage source	Current source
Series branch	Parallel branch
Voltage	Current
Mesh analysis	Node analysis
Open circuit	Short circuit
Loop analysis	Cut set analysis

FIGURE 8.1   (a) A network containing two meshes and (b) its dual containing two nodes

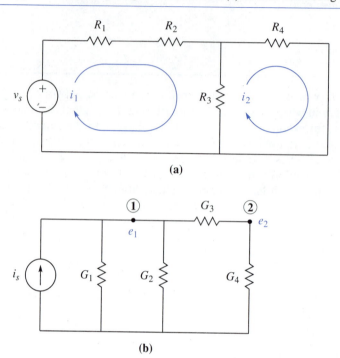

SECTION 8.3   CONTOURS

Any simple, closed contour placed in the xy plane separates the plane into two regions or domains, each having the contour as its boundary. One of these domains is called the interior domain and is bounded. The other, called the exterior domain, is unbounded. The network shown in Fig. 8.2a possesses the connected, yet unoriented, graph shown in Fig. 8.2b. The closed contour consisting of branches 1, 5, and 6 divides the graph into two domains: the interior domain, which is the network itself, and the exterior domain, which contains the entire xy plane except for the network.

The positive direction of a closed contour can be specified by considering the interior of the contour. To determine direction, some people imagine a leprechaun bicycling strenuously around the contour and tracing its path. If the bicycle has training wheels, then one of the training wheels will be in the interior domain and one will be in the exterior domain of the contour. The direction of the contour may then be specified by a consideration of whether the interior or exterior domains of the contour lie to the right or to the left of the direction of contour traversal. Here, a positive direction will be considered as the direction demanded when the interior domain or region lies to the right of the contour when the direction around the contour is clockwise.

With the foregoing in mind, it is observed that the meshes labeled 1, 2, and 3 in the graph shown in Fig. 8.5c are to be oriented in the clockwise direction. This, however, is not case for the closed contour consisting of branches 1, 5, and 6. This contour defines the entire xy plane exterior to the network, with the exception of the network in its interior. This mesh, called an outer mesh, truly satisfies the definition

FIGURE 8.2

(a) A network, (b) its unoriented graph, and (c) its graph showing n_t nodes and m_t meshes

One of the meshes is the outer mesh with counter-clockwise orientation.

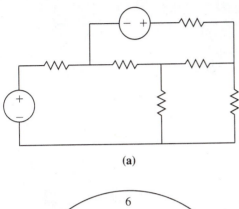

(a)

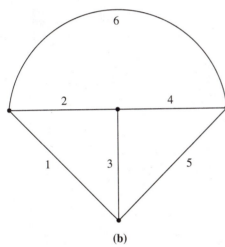

(b)

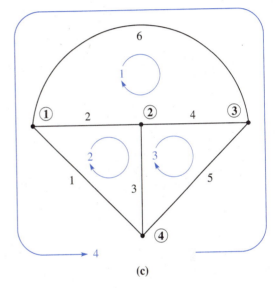

(c)

of a mesh (no branches or loops in its interior), and it must bear a counterclockwise orientation.

Thus, the graph shown in Fig. 8.5c contains a total of $n_t = 4$ nodes and $m_t = 4$ meshes, and it is the outer mesh that corresponds to the ground or datum node.

CONSTRUCTING A DUAL NETWORK

SECTION 8.4

A dual network may be constructed from any planar network. For example, consider the network shown in Fig. 8.3a containing four meshes, including the outer mesh, four nodes, including the ground or datum node, and a potpourri of R's, L's, and C's.

Because each mesh in a network locates a node in its dual, dots may be placed within each mesh, as indicated in Fig. 8.3b. The dots are connected as shown by the dashed lines to yield the branches of the dual network, and the elements in these branches are the duals of the elements in the original network (Table 8.1 may be consulted).

FIGURE 8.3

(a) A network containing a total of four nodes and four meshes, with mesh 4 as the outer mesh, (b) the procedure for constructing its dual, and (c) the actual dual of the network in (a)

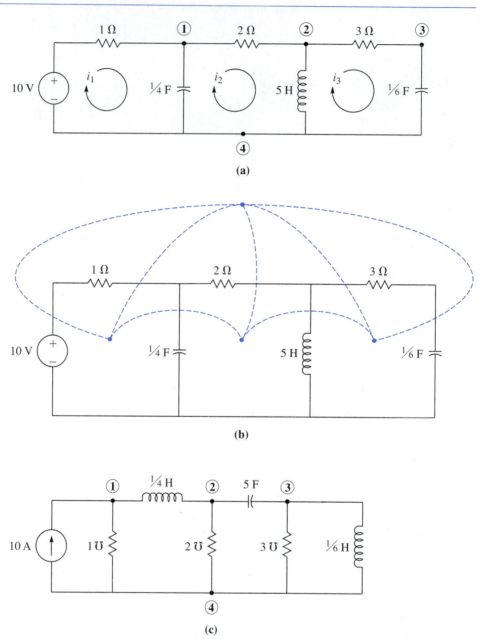

(a)

(b)

(c)

An actual dual of the network in Fig. 8.3a is shown in Fig. 8.3c. It was formed by proceeding from Fig. 8.3a through Fig. 8.3b. Observe that because node 1 in Fig. 8.3c replaces mesh 1 in Fig. 8.3a, the current source is oriented with the current flowing into the node. This is consistent with the positive direction of the mesh. Also observe that the ground node is the dual of the outer mesh.

A quick check of the validity of the foregoing procedure is in order. For the network in Fig. 8.3c, the integro-differential equations are

$$e_1 + 4 \int e_1 \, dt - 4 \int e_2 \, dt = 10$$

$$-4 \int e_1 \, dt + 5 \frac{de_2}{dt} + 2e_2 + 4 \int e_2 \, dt - 5 \frac{de_3}{dt} = 0$$

$$-5 \frac{de_2}{dt} + 5 \frac{de_3}{dt} + 3e_3 + 6 \int e_3 \, dt = 0$$

For the network in Fig. 8.3a, the three integro-differential node equations are

$$i_1 + 4 \int i_1 \, dt - 4 \int i_2 \, dt = 10$$

$$-4 \int i_1 \, dt + 5 \frac{di_2}{dt} + 2i_2 + 4 \int i_2 \, dt - 5 \frac{di_3}{dt} = 0$$

$$-5 \frac{di_2}{dt} + 5 \frac{di_3}{dt} + 3i_3 + 6 \int i_3 \, dt = 0$$

The form of each of the corresponding equations is identical, and all i's in the mesh equations are replaced by e's in the node equations.

SOME WORDS OF CAUTION

The method of constructing a dual network discussed in the previous section will not apply in most cases, if mutual inductance (discussed in Section 14.2) is present in the network. It will also not apply for a nonplanar network that does not have a dual.

A nonplanar graph is defined as a graph in which at least two branches possess an intersection (in the visual sense) that is not a node. Such a graph is indicated in Fig. 8.4. Observe that it is impossible to stretch the branches that intersect at the center of the graph to eliminate the crossing.

An unoriented graph of a nonplanar network

FIGURE 8.4

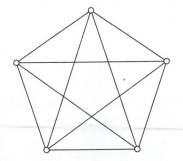

The allocation of meshes in a nonplanar graph may result in three or more mesh currents passing through a single branch. In this event, the single branch so involved would require three nodes at its extremities. This, of course, is clearly impossible, and it must be concluded that a nonplanar network does not possess a dual. Of even more importance is the fact that because of this, the method of mesh analysis must be generalized to loop analysis in order to be applied to nonplanar networks.

SECTION 8.6   OPERATIONAL AMPLIFIERS

8.6.1 Introduction

The *operational amplifier*, or *op-amp*, is an electronic device that has become a versatile network element. Operational amplifiers are commonly available in integrated circuit packages at minimum cost in several forms, ranging from configurations with 8 to 14 terminals and 1 to 4 op-amps. Typical of these are the 8-pin metal package, the 10-pin *flatpack*, and the 8-pin *dual in-line package*, or *DIP*. An artist's conception of the 8-pin DIP is shown in Fig. 8.5a and Fig. 8.5b indicates pin or lead connections.

By focusing on the terminal behavior of the op-amp, one can appreciate its use as a network element without a knowledge of its internal behavior. Consequently, the interest here is confined to the input/output characteristics of the op-amp, and in this way, an appreciation for the use of the device as a network element can be developed.

8.6.2 Exterior Connections

The voltage signal connections to the op-amp are indicated by the numerals 2, 3, and 6 in Figs. 8.6a and 8.6b. External sources of power that are required for the operation of the device are indicated at terminals 4 and 7. Other terminals that are used for other functions, such as frequency compensation and offset nulling, are not shown, nor are the terminals that have no connection whatsoever.

In Fig. 8.6c, v_o is designated as the output voltage, and v_1 and v_2 are the input voltages at the op-amp terminals. Notice in Figs. 8.6a and 8.6b that terminal 2 is labeled with a *minus* sign. This terminal is called the *inverting input*. Terminal 3 is labeled with a *plus* sign and is referred to as the *noninverting input*. The reference potential for all terminals is zero volts or ground potential.

The circuit symbol for the ideal op-amp is shown in Fig. 8.7a, and the equivalent circuit for the ideal op-amp is shown in Fig. 8.7b. Notice in Fig. 8.7b that the output v_o is a function of the difference between v_2 and v_1 and that this function is linear. The proportionally constant, A, is called the *open-loop voltage gain*, or just the *voltage gain*.

The signal range of the output v_o is dependent on and is limited by the voltages provided by the power supplies connected at points 4 and 7 in Figs. 8.6a and 8.6b. In the ideal op-amp, it is assumed, as indicated in Fig. 8.8, that operation in the linear region is without distortion. If the output v_o tends to go beyond the limits provided by the power supply (the limits are sometimes referred to as the *power supply rails*), the output is in the *saturation* region and the op-amp is said to *saturate*.

(a) Artist's conception of 8-pin dual-in-line package (DIP) and (b) pin or lead connections

FIGURE 8.5

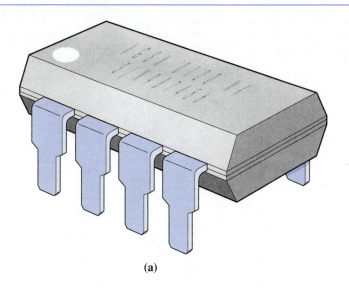

(a)

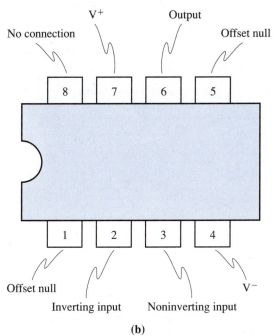

(b)

FIGURE 8.6   (a) and (b) Alternative representatives for the op-amp, showing signal voltages
at terminals 2, 3, and 6 and power supply voltages at terminals 4 and 7, and (c)
the representation showing input and output voltages. The numerals within circles
correspond to the 8-pin DIP connections in Fig. 8.5b.

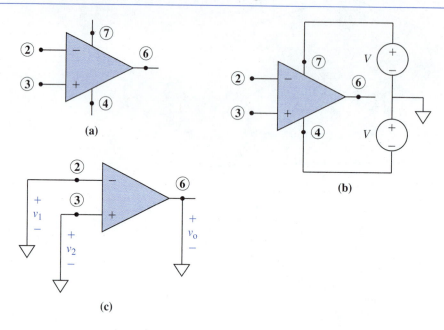

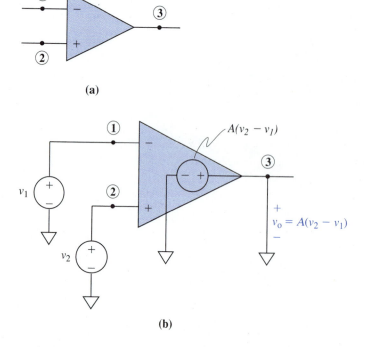

FIGURE 8.7   (a) Circuit symbol for the ideal op-amp and (b) the equivalent circuit
for the ideal op-amp

FIGURE 8.8

Transfer characteristics of a typical operational amplifier

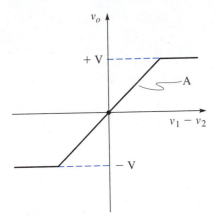

8.6.3 Characteristics of the Ideal Op-Amp

The op-amp senses the difference between two applied signals, multiples this difference by the open-loop gain, and provides the resulting voltage at the output terminal. In order to do this effectively, the ideal op-amp must possess three characteristics:

1. *The op-amp must have infinite-input resistance*; the op-amp must not draw current at terminals 1 and 2. Thus, the input impedance or resistance of the ideal op-amp is enormous, $R_i \rightarrow \infty$.
2. *The op-amp must have zero-output resistance*; because the output terminal of the op-amp must be treated as an ideal voltage source, the output voltage must be independent of any current drawn by a load impedance. Thus, the output impedance or resistance of the ideal op-amp is zero, $R_o = 0$.
3. *The op-amp must have very high (ideally infinite) gain*; high value of the open-loop gain (typically 10^5 or 10^6) guarantees the accuracy of the op-amp and ensures that the actual gain of the op-amp in a network application is not dependent on the open-loop gain. For the ideal op-amp, the open-loop gain is $A \rightarrow \infty$.

THE INVERTING CONFIGURATION   SECTION 8.7

The inverting configuration is displayed in Fig. 8.9, where it is to be noted that R_f, called the *feedback resistor*, connects terminals 1 and 3, terminal 2 is grounded and R_1 is connected between the voltage source v_i and terminal 1. The output voltage, as indicated in Fig. 8.7b is

$$v_o = A(v_2 - v_1)$$

and hence,

$$v_2 - v_1 = \frac{v_o}{A}$$

FIGURE 8.9   The inverting configuration

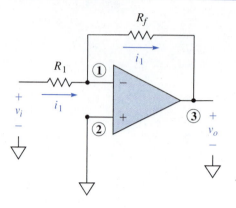

Because the amplifier circuit is designed to produce an output voltage that lies in the range $-V \le v_o \le V$, and because the magnitude of V is some moderate number (like 20 V), it is observed that if A is very large (like 10^6),

$$v_2 - v_1 \approx 0$$

or

$$v_2 \approx v_1$$

Here, in Fig. 8.9, v_2 is at ground potential, so that as long as A is very high, little error occurs if it is assumed that

$$v_1 = v_2 = 0$$

and it is said that the input terminals of the ideal op-amp are virtually at ground potential. The terminology *virtual ground* or *virtual short circuit* is often used.

In Fig. 8.9,

$$i_1 = \frac{v_i - v_1}{R_1} = \frac{v_i}{R_1}$$

because the assumption is made that $v_1 = 0$. Because the ideal op-amp does not draw current ($R_i = \infty$), KCL shows that the current through R_f is i_1, and

$$v_o = v_1 - i_1 R_f = 0 - i_1 R_f$$

or

$$v_o = -\frac{v_i}{R_1} R_f$$

The ratio of v_o to v_i, called the *closed-loop voltage gain*, is

$$\frac{v_o}{v_i} = -\frac{R_f}{R_1} \tag{8.1}$$

and it is noted that this closed-loop amplifier provides a signal inversion. This is why it called an *inverting amplifier*. In analog computer parlance, the inverting amplifier is often called an *inverter*.

8.7.1 The Summing Inverter

The inverting configuration can handle multiple inputs as shown in Fig. 8.10. Here, v_1 is a virtual ground, and KCL at point 1 provides

$$i_1 + i_2 + i_3 + \cdots + i_n = -i_o$$

or

$$\frac{v_{i1} - v_1}{R_1} + \frac{v_{i2} - v_1}{R_2} + \frac{v_{i3} - v_1}{R_3} + \cdots + \frac{v_{in} - v_1}{R_n} = \frac{v_1 - v_o}{R_f}$$

Because $v_1 \approx 0$,

$$\frac{v_{i1}}{R_1} + \frac{v_{i2}}{R_2} + \frac{v_{i3}}{R_3} + \cdots + \frac{v_{in}}{R_n} = -\frac{v_o}{R_f}$$

and this may be solved for v_o:

$$v_o = -\left(\frac{R_f}{R_1} v_{i1} + \frac{R_f}{R_2} v_{i2} + \frac{R_f}{R_3} v_{i3} + \cdots + \frac{R_f}{R_n} v_{in} \right) \tag{8.2}$$

Equation (8.2) shows why the configuration illustrated in Fig. 8.10 is called *the summing inverter*, the *weighted summer*, or merely the *summer*.

8.7.2 The Integrator

If a capacitor is used instead of the feedback resistor, as indicated in Fig. 8.11, an integrating circuit is obtained. Here, because of the assumption that $v_1 = 0$,

$$i_1 = \frac{v_i}{R_1}$$

The summing inverter

FIGURE 8.10

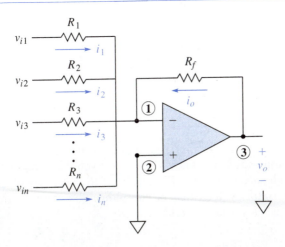

FIGURE 8.11   The integrator

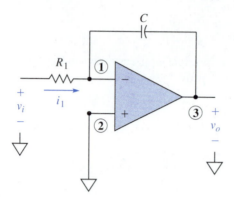

and because this current must flow through the capacitor, due to the almost infinite input impedance of the op-amp, the voltage across the capacitor is

$$v_o = -\frac{1}{C} \int_{-\infty}^{t} i_1 \, dz$$

From these simple relationships, the integrating action is obtained as

$$v_o = -\frac{1}{R_1 C} \int_{0}^{t} v_i \, dz + v_C(0) \tag{8.3}$$

where $v_C(0)$ is the initial voltage across the capacitor.

Care must be exercised, however. If the input is a unit step, $v_i = u(t)$, the integration produces an inverted unit ramp, $v_o = r(t)$, which has an initial value at $t = 0$ of $v_o(0)$ and continues until the output saturates, as indicated in Fig. 8.8. This effect can be troublesome in a variety of real-world situations. For example, in using an analog computer, the analyst must scale the problem so that the required information is obtained before the integrator saturates.

The integrator can handle multiple inputs. When it does, it becomes the summing integrator shown in Fig. 8.12. In this case, a development similar to the one given in the previous discussion of the summing inverter will indicate that

$$v_o = -\left(\frac{1}{R_1 C} \int_{0}^{t} v_{i1} \, dz + \frac{1}{R_2 C} \int_{0}^{t} v_{i2} \, dz + \frac{1}{R_3 C} \int_{0}^{t} v_{i3} \, dz + \cdots \right.$$
$$\left. + \frac{1}{R_n C} \int_{0}^{t} v_{in} \, dz \right) + v_C(0) \tag{8.4}$$

8.7.3 The Differentiator

The ideal differentiating op-amp is rarely employed in practice because the input voltage signal always carries a spurious signal known as *noise*. Even if the manitude of the noise is low so that a high *signal-to-noise ratio* results, if the noise waveform has a large slope, its effect on the output voltage may be enormous. This configuration is called a differentiator, and it is shown in Fig. 8.13.

The summing integrator

FIGURE 8.12

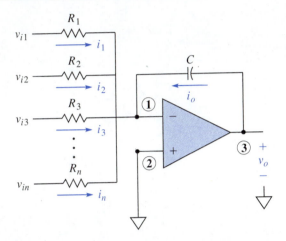

In this case, again with the assumption that $v_1 = 0$,

$$i_f = -\frac{v_o}{R_f}$$

and because the ideal op-amp does not draw current, KCL shows that i_f is also given by

$$i_c = C\frac{dv_i}{dt}$$

When these are equated, the differentiating action is observed:

$$v_o = -R_fC\frac{dv_i}{dt} \tag{8.5}$$

The differentiator

FIGURE 8.13

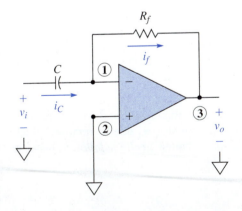

FIGURE 8.14   The noninverting configuration

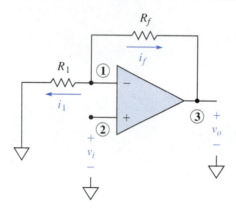

SECTION 8.8   THE NONINVERTING CONFIGURATION

Observe that in the ideal op-amp in Fig. 8.14, the input is applied at terminal 2 and
that R_1 is connected to ground. Application of KCL at terminal 1 gives

$$\frac{v_1}{R_1} + \frac{v_1 - v_o}{R_f} = 0$$

With $v_1 = v_2 = v_i$,

$$v_i R_f + v_i R_1 - v_o R_1 = 0$$

$$v_i(R_1 + R_f) = v_o R_1$$

$$v_i\left(1 + \frac{R_f}{R_1}\right) = v_o$$

and the closed-loop voltage gain for the noninverting configuration is therefore

$$\frac{v_o}{v_i} = 1 + \frac{R_f}{R_1} \qquad\qquad (8.6)$$

SECTION 8.9   EXAMPLES OF OP-AMP CIRCUITS

8.9.1 The Voltage Follower

■ EXAMPLE 8.1

In the noninverting configuration shown in Fig. 8.14, the feedback resistor
can be replaced by a short circuit, and the resistor R_1 can be removed. Such
an arrangement is shown in Fig. 8.15 and is called a *voltage follower*, because
with $R_f = 0$ and $R_1 = \infty$, eq. (8.6)

$$\frac{v_o}{v_i} = 1 + \frac{R_f}{R_1} \qquad\qquad (8.6)$$

FIGURE 8.15

The voltage follower

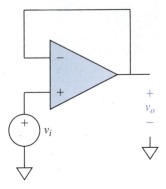

reduces to

$$\frac{v_o}{v_i} = 1$$

8.9.2 A Strain Gage

EXAMPLE 8.2

A strain gage is a device that measures mechanical strain (elongation, compression, or deflection) by means of a resistor whose resistance value changes slightly when it is bent or twisted. Figure 8.16 shows such a resistor in the feedback loop of an op-amp. In Fig. 8.16, a reference input voltage v_i is connected to both the positive and the negative terminals of the op-amp, and by voltage division.

$$v_1 = v_i + \frac{R_a}{R_a + R_b + \Delta R}(v_o - v_i)$$

$$v_2 = \frac{R_b}{R_a + R_b}v_i$$

The op-amp configuration for a strain gage

FIGURE 8.16

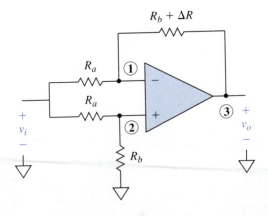

Because $v_1 = v_2$, these may be equated:

$$v_i + \frac{R_a}{R_a + R_b + \Delta R} v_o - \frac{R_a}{R_a + R_b + \Delta R} v_i = \frac{R_b}{R_a + R_b} v_i$$

A little algebra can show that

$$\left(\frac{R_b}{R_a + R_b} + \frac{R_a}{R_a + R_b + \Delta R} - 1 \right) v_i = \frac{R_a}{R_a + R_b + \Delta R} v_o$$

and then

$$-R_a \Delta R\, v_i = R_a(R_a + R_b) v_o$$

or

$$v_o = -\frac{\Delta R}{R_a + R_b} v_i$$

so ΔR can be written as a function of v_o for known values of R_a, R_b, and v_i:

$$\Delta R = -\left(\frac{R_a + R_b}{v_i} \right) v_o$$

■

8.9.3 A More Practical Arrangement

The next example shows how an inverter may be put together by using resistors that have relatively small resistance values.

■ **EXAMPLE 8.3**

Find the value of R_a if the op-amp arrangement shown in Fig. 8.17 is to have a closed-loop gain of -1200, $R_1 = 2000\ \Omega$, and $R_b = 50\ \Omega$.

FIGURE 8.17   A more practical inverting configuration

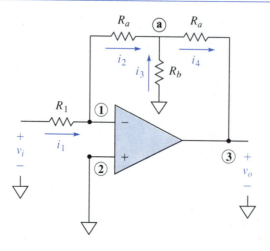

Solution   Observe that if $R_1 = 2000\ \Omega$, eq. (8.1) shows that the value of a single feedback resistor to yield a gain of -1200 would be

$$R_f = -R_1 \frac{v_o}{v_i} = -2000(-1200) = 2.4 \times 10^6 = 2.4\ \text{M}\Omega$$

In Fig. 8.17 with $v_1 = 0$,

$$i_1 = \frac{v_i - v_1}{R_1} = \frac{v_i}{R_1}$$

and

$$i_2 = i_1 = \frac{v_i}{R_1}$$

The voltage at node a will be

$$v_a = v_1 - i_2 R_a = -i_2 R_a = -\frac{v_i}{R_1} R_a = -\frac{R_a}{R_1} v_i$$

Then,

$$i_3 = \frac{0 - v_a}{R_b} = \frac{R_a}{R_1 R_b} v_i$$

KCL at node a gives

$$i_4 = i_2 + i_3 = \frac{v_i}{R_1} + \frac{R_a}{R_1 R_b} v_i = \frac{R_a + R_b}{R_1 R_b} v_i$$

so that

$$v_o = v_a - R_a i_4 = -\frac{R_a}{R_1} v_i - \frac{R_a(R_a + R_b)}{R_1 R_b} v_i$$

or

$$v_o = -\frac{R_a}{R_1}\left(1 + \frac{R_a + R_b}{R_b}\right) v_i$$

If $R_1 = 2000\ \Omega$, $R_b = 50\ \Omega$, and $v_o/v_i = -1200$, then

$$-1200 = -\frac{R_a}{2000}\left(1 + \frac{R_a + 50}{50}\right)$$

and the reader may verify that $R_a = 10{,}900\ \Omega$. Notice that four resistors with values less than $12{,}000\ \Omega$ are required to take the place of one resitor of $2.4\ \text{M}\Omega$.

FIGURE 8.18   The negative impedance converter

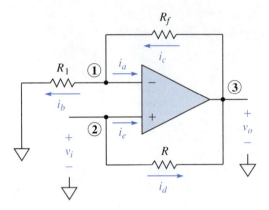

8.9.4 The Negative Impedance (Resistance) Converter

■ **EXAMPLE 8.4**

Figure 8.18 shows the use of an ideal op-amp in the noninverting configuration as a *negative impedance converter*. Notice that $i_a = i_e = 0$ and that $v_1 = v_2 = v_i$. Ohm's law gives

$$i_b = \frac{v_i}{R_1}$$

and KCL at node 1 indicates that

$$i_c = i_b = \frac{v_i}{R_1}$$

By KVL,

$$v_o = v_i + i_c R_f = v_i + \frac{R_f}{R_1} v_i$$

or

$$v_o = \left(1 + \frac{R_f}{R_1}\right) v_i$$

and by Ohm's law,

$$i_d = \frac{v_i - v_o}{R} = \frac{v_i}{R} - \left(1 + \frac{R_f}{R_1}\right)\left(\frac{v_i}{R}\right)$$

or

$$i_d = -\frac{R_f}{R_1 R} v_i$$

The input resistance is defined as v_i/i_d, so that here

$$R_{\text{in}} = -\frac{R_1}{R_f} R$$

It can be shown that when $R_1 = R_f$, the circuit acts as a voltage-to-current converter that can supply a current that is independent of the load resistance. ■

8.9.5 Programming the Analog Computer

■ **EXAMPLE 8.5**

The analog computer can be used to look at the form of the solution to a pair of simultaneous differential equations such as

$$\frac{di}{dt} + 4i - 2v = 2$$

$$-2i + \frac{dv}{dt} + 5v = 0$$

with initial conditions, for simplicity, taken as $i(0) = v(0) = 0$. It is then a matter of procedure to connect summing integrators and inverters to provide signals representing i and v to recording devices for viewing purposes. The connections for the pair of equations under consideration are indicated in Fig. 8.19. They are based on the fact that the input to each op-amp represents a sum and that the derivatives are

$$\frac{di}{dt} = -4i + 2v + 2$$

Operational amplifier connections for the analog solution of a pair of simple, first-order, simultaneous, linear differential equations

FIGURE 8.19

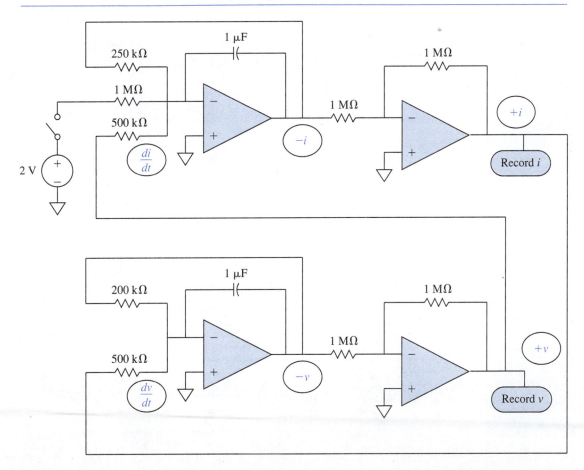

and

$$\frac{dv}{dt} = 2i - 5v$$

■

SECTION 8.10   SPICE EXAMPLE

Reading: In addition to reading Sections C.1 through C.6 in Appendix C, which are recommended for reading in previous chapters, the reader should read and understand Section C.7 before proceding to the SPICE example that follows. In this example, consider that both operational amplifiers have an input resistance of $R_i =$ 500 kΩ, an open-loop gain of $A = 10^6$, and an output resistance of $R_o = 2000$ Ω.

EXAMPLE S8.1

The output voltage in the operational amplifier arrangement in Fig. S8.1 is to be found. The network made ready for analysis with PSPICE is shown in Fig. S8.2, the input file is reproduced in Fig. S8.3, and pertinent extracts from the output file are presented in Fig. S8.4. The reader may wish to verify that if the op-amps were ideal, the voltage appearing at the 10-kΩ resistor would be 20 V.

Figure S8.1   Arrangement of operational amplifiers

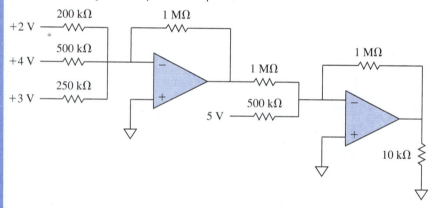

Figure S8.2   The arrangement of Fig. S8.1 ready for PSPICE analysis

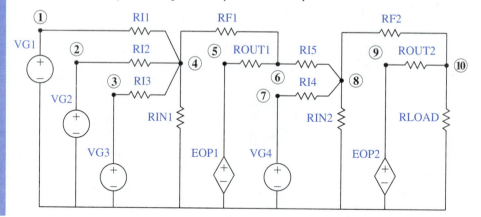

Figure S8.3 Input PSPICE file for analysis of the arrangement of Fig. S8.1

```
SPICE EXAMPLE - CHAPTER 8 - NUMBER 1 - OPAMP SUMMING INVERTER
*FIRST ENTER DATA FOR BOTH OPAMPS
*****************************************************************
RIN1    4       0       500K
RIN2    8       0       500K
ROUT1   5       6       2000
ROUT2   9       10      2000
EOP1    5       0       4       0       -1E6
EOP2    9       0       8       0       -1E6
*****************************************************************
*THEN THE BALANCE OF THE NETWORK
VG1     1       0       DC      2
VG2     2       0       DC      4
VG3     3       0       DC      3
VG4     7       0       DC      5
RI1     1       4       200K
RI2     2       4       500K
RI3     3       4       250K
RI4     7       8       500K
RI5     6       8       1MEG
RLOAD   10      0       10K
RF1     4       6       1MEG
RF2     8       10      1MEG
*****************************************************************
*HERE IS THE SINGLE CONTROL STATEMENT
.END
```

Figure S8.4 Pertinent output for the PSPICE analysis of the arrangement in Fig. S8.1

```
SPICE EXAMPLE - CHAPTER 8 - NUMBER 1 - OPAMP SUMMING INVERTER


****      SMALL SIGNAL BIAS SOLUTION         TEMPERATURE =     27.000 DEG C


*********************************************************************************


NODE    VOLTAGE     NODE    VOLTAGE     NODE    VOLTAGE     NODE    VOLTAGE


(   1)    2.0000  (    2)     4.0000  (    3)     3.0000  (     4) 30.12E-06

(   5)  -30.1200  (    6)   -30.0000  (    7)     5.0000  (     8)-24.04E-06

(   9)   24.0390  (   10)    19.9990
```

CHAPTER 8

SUMMARY

- Dual quantities in electric networks are displayed in Table 8.1.

- The ideal operational amplifier possesses the following characteristics:

    —It has *infinite-input resistance*; the op-amp must not draw current at its inverting and noninverting terminals. Thus, the input impedance or resistance of the ideal op-amp is infinite, $R_i \rightarrow \infty$.

    —It has *zero-output resistance*; because the output terminal of the op-amp must be treated as an ideal voltage source, the output voltage must be independent of any current drawn by a load impedance. Thus, the output impedance or resistance of the ideal op-amp is zero, $R_0 = 0$.

    —It has *very high (Ideally infinite) gain*; the high value of the open-loop gain (typically 10^5 or 10^6) guarantees the accuracy of the op-amp and ensures that the actual gain of the op-amp in a network application is not dependent on the open-loop gain. For the ideal op-amp, the open-loop gain is $A \rightarrow \infty$.

- The open-loop voltage gain for the inverting configuration is

$$\frac{v_o}{v_i} = -\frac{R_f}{R_i}$$

and the open-loop voltage gain for the noninverting configuration is

$$\frac{v_o}{v_i} = 1 + \frac{R_f}{R_i}$$

- The noninverting configuration can be employed as a summing inverter, which gives the output as the sum of n inputs:

$$v_o = -\left(\frac{R_f}{R_1}v_{i1} + \frac{R_f}{R_2}v_{i2} + \frac{R_f}{R_3}v_{i3} + \cdots + \frac{R_f}{R_n}v_{in}\right)$$

- The noninverting configuration can be employed as an integrator, which gives the output as a single input,

$$v_o = -\frac{1}{R_1 C}\int_0^t v_i \, dz + v_C(0)$$

or as the sum of n inputs,

$$v_o = -\left(\frac{1}{R_1 C}\int_0^t v_{i1} \, dz + \frac{1}{R_2 C}\int_0^t v_{i2} \, dz + \frac{1}{R_3 C}\int_0^t v_{i3} \, dz + \cdots \right.$$
$$\left. + \frac{1}{R_n C}\int_0^t v_{in} \, dz\right) + v_C(0)$$

- The operational amplifier can be used in many applications. Five that are discussed in detail in this chapter are listed here:

—The voltage follower

—A strain gage

—A practical arrangement of resistors

—The negative impedance converter

—The analog computer

Additional Readings

Blackwell, W.A., and L.L. Grigsby. *Introductory Network Theory*. Boston: PWS Engineering, 1985, pp. 217–231, 282–284.

Bobrow, L.S. *Elementary Linear Circuit Analysis*. 2d ed. New York: Holt, Rinehart and Winston, 1987, pp. 104–114, 213–233, 303–305.

Del Toro, V. *Engineering Circuits*. Englewood Cliffs, N.J.: Prentice-Hall, 1987, pp. 47–59, 308, 425–430.

Dorf, R.C. *Introduction to Electric Circuits*. New York: Wiley, 1989, pp. 126, 425–442.

Hayt, W.H., Jr., and J.E. Kemmerly. *Engineering Circuit Analysis*. 4th ed. New York: McGraw-Hill, 1986, pp. 45–48, 135–139, 221, 222, 365–367.

Irwin, J.D. *Basic Engineering Circuit Analysis*. 3rd ed. New York: Macmillan, 1989, pp. 121–132, 395, 396.

Johnson, D.E., J.L. Hilburn, and J.R. Johnson. *Basic Electric Circuit Analysis*. 4th ed. Englewood Cliffs, N.J.: Prentice-Hall, 1989, pp. 64–70, 89, 90, 99–102, 194.

Karni, S. *Applied Circuit Analysis*. New York: Wiley, 1988, pp. 34, 86–93, 174, 175, 183, 295.

Madhu, S. *Linear Circuit Analysis*. Englewood Cliffs, N.J.: Prentice-Hall, 1988, pp. 74, 144–153, 176, 177, 203–205, 268, 307, 308, 372, 373.

Nilsson, J.W. *Electric Circuits*. 3rd ed. Reading, Mass.: Addison-Wesley, 1990, pp. 145–168, 245–253, 297–305.

Paul, C.R. *Analysis of Linear Circuits*. New York: McGraw-Hill, 1989, pp. 59–64.

CHAPTER 8 — PROBLEMS

Section 8.4

8.1   Draw the dual of the network shown in Fig. P8.1.

Figure P8.1

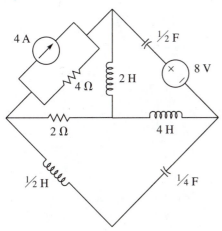

8.3   Draw the dual of the network shown in Fig. P8.3.

Figure P8.3

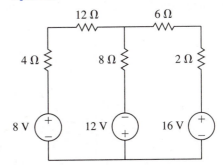

8.2   Draw the dual of the network shown in Fig. P8.2.

Figure P8.2

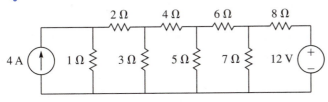

8.4   Draw the dual of the network shown in Fig. P8.4.

Figure P8.4

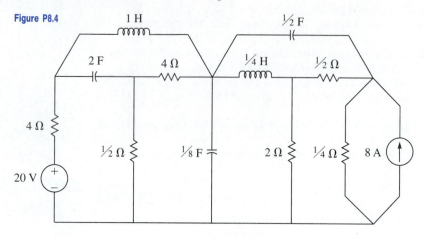

8.5   Draw the dual of the network shown in Fig. P8.5.

Figure P8.5

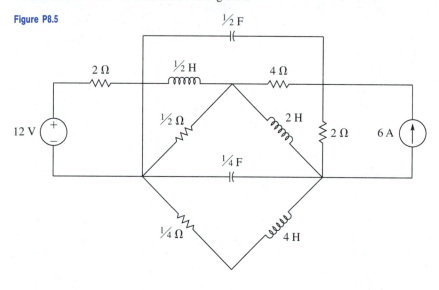

8.6   Draw the dual of the network shown in Fig. P8.6.

Figure P8.6

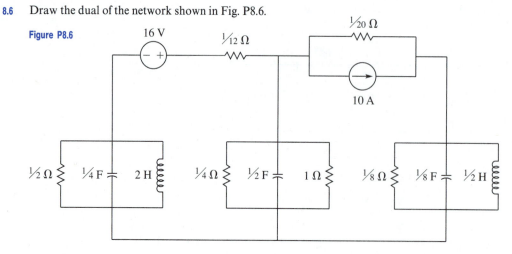

Section 8.7

8.7 Find the voltage gain v_o/v_i and the input resistance R_{in} for each of the op-amp configurations in Fig. P8.7.

Figure P8.7

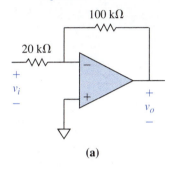
(a)

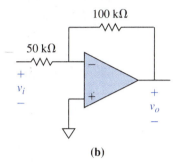
(b)

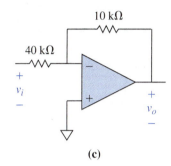
(c)

8.8 Find the voltage gain v_o/v_i and the input resistance R_{in} for each of the op-amp configurations in Fig. P8.8.

Figure P8.8

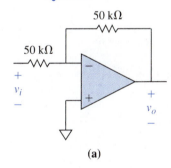
(a)

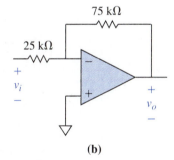
(b)

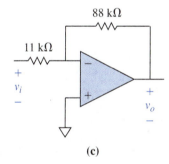
(c)

8.9 An op-amp in the inverting configuration is to have a gain of -125 and an input resistance as high as possible. If no resistor used in the op-amp circuit is to have a value larger than 5 MΩ, how is this design to be achieved by using two resistors? What is the input resistance?

8.10 In the op-amp circuit of Fig. P8.9, what is v_o?

Figure P8.9

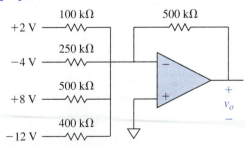

8.11 In the op-amp circuit of Fig. P8.10, what is the value of v_3 necessary to make $v_o = 18.5$ *V*?

Figure P8.10

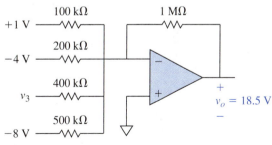

8.12   What is the gain of the op-amp circuit in Fig. P8.11?

Figure P8.11

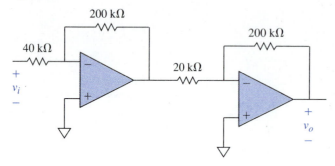

8.13   What is the value of the capacitor in the op-amp circuit of Fig. P8.12 if $v_i(t) = -12u(t)$ volts and $v_o(t) = 48r(t)V$? Note that $v_C(0) = 0$ V.

Figure P8.12

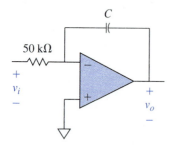

8.14   In the op-amp circuit show in Fig. P8.13, if v_i is a 4-V step, what is the output voltage?

Figure P8.13

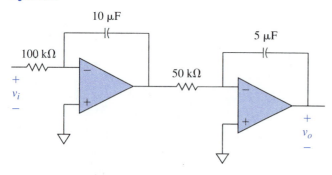

8.15 Figure 8.14 shows an op-amp circuit with the op-amp connected as a differential amplifier. By evaluating v_1 and v_2 in terms of v_{i1} and v_{i2}, find an expression for $v_0 = f(v_{i1}, v_{i2})$.

Figure P8.14

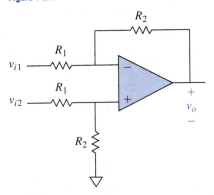

8.16 Determine the ratio v_o/v_i in the op-amp configuration of Fig. P8.15

Figure P8.15

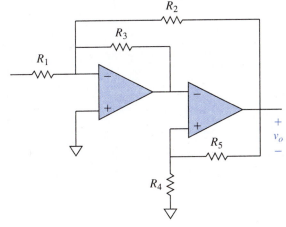

Section 8.8

8.17 Find v_o in the op-amp configuration of Fig. P8.16.

Figure P8.16

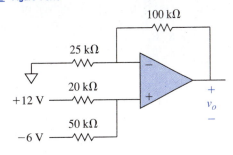

8.18 In the op-amp circuit of Fig. P8.17, if $v_{i1} = 2$ V, $v_{i2} = -4$ V, and $v_o = 12$ V, what is the value of v_3?

Figure P8.17

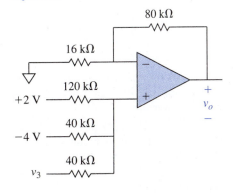

8.19 For the op-amp arrangement shown in Fig. P8.18, if $v_2 = 6$ V, select the value of R to make $v_o = 24$ V.

Figure P8.18

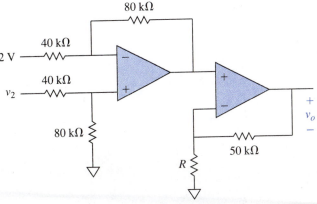

8.20 For the op-amp arrangement shown in Fig. P8.18, it $R = 40$ kΩ, what value of v_2 is required to make $v_o = 11.25$ V?

8.21 For the op-amp arrangement shown in Fig. P8.19, show that the output voltage is given by

$$v_o = \frac{1 + R_2/R_1}{1 + R_3/R_4} v_2 - \frac{R_2}{R_1} v_1$$

Figure P8.19

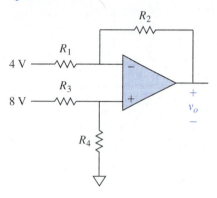

Section 8.9

8.22 In Fig. P8.19, $v_1 = 4$ V, $v_2 = 0$ V, $R_2 = 80$ kΩ, and $R_4 = 40$ kΩ. Find the values of R_1 and R_3 required to make $v_o = 4$ V subject to the constraint that $R_1 = 1.25R_3$.

8.23 What is v_o in the op-amp configuration of Fig. P8.20 if $v_2 = -1$ V and $R = 25$ kΩ?

Figure P8.20

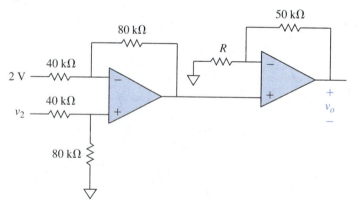

8.24 The set of simultaneous differential equations

$$\frac{dv_1}{dt} + 5v_1 - 2v_2 = 18$$

$$-2v_1 + 2\frac{dv_2}{dt} + 2v_2 = 0$$

is to be programmed into an analog computer. Draw the patch diagram showing recorder outputs for v and i. Provision need not be made for initial conditions.

8.25 The differential equation

$$\frac{d^2i}{dt^2} + 10\frac{di}{dt} + 16i = 40t$$

is to be programmed into an analog computer. Draw the patch diagram showing recorder output for i. Provision need not be made for initial conditions.

ANALYSIS OF AC NETWORKS

COMPLEX FREQUENCY
AND THE PHASOR

9

OBJECTIVES

The objectives for this chapter are to:

- Develop the concept of complex frequency by considering the forms that s may take in the exponential, $f(t) = Fe^{st}$.

- Introduce the concept of the s-domain and the complex plane.

- Consider the concepts of impedance and admittance to exponentials and show how these may be evaluated for both dc and ac signals.

- Show how alternating current impedance and admittance may be determined.

- Develop the concept of the phasor, the phasor domain, and the relationship between the phasor and the amplitude-phase angle form of the sinusoid.

INTRODUCTION

SECTION 9.1

The sinusoid is the electrical waveform that is most easily generated by physical devices. Most of the electric power systems throughout the world produce, transmit, and use electric energy that varies sinusoidally and a study of electrical networks that are excited by sinusoids can be justified on just this basis alone.

But there is much more. The world is full of sinusoidal phenomena, from the motion of the stars, planets, and tides to the most sophisticated of communications systems. Moreover, Fourier analysis (to be considered in detail in a later chapter) permits arbitrary-signal waveforms to be decomposed and expressed as sinusoids, thereby extending the range of sinusoidal analysis to more general cases.

In this chapter, some analytical tools are introduced so that in subsequent chapters, they may be used as the basis for steady-state, alternating-current (ac) network analysis. In all of this, the reader will be able to observe and continuously reflect on the heritage that derives from a study of direct-current resistive networks.

The reader may wish to review the algebra of complex numbers before proceeding. A summary of this subject is presented in Appendix A.

SECTION 9.2   **COMPLEX FREQUENCY**

Consider the function

$$f(t) = Fe^{st} \tag{9.1}$$

where F and s are complex constants. Here, $s = \sigma + j\omega$ and is known as the *complex frequency*. If $s = 0$, then $\sigma = \omega = 0$, and $f(t)$ is a constant:

$$f(t) = F$$

If $\sigma \neq 0$ but $\omega = 0$, then $s = \sigma$, and $f(t)$ is a rising (*ascending*) or falling (*descending*) exponential function:

$$f(t) = Fe^{\sigma t}$$

If $\sigma = 0$ but $\omega \neq 0$, then $s = j\omega$, and $f(t)$ is a complex sinusoid without any exponential enhancement either in the form of a rise or a decay,

$$f(t) = Fe^{j\omega t} \tag{9.2a}$$

or by one of Euler's equations (see Appendix A),

$$f(t) = F(\cos \omega t + j \sin \omega t) \tag{9.2b}$$

Because the cosine function can be represented by a pair of exponentials,

$$\cos x = \frac{1}{2}(e^{jx} + e^{-jx})$$

it is observed that the cosine function

$$f(t) = \cos(\omega t + \phi)$$

can be represented by

$$f(t) = \frac{1}{2}Fe^{j(\omega t + \phi)} + \frac{1}{2}Fe^{-j(\omega t + \phi)} = \frac{1}{2}Fe^{j\phi}e^{j\omega t} + \frac{1}{2}Fe^{-j\phi}e^{-j\omega t}$$

or

$$f(t) = F_1 e^{j\omega t} + F_2 e^{-j\omega t}$$

where F_1 and F_2 are complex conjugates,

$$F_1 = \frac{1}{2}Fe^{j\phi} \quad \text{and} \quad F_2 = \frac{1}{2}Fe^{-j\phi}$$

and where in

$$f(t) = F_1 e^{s_1 t} + F_2 e^{s_2 t}$$

both $s_1 = j\omega$ and $s_2 = -j\omega$ are also complex conjugates.

If $s = \sigma + j\omega$, where neither σ nor ω are equal to zero, then

$$f(t) = Fe^{(\sigma + j\omega t)}$$

or

$$f(t) = Fe^{\sigma t}e^{j\omega t} = Fe^{\sigma t}(\cos \omega t + j \sin \omega t)$$

This case is called the *damped sinusoid* whether or not σ is negative. A damped cosine with phase angle ϕ,

$$Fe^{\sigma t} \cos(\omega t + \phi)$$

can be written as

$$f(t) = \frac{1}{2} Fe^{\sigma t}(e^{j(\omega t + \phi)} + e^{-j(\omega t + \phi)}) = \frac{1}{2} Fe^{\sigma t}e^{j\phi}e^{j\omega t} + \frac{1}{2} Fe^{\sigma t}e^{-j\phi}e^{-j\omega t}$$

or

$$f(t) = F_1 e^{j\omega t} + F_2 e^{-j\omega t} \qquad\qquad (9.3)$$

where in this case,

$$F_1 = \frac{1}{2} Fe^{\sigma t}e^{j\phi} \qquad \text{and} \qquad F_2 = \frac{1}{2} Fe^{\sigma t}e^{-j\phi}$$

are also complex conjugates and where, because $e^{j\omega t}$ is the complex conjugate of $e^{-j\omega t}$, both terms in eq. (9.3) are complex conjugates.

There is a threefold significance to the foregoing:

1. With $s = \sigma + j\omega$ in

$$f(t) = Fe^{st} = Fe^{(\sigma + j\omega)t}$$

   both σ and $j\omega$ multiply t. The angular frequency is ω in *radians per unit time* (usually in radians per second) and is the imaginary part of s. The radian has no units, so that ω has the units of reciprocal seconds (s^{-1}). The real part of s, which is σ, must also possess the units s^{-1} and is also a frequency. It is called the *neper frequency*; the neper is a measure of the attenuation in a system and has no units. Thus, σ is in nepers per second.

2. There is a relation between the value s and actual dc or ac signals. Direct current implies, as indicated by eq. (9.1), that $s = 0$, and eqs. (9.2) indicate that for ac, $s = j\omega$.

3. The form of eq. (9.1) as both σ and ω in $s = \sigma + j\omega$ take on various values is of considerable importance. Consider the s plane diagram of Fig. 9.1, and observe that the imaginary axis (called the $j\omega$ axis) divides the s-plane into what are called the left-half and right-half planes (the LHP and RHP). The function $f(t)$ will decay if $\sigma < 0$, will have constant amplitude if $\sigma = 0$, and will increase without bound if $\sigma > 0$.

FIGURE 9.1   The behavior of $f(t) = e^{(\sigma + j\omega)t}$ in the s plane

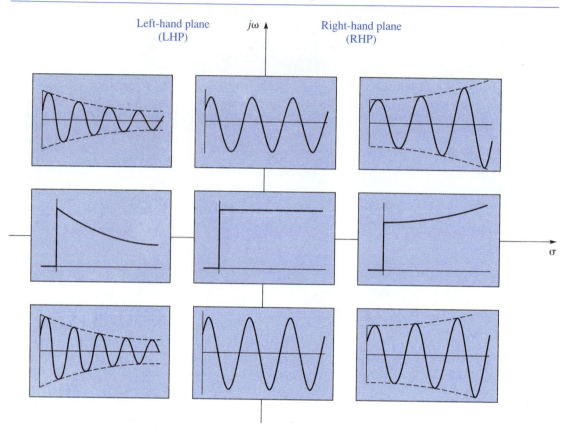

SECTION 9.3   **IMPEDANCE TO EXPONENTIALS**

The ratio of the exponential voltage function Ve^{st} to the exponential current function Ie^{st} is called the *impedance Z*,

$$Z = \frac{Ve^{st}}{Ie^{st}} \quad \text{(ohms)} \tag{9.4}$$

and the reciprocal of the impedance is the *admittance*,

$$Y = \frac{1}{Z} = \frac{Ie^{st}}{Ve^{st}} \quad \text{(mhos)} \tag{9.5}$$

Networks composed of resistors, inductors, and capacitors all possess both an impedance and an admittance. For the resistor, the elemental equation that relates voltage and current is Ohm's law, $v = Ri$, so that

$$Z_R = \frac{Ve^{st}}{Ie^{st}} = \frac{RIe^{st}}{Ie^{st}} = R \qquad \text{(ohms)} \tag{9.6}$$

For the inductor, where $v(t) = L\, di/dt$,

$$Z_L = \frac{Ve^{st}}{Ie^{st}} = \frac{(Ls)Ie^{st}}{Ie^{st}} = Ls \qquad \text{(ohms)} \tag{9.7}$$

and for the capacitor, where $i = C\, dv/dt$,

$$Z_C = \frac{Ve^{st}}{Ie^{st}} = \frac{Ve^{st}}{(Cs)Ve^{st}} = \frac{1}{Cs} \qquad \text{(ohms)} \tag{9.8}$$

The admittances are merely the reciprocals of the impedances. Thus,

$$Y_R = \frac{1}{R} = G \qquad \text{(mhos)} \tag{9.9}$$

$$Y_L = \frac{1}{Ls} \qquad \text{(mhos)} \tag{9.10}$$

$$Y_C = Cs \qquad \text{(mhos)} \tag{9.11}$$

In Section 9.2, it was observed that in the exponential function Fe^{st}, if $s = 0$, the function becomes the constant F. This is a condition of direct current, and for this condition, with $s = 0$, $Z_R = R$, $Z_L = 0$, and $Z_C = \infty$. This indicates that in the steady state, under dc conditions, the inductor behaves as a short circuit and the capacitor acts as an open circuit.

It was also noted in Section 9.2 that if $s = j\omega$, a complex sinusoid results, and this represents the case of alternating current. This is most important and will be considered in more detail in Section 9.4. Here, however, it may be noted that when $s = j\omega$, the impedances become $Z_R = R$, $Z_L = j\omega L$, and $Z_C = -j(1/\omega C)$. The admittances when $s = j\omega$ are the reciprocal of the impedances, $Y_R = 1/R = G$, $Y_L = -j/\omega L$, and $Y_C = j\omega C$.

All of the foregoing is summarized in Table 9.1, where Z and Y are now shown as $Z(s)$ and $Y(s)$.

The concept of impedance to exponentials (or merely impedance) has great merit. All impedances are in ohms, and all network elements may be combined in what is called the *s domain*, just as resistances are combined in resistive networks. The alert reader may have already deduced that the combination of resistances in series and parallel in dc analysis may be considered a combination in the *s* domain, because with $s = 0$, $Z_R(s) = R$ and $Y_R(s) = G$. This can be extended to a purely resistive network operating under alternating-current conditions. Here, the laws of combination also hold, because this case is but a specific application of the more general case with $s = j\omega$, $Z_R(j\omega) = R$, and $Y_R(j\omega) = G$.

When inductors and capacitors are present, the total impedance will be a function of *s*.

TABLE 9.1 Summary of impedance and admittance

Element	$Z(s)$	$s = 0$ (dc)	$s = j\omega$ (ac)	Element	$Y(s)$	$s = 0$ (dc)	$s = j\omega$ (ac)
R	R	R	R	R	G	G	G
L	Ls	0	$j\omega L = jX_L$	L	$\dfrac{1}{Ls}$	∞	$-j\dfrac{1}{\omega L}$
C	$\dfrac{1}{Cs}$	∞	$-j\dfrac{1}{\omega C} = -jX_C$	C	Cs	0	$j\omega C$

■ **EXAMPLE 9.1**

For the network shown in Fig. 9.2, what is the driving-point impedance $Z(s)$?

Solution For the parallel combination of R and C,

$$Z_p(s) = \frac{Z_R(s)Z_C(s)}{Z_R(s) + Z_C(s)} = \frac{R(1/Cs)}{R + (1/Cs)} = \frac{R}{RCs + 1}$$

This makes the driving-point impedance

$$Z(s) = Ls + \frac{R}{RCs + 1} = \frac{RLCs^2 + Ls + R}{RCs + 1}$$

or

$$Z(s) = L\frac{s^2 + (1/RC)s + (1/LC)}{s + (1/RC)}$$

With $R = \frac{5}{12}\,\Omega$, $L = \frac{1}{2}\,$H, and $C = \frac{2}{5}\,$F, the parameters in $Z(s)$ can be computed as $RC = \frac{1}{6}$, $1/RC = 6$, $LC = \frac{1}{5}$, and $1/LC = 5$; so that

$$Z(s) = \frac{1}{2}\frac{s^2 + 6s + 5}{s + 6}$$

■

FIGURE 9.2 A network for which the driving-point impedance is to be determined

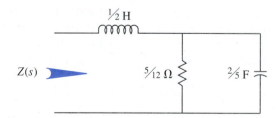

FIGURE 9.3

Network (Exercise 9.1)

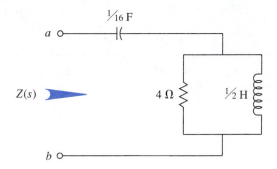

$Z(s)$

EXERCISE 9.1

In Chapter 2, it was shown that the usual strategy for the analysis of resistive ladder networks is to first determine the *driving-point impedance*, or the resistance *looking into* the input terminals of the network. What is the impedance (the driving-point impedance) looking into the terminals $a-b$ in the network shown in Fig. 9.3?

Answer

$$Z(s) = 4\,\frac{s^2 + 4s + 32}{s(s + 8)}\ \text{ohms}$$

ALTERNATING-CURRENT IMPEDANCE

If eq. (9.1) represents a forcing function that is applied to a network composed of an arrangement of resistors, inductors, and capacitors, one may write, as shown in Section 9.3, an expression for the impedance of the network, which is now referred to as $Z(s)$. Reference to Table 9.1 shows that, in general,

$$Z_R(s) = R \qquad Z_L(s) = Ls \qquad Z_C(s) = \frac{1}{Cs}$$

In the case of alternating current, where $s = j\omega$, Z_L is

$$Z_L(j\omega) = j\omega L = jX_L$$

where X_L is the *inductive reactance,*

$$X_L = \omega L \tag{9.12}$$

and Z_C is

$$Z_C(j\omega) = \frac{1}{j\omega C} = -j\,\frac{1}{\omega C} = -jX_C$$

where X_C is the *capacitive reactance*,

$$X_C = \frac{1}{\omega C} \tag{9.13}$$

Alternating-current impedance functions may be written in terms of the complex frequency s and then evaluated by letting $s = j\omega$. The same result is achieved if the impedance of each network element is evaluated and then combined in accordance with the rules developed in Chapter 2 for the addition of resistances in series and the combination of resistances in parallel.

■ **EXAMPLE 9.2**

In Example 9.1, the driving-point impedance of the network in Fig. 9.2, repeated here as Fig. 9.4, was shown to be

$$Z(s) = \frac{1}{2}\frac{s^2 + 6s + 5}{s + 6}$$

What is the value of the ac impedance if $\omega = 4$ rad/s?

Solution   If $s = j\omega$,

$$Z(j\omega) = \frac{1}{2}\frac{-\omega^2 + j(6\omega) + 5}{j\omega + 6}$$

and if $\omega = 4$ rad/s,

$$Z(j4) = \frac{1}{2}\frac{-16 + 5 + j24}{j4 + 6} = \frac{-\frac{11}{2} + j12}{6 + j4}$$

This may be put into polar and rectangular form after an exercise in complex number algebra (Appendix A):

$$Z = \frac{-5.5 + j12}{6 + j4} = \frac{13.20\underline{/114.62°}}{7.21\underline{/33.69°}} = 1.83\underline{/80.93°} = 0.29 + j1.81 \ \Omega$$

The need for writing the driving-point impedance in terms of s is eliminated by working with each individual element. If $C = \frac{2}{5}$ F, then $Z_C(s) = \frac{5}{2s}$

A network containing three elements

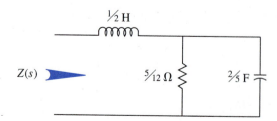

and $X_C = 1/\omega C = 5/(2\omega)$. With $Z_C = -jX_C$ and $\omega = 4$ rad/s,

$$Z_C(j4) = -j\frac{5}{8}\,\Omega$$

This impedance may be combined with that of the resistor,

$$Z_R = \frac{5}{12}\,\Omega$$

to yield the impedance of the parallel combination,

$$Z_p(j4) = \frac{Z_R Z_C}{Z_R + Z_C} = \frac{\frac{5}{12}(-j\frac{5}{8})}{\frac{5}{12} - j\frac{5}{8}} = \frac{-j\frac{25}{96}}{(\frac{1}{96})(40 - j60)} = \frac{-j5}{8 - j12}$$

$$= \frac{5\underline{/-90°}}{14.42\underline{/-56.31°}} = 0.35\underline{/-33.69°}$$

or

$$Z_p = 0.29 - j0.19\,\Omega$$

Then with $L = \frac{1}{2}$ H and $Z_L(s) = \frac{1}{2}s$, $X_L = \omega L = \frac{1}{2}\omega$. With $Z_L = jX_L$ at $\omega = 4$ rad/s,

$$Z_L = j2\,\Omega$$

This makes the driving-point impedance

$$Z = Z_p + Z_L = (0.29 - j0.19) + j2 = 0.29 + j1.81\,\Omega$$

which confirms the previous result. ∎

EXERCISE 9.2

What is the impedance looking into terminals a–b of the network in Fig. 9.3 if $\omega = 8$ rad/s?

Answer $Z = 2 + j0 = 2\underline{/0°}\,\Omega$.

THE AMPLITUDE–PHASE ANGLE FORM FOR SINUSOIDS

SECTION 9.5

The form of one of the Euler relationships,

$$e^{j\omega t} = \cos \omega t + j \sin \omega t$$

suggests that an ac signal or a response such as a branch voltage or branch current is composed of two sinusoids at the same frequency:

$$f(t) = A \cos \omega t + B \sin \omega t \tag{9.14}$$

which can be represented in what is called the *amplitude–phase angle form,*

$$f(t) = F \cos(\omega t + \phi) \tag{9.15}$$

or

$$f(t) = F \sin(\omega t + \theta) \tag{9.16}$$

Suppose that it is desired to represent eq. (9.14) in the form of eq. (9.15). In this event, the familiar trigonometric identity

$$\cos(x + y) = \cos x \cos y - \sin x \sin y$$

can be used with $x = \omega t$ and $y = \phi$. The identity can be applied to eq. (9.15):

$$f(t) = F \cos(\omega t + \phi) = F(\cos \omega t \cos \phi - \sin \omega t \sin \phi)$$

A comparison of this with eq. (9.14) indicates that

$$F \cos \phi = A \quad \text{and} \quad F \sin \phi = -B$$

The amplitude F can be expressed in terms of A and B. If the squares of each of these are added,

$$A^2 + B^2 = F^2(\cos^2 \phi + \sin^2 \phi)$$

one obtains

$$F = \sqrt{A^2 + B^2} \tag{9.17}$$

because $\cos^2 \phi + \sin^2 \phi = 1$. A division gives

$$\frac{F \sin \phi}{F \cos \phi} = \frac{-B}{A}$$

so that

$$\phi = \arctan \frac{-B}{A} \tag{9.18}$$

■ **EXAMPLE 9.3**
Put

$$f(t) = 8 \cos 400t - 6 \sin 400t$$

into the amplitude—phase angle form

$$f(t) = F \cos(400t + \phi)$$

Solution Here, $A = 8$ and $B = -6$. Thus,

$$F = \sqrt{8^2 + (-6)^2} = \sqrt{64 + 36} = \sqrt{100} = 10$$

$$\phi = \arctan \frac{-(-6)}{8} = \arctan 0.75 = 36.87°$$

The result is

$$f(t) = 10 \cos(400t + 36.87°)$$

or

$$f(t) = 10 \cos(400t + 0.644)$$

Observe that when the angle is in degrees, the degree sign must be shown. ■

EXERCISE 9.3

Put

$$f(t) = -12 \cos 377t + 5 \sin 377t$$

into the amplitude—phase angle form

$$f(t) = F \cos(377t + \phi)$$

Answer $f(t) = 13 \cos(377t - 157.38°)$.

THE PHASOR

The notion that the sinusoid can be represented by a rotating line was discussed in Section 5.6, where the horizontal motion of a pendulum was represented as the horizontal projection of a rotating line. This may be extended to any alternating sinusoidal quantity such as a current or a voltage, and there is no need to confine the discussion to the horizontal projection. In fact, if the rotating line, which is to be called a *phasor*, is to rotate at an angular velocity ω, the line may be hinged at the origin of the complex plane. If the line is at angle ϕ, measured counterclockwise from the positive real axis at $t = 0$, the functions $F \cos(\omega t + \phi)$ and $F \sin(\omega t + \phi)$ may be obtained from projections on the real and imaginary axes, respectively, as shown in Fig. 9.5.

Thus, a rotating phasor is a rotating line in the complex plane, and it may be used to represent a sinusoidally varying quantity. The length of the line represents the magnitude of the quantity, its angular velocity around the origin of the complex plane corresponds to the angular velocity of the sinusoidal function, and its angle with respect to the real axis is its phase at any instant. Some authors use the words *phasor* and *vector* interchangeably, but here this practice will be resisted. Although phasors behave like vectors, which are directed quantities in space, they will not be referred to as vectors.

FIGURE 9.5

A line of length F, hinged at the origin of the complex plane and rotating at an angular velocity ω

The line is at angle ϕ with the positive real axis at $t = 0$ and projections of the line on the real and imaginary axes trace its cosinusoidal and sinusoidal components respectively. In a sense, this is a snapshot of a rotating phasor at $t = 0$.

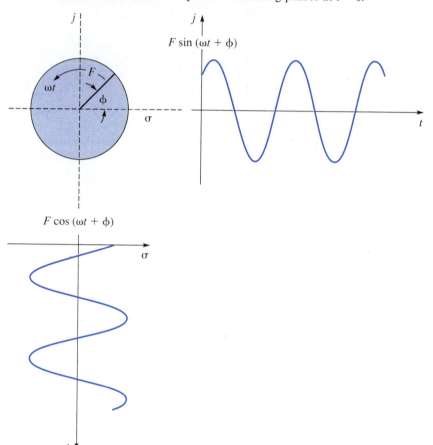

FIGURE 9.6

An *RLC* series circuit excited by a sinusoidal voltage source

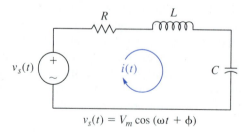

$$v_s(t) = V_m \cos(\omega t + \phi)$$

9.6.1 The Phasor from the Exponential Form

The network shown in Fig. 9.6 is to be excited by a voltage

$$v_s(t) = V_m \cos(\omega t + \phi) \text{ volts}$$

and possesses, at an angular velocity of $\omega = 2\pi f$, an impedance

$$Z = R + j\left(\omega L - \frac{1}{\omega C}\right)$$

Because sinusoidal voltages, when applied to a network, yield sinusoidal currents, it is desired to find the current flowing through all of the elements in series,

$$i(t) = I_m \cos(\omega t + \theta)$$

where θ is not to be assumed to be equal to ϕ.

It was shown in Section 9.2 that $v_s(t)$ may be written as

$$v_s(t) = \frac{V_m}{2} e^{j\phi} e^{j\omega t} + \frac{V_m}{2} e^{-j\phi} e^{-j\omega t} \tag{9.19}$$

where the two terms are complex conjugates. This is the exponential form, and it is observed that $i(t)$ can be represented in the same manner,

$$i(t) = \frac{I_m}{2} e^{j\theta} e^{j\omega t} + \frac{I_m}{2} e^{-j\theta} e^{-j\omega t} \tag{9.20}$$

In both eqs. (9.19) and (9.20), two of the terms represent a complex number rotating in the counterclockwise direction at an angular velocity ω. The other two also represent a complex number rotating at the angular velocity ω but in the *clockwise* direction. All four terms therefore fit the definition of a rotating phasor, and four phasors (two pairs of complex conjugates) may be defined:[1]

$$\hat{V} = V_m e^{j\phi} \quad = V_m\underline{/\phi} \tag{9.21a}$$

$$\hat{V}* = V_m e^{-j\phi} = V_m\underline{/-\phi} \tag{9.21b}$$

$$\hat{I} = I_m e^{j\theta} \quad = I_m\underline{/\theta} \tag{9.21c}$$

$$\hat{I}* = I_m e^{-j\theta} = I_m\underline{/-\theta} \tag{9.21d}$$

so that

$$v_s(t) = \frac{1}{2}(\hat{V}e^{j\omega t} + \hat{V}*e^{-j\omega t}) \tag{9.22}$$

$$i(t) = \frac{1}{2}(\hat{I}e^{j\omega t} + \hat{I}*e^{-j\omega t}) \tag{9.23}$$

---

[1] Phasors will be represented by symbols wearing hats, such as $\hat{V}$ and $\hat{I}$.

KVL may be applied to the network of Fig. 9.6 to yield an integro-differential equation,

$$L\frac{di}{dt} + Ri + \frac{1}{C}\int i\,dt = v_s(t)$$

With eqs. (9.22) and (9.23) substituted, the result, after complex algebra simplification, is

$$\left[R + j\left(\omega L - \frac{1}{\omega C}\right)\right]\hat{I}e^{j\omega t} - \hat{V}e^{j\omega t} + \left[R - j\left(\omega L - \frac{1}{\omega C}\right)\right]\hat{I}*e^{-j\omega t} - \hat{V}*e^{-j\omega t} = 0$$

or with

$$Z(j\omega) = R + j\left(\omega L - \frac{1}{\omega C}\right)$$

$$Z(-j\omega) = R - j\left(\omega L - j\frac{1}{\omega C}\right)$$

a more compact expression can be written,

$$[Z(j\omega)\hat{I} - \hat{V}]e^{j\omega t} + [Z(-j\omega)\hat{I}* - \hat{V}*]e^{-j\omega t} = 0 \qquad (9.24)$$

For eq. (9.24) to be satisfied for all t, one must have

$$Z(j\omega)\hat{I}\ -\ \hat{V}\ = 0$$
$$Z(-j\omega)\hat{I}* - \hat{V}* = 0$$

or

$$\hat{I} = \frac{\hat{V}}{Z(j\omega)} \qquad (9.25)$$

$$\hat{I}* = \frac{\hat{V}*}{Z(-j\omega)} \qquad (9.26)$$

These important equations lie at the core of the validity of the phasor concept. They are complex conjugates, and this can be verified by taking the conjugate of each term of the first equation, thereby obtaining the second. Thus, either equation is satisfied if one is satisfied, and the network may be analyzed by considering only $\hat{V}$ to obtain $\hat{I}$ or by considering $\hat{V}*$ to obtain $\hat{I}*$.

But $\hat{V}$, $\hat{I}$, $\hat{V}*$, and $\hat{I}*$ are all phasors, so that with $Z(j\omega) = |Z|\underline{/\alpha}$,

$$I_m\underline{/\theta} = \frac{V_m\underline{/\phi}}{|Z|\underline{/\alpha}} \qquad (9.27)$$

$$I_m\underline{/-\theta} = \frac{V_m\underline{/-\phi}}{|Z|\underline{/-\alpha}} \qquad (9.28)$$

It is seen from either of these that

$$I_m = \frac{V_m}{|Z|} \qquad (9.29)$$

$$\theta = \phi - \alpha \qquad (9.30)$$

It is expected that $i(t)$ is a totally real quantity,

$$i(t) = I_m \cos(\omega t + \theta)$$

and this will now be confirmed. The total current is the sum of two current components, as indicated by eq. (9.23). Putting eqs. (9.25) and (9.26) into eq. (9.23) yields

$$i(t) = \frac{1}{2}(\hat{I}e^{j\omega t} + \hat{I}^* e^{-j\omega t}) = \frac{1}{2}\left[\frac{\hat{V}}{Z(j\omega)} e^{j\omega t} + \frac{\hat{V}^*}{Z(-j\omega)} e^{-j\omega t}\right]$$

and by eqs. (9.21), with

$$Z(j\omega) = R + j\left(\omega L - \frac{1}{\omega C}\right)$$

it is seen that

$$i(t) = \frac{V_m/2}{|Z|} e^{j(\omega t + \theta)} + \frac{V_m/2}{|Z|} e^{-j(\omega t + \theta)}$$

or

$$i(t) = I_m \cos(\omega t + \theta)$$

where $I_m = V_m/|Z|$ and where $|Z|$ is the magnitude of $Z(j\omega)$.

9.6.2 The Phasor from the Real Parts of $v(t)$ and $i(t)$

Once again, consider the RLC series network shown in Fig. 9.6, and observe that one of the Euler equations shows that the forcing voltage is the real part of an exponential voltage,

$$v_s(t) = V_m \cos(\omega t + \phi) = \text{Re}(V_m e^{j(\omega t + \phi)}) \tag{9.31}$$

By the same reasoning,

$$i(t) = I_m \cos(\omega t + \theta) = \text{Re}(I_m e^{j(\omega t + \theta)}) \tag{9.32}$$

It can be shown that

$$\frac{d}{dt} \text{Re}(x) = \text{Re}\left(\frac{dx}{dt}\right)$$

and

$$\int \text{Re}(x)\,dt = \text{Re}\left(\int x\,dt\right)$$

With these facts, eqs. (9.31) and (9.32) may be put into the governing integro-differential equation,

$$L\frac{di}{dt} + Ri + \frac{1}{C}\int i\,dt = v_s(t)$$

to yield

$$L \frac{d}{dt} \mathrm{Re}(I_m e^{j(\omega t + \theta)}) + R \, \mathrm{Re}(I_m e^{j(\omega t + \theta)}) + \frac{1}{C} \int \mathrm{Re}(I_m e^{j(\omega t + \theta)}) \, dt = \mathrm{Re}(V_m e^{j(\omega t + \phi)})$$

or

$$\mathrm{Re}\left\{ \left[R + j\left(\omega L - \frac{1}{\omega C} \right) \right] I_m e^{j(\omega t + \theta)} \right\} = \mathrm{Re}(V_m e^{j(\omega t + \phi)})$$

which says that for all values of time, the real component of the current is a function of the real component of the voltage. Thus, the symbolism Re may be suppressed, and with

$$Z(j\omega) = R + j\left(\omega L - \frac{1}{\omega C} \right) = |Z| \underline{/\alpha}$$

one can obtain

$$I_m e^{j(\omega t + \theta)} = \frac{V_m e^{j(\omega t + \phi)}}{|Z| \underline{/\alpha}}$$

or

$$I_m e^{j\theta} e^{j\omega t} = \frac{V_m e^{j\phi} e^{j\omega t}}{|Z| \underline{/\alpha}}$$

But by eqs. (9.21), $I_m e^{j\theta}$ and $V_m e^{j\phi}$ are phasors, so that once again it is seen that

$$\hat{I} = \frac{\hat{V}}{|Z| \underline{/\alpha}} \qquad \text{or} \qquad I \underline{/\theta} = \frac{V \underline{/\phi}}{|Z| \underline{/\alpha}}$$

which confirms that

$$I_m = \frac{V_m}{|Z|} \qquad \theta = \phi - \alpha$$

The beauty of network analysis using phasor methods lies in the fact that all voltages and currents are referenced, in the phasor domain, to the horizontal real axis of the complex plane. The voltage and current phasors always bear a fixed relationship to one another at any particular instant of time. The frequency and time involvement is inherent in the rotation of the *entire* complex plane at an angular velocity ω. To be sure, the value of ω governs the values of the network impedances and admittances that relate the voltages and the currents. But once the impedances and admittances are evaluated, the voltages and currents are fixed with respect to a common reference, which can be chosen by the analyst.

■ **EXAMPLE 9.4**

Notice that the voltages in Fig. 9.7 are given in instantaneous form. What is the instantaneous form of the voltage across impedance Z_2?

Series network of impedances

FIGURE 9.7

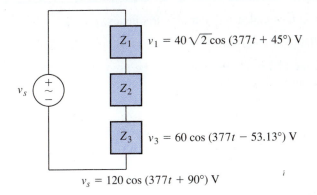

$v_1 = 40\sqrt{2}\cos(377t + 45°)$ V

$v_3 = 60\cos(377t - 53.13°)$ V

$v_s = 120\cos(377t + 90°)$ V

Solution First, recognize that KVL must apply. Then, via KVL, the relationship between the voltages,

$$\hat{V}_2 = \hat{V}_s - \hat{V}_1 - \hat{V}_3$$

is obtained. The rest is straightforward:

$$\hat{V}_2 = 120\underline{/90°} - 40\sqrt{2}\underline{/45°} - 60\underline{/-53.13°}$$
$$= j120 - (40 + j40) - (36 - j48) = -76 + j128$$
$$= 148.86\underline{/120.70°} \text{ volts}$$

A transformation back into the time domain provides the sought-after result:

$$v_2(t) = 148.86\cos(377t + 120.70°) \text{ volts} \qquad \blacksquare$$

EXERCISE 9.4

What is the value of $i_3(t)$ at point n shown in Fig. 9.8. The values of $i_1(t)$, $i_2(t)$, and $i_4(t)$ are

$$i_1(t) = 4\sqrt{2}\cos(400t + 135°) \text{ amperes}$$

$$i_2(t) = 5\cos(400t + 126.87°) \text{ amperes}$$

$$i_4(t) = 6.50\cos(400t - 67.38°) \text{ amperes}$$

Answer $i_3(t) = 6.95\cos(400t - 59.74°)$ amperes.

A point in a network (Exercise 9.4)

FIGURE 9.8

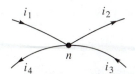

CHAPTER 9

SUMMARY

- In $f(t) = Fe^{st}$, $s = \sigma + j\omega$ is called the complex frequency.

- If $\sigma = \omega = 0$, then $f(t) = F$, a constant.

- If $\sigma = 0$ but $\omega \neq 0$, one of Euler's equations can be used to show that $f(t) = Fe^{j\omega t}$ will yield the sinusoid

$$f(t) = F_1 \cos \omega t + F_2 \sin \omega t$$

This can be represented in amplitude—phase angle form by

$$f(t) = F \cos(\omega t + \phi) \quad \text{or} \quad f(t) = F \sin(\omega t + \theta)$$

where

$$F = \sqrt{F_1^2 + F_2^2} \qquad \phi = \arctan \frac{-F_2}{F_1} \qquad \theta = \arctan \frac{F_2}{F_1}$$

- Alternating-current impedances and admittances, with inductive reactance defined as $X_L = \omega L$ and capacitive reactance defined as $1/\omega C$, are

$$Z_R = R \qquad\qquad\qquad Y_R = G = \frac{1}{R}$$

$$Z_L = jX_L = j\omega L \qquad\qquad Y_L = \frac{-j}{X_L} = \frac{-j}{\omega L}$$

$$Z_C = -jX_C = \frac{-j}{\omega C} \qquad\qquad Y_C = \frac{j}{X_C} = j\omega C$$

- A sinusoidal function of the form

$$f(t) = F \cos(\omega t + \phi)$$

is related to a phasor

$$\hat{F} = F \underline{/\phi} = Fe^{j\phi}$$

- Voltage and current phasors always bear a fixed relationship to one another at any particular instant of time. The frequency and time involvement is inherent in the rotation of the *entire* complex plane at an angular velocity ω.

Additional Readings

Blackwell, W.A., and L.L. Grigsby. *Introductory Network Theory*. Boston: PWS Engineering, 1985, pp. 326–336, 342–350.

Bobrow, L.S. *Elementary Linear Circuit Analysis*. 2d ed. New York: Holt, Rinehart and Winston, 1987, pp. 361–375, 471–477.

Del Toro, V. *Engineering Circuits*. Englewood Cliffs, N.J.: Prentice-Hall, 1987, pp. 193–198, 201–224, 259–261.

Dorf, R.C. *Introduction to Electric Circuits*. New York: Wiley, 1989, pp. 326–343, 529.

Hayt, W.H., Jr., and J.E. Kemmerly. *Engineering Circuit Analysis.* 4th ed. New York: McGraw-Hill, 1986, pp. 243–259, 334–344.

Irwin, J.D. *Basic Engineering Circuit Analysis.* 3d ed. New York: Macmillan, 1989, pp. 362–383.

Johnson, D.E., J.L. Hilburn, and J.R. Johnson. *Basic Electric Circuit Analysis.* 4th ed. Englewood Cliffs, N.J.: Prentice-Hall, 1989, pp. 306–308, 316–330, 433–444.

Karni, S. *Applied Circuit Analysis.* New York: Wiley, 1988, pp. 255–272, 429.

Madhu, S. *Linear Circuit Analysis.* Englewood Cliffs, N.J.: Prentice-Hall, 1988, pp. 347–354, 360–367, 369–375, 377, 378, 573–577.

Nilsson, J.W. *Electric Circuits.* 3d ed. Reading, Mass.: Addison-Wesley, 1990, pp. 270, 317–335, 580.

Paul, C.R. *Analysis of Linear Circuits.* New York: McGraw-Hill, 1989, pp. 256–275, 283, 333, 334, 565, 634

CHAPTER 9 · PROBLEMS

Section 9.1

Problems 9.1 through 9.10 involve the three complex numbers $N_1 = 4 - j3$, $N_2 = 2\sqrt{2}\underline{/45°}$, and $N_3 = 5e^{j90°}$. Perform the indicated operations and give the result in rectangular, polar, and exponential form.

9.1 $N = N_1 + N_2 + N_3$

9.2 $N = \dfrac{N_1 N_2}{N_3}$

9.3 $N = (N_1 - N_2)^2 N_3$

9.4 $N = 2N_1 N_2 N_3$

9.5 $N = \dfrac{N_1}{N_2} + \dfrac{N_1}{N_3}$

9.6 $N = (\sqrt{N_1 N_2} - N_3)N_1$

9.7 $N = \dfrac{N_1 + N_2}{N_1 - N_2}$

9.8 $N = (N_1 + N_2 + N_3)\sqrt{N_1 N_2 N_3}$

9.9 $N = N_1^3 + N_2^2 + N_3$

9.10 $N = N_1 N_2 + N_2 N_3 + N_1 N_3$

Section 9.2

In Problems 9.11 through 9.16, complex frequencies are given. Write expressions in the general form $Ae^{-\sigma t}\cos(\omega t + \phi)$.

9.11 $s = -3 + j4$

9.12 $s = -12 + j12$

9.13 $s = -2$

9.14 $s = -4 + j3$

9.15 $s = -j6$

9.16 $s = -6 - j8$

Section 9.4

9.17 If $\omega = 400$ rad/s, find the driving-point impedance at terminals a and b in the network in Fig. P9.1.

Figure P9.1

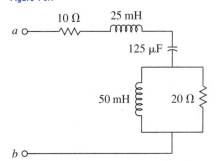

9.18 If $\omega = 200$ rad/s, find the driving-point impedance at terminals a and b in the network in Fig. P9.2.

Figure P9.2

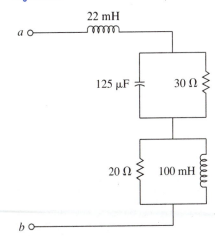

9.19 If $\omega = 100$ rad/s, find the driving-point admittance at terminals a and b in the network in Fig. P9.3.

Figure P9.3

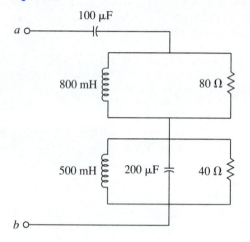

9.20 If $\omega = 500$ rad/s, find the driving-point admittance at terminals a and b in the network in Fig. P9.4.

Figure P9.4

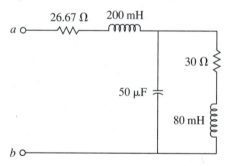

9.21 If $\omega = 1000$ rad/s, find the driving-point impedance at terminals a and b in the network in Fig. P9.5.

Figure P9.5

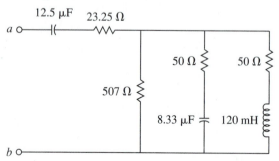

Section 9.5

In Problems 9.22 through 9.26, A and B, in the form $f(t) = A \cos \omega t + B \sin \omega t$, are given. Find F and ϕ in the form $f(t) = F \cos(\omega t + \phi)$.

9.22 $A = -4$, $B = 4$

9.23 $A = -64$, $B = 120$

9.24 $A = -60$, $B = -45$

9.25 $A = -70$, $B = -168$

9.26 $A = 28$, $B = -96$

In Problems 9.27 through 9.31, F and ϕ, in the form $f(t) = F \cos(\omega t + \phi)$, are given. Find A and B in the form $f(t) = A \cos \omega t + B \sin \omega t$

9.27 $F = 275$, $\phi = -143.13°$

9.28 $F = 390$, $\phi = 112.62°$

9.29 $F = 800$, $\phi = -196.26°$

9.30 $F = 120$, $\phi = 150°$

9.31 $F = 1000$, $\phi = -135°$

Section 9.6

In Problems 9.32 through 9.36, either a voltage or a current is expressed in its instantaneous form. Find the voltage or current phasor, and express the phasor in rectangular, polar, and exponential form.

9.32 $v(t) = 208 \cos(377t + 60°)$ volts

9.33 $i(t) = 4\sqrt{2} \cos(400t - 135°)$ amperes

9.34 $i(t) = 120\sqrt{3} \cos(600t + 15°)$ amperes

9.35 $v(t) = 1750 \cos(200t - 163.74°)$ volts

9.36 $v(t) = 400 \cos(1200t + 90°)$ volts

9.37 Two elements are connected in parallel, as shown in Fig. P9.6. The line current is $i(t) = 10\sqrt{2} \cos(377t + 135°)$ amperes and $i_1(t) = 6 \cos(377t + 30°)$ amperes. Express i_2 in instantaneous form.

Figure P9.6

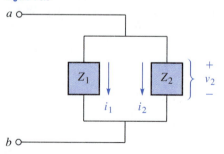

9.38 In Fig. P9.6, $\hat{I}_1 = 2\sqrt{2}\underline{/45°}$ A and $\hat{V}_2 = 120\underline{/53.13°}$ V. If the line current is $i = 4\cos(\omega t + 90°)$ amperes, determine the four components that form the parallel combination if $f = 250/2\pi$ Hz.

9.39 Represent the three sources shown in Fig. P9.7 as a single voltage source connected across terminals a and b. In Fig. P9.7, $i_s(t) = 8\cos(400t + 90°)$ amperes, $v_{s1}(t) = 200\cos(400t + 36.87°)$ volts, and $v_{s2}(t) = 200\sqrt{2}\cos(400t + 135°)$ volts.

Figure P9.7

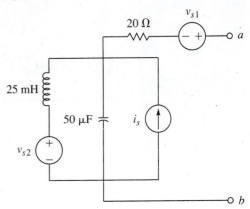

10

TECHNIQUES IN AC ANALYSIS

OBJECTIVES

The objectives of this chapter are to:

- Show how networks may be transformed from the time domain to the phasor domain.

- Provide examples pertaining to the solution of ladder networks.

- Demonstrate ac analysis by using node, mesh and loop analysis.

- Illustrate the use of the superposition principle and Thévenin's theorem in an ac network.

SECTION 10.1   **INTRODUCTION**

The impedance and phasor concepts allow a network that is driven by sinusoidal voltage and current sources to be analyzed by the same techniques used for the analysis of resistive networks. Once certain preliminaries are considered, the analysis, which is admittedly somewhat more difficult due to the calculations using complex numbers, is conducted in exactly the same manner.

The preliminary considerations involve the determination of the angular frequency and then the calculation of each and every element impedance or admittance in the network. When this and the conversion of all sources from instantaneous to phasor form are accomplished, the network may be redrawn in what is called the *phasor domain.* But it must be emphasized that all voltages and currents in the phasor domain must be consistent in their employment of either maximum or rms magnitudes[1]. If in

$$\hat{V} = V \underline{/\phi}$$

---

[1] Effective or root-mean-square (rms) values for a periodic function are developed in Section 5.7.2.

the magnitude V is the maximum amplitude of the sinusoid, then in

$$\hat{I} = I \underline{/\theta}$$

the magnitude I is also the maximum amplitude. Either V or I can be rms values, but if one is, so is the other. This important message regarding consistency cannot be overemphasized.

In the sections that follow, examples will be provided that demonstrate how ac network analysis in the phasor domain may be conducted. In these examples, phasors will be indicated by a capital letter bearing a hat.

LADDER NETWORKS

SECTION 10.2

The analysis of the ladder network shown in Fig. 10.1a will demonstrate the use of the phasor method. Here, the branch currents are designated, in phasor form, by $\hat{J}$ (an uppercase letter) so that branch currents may be distinguished from mesh or loop currents, which bear the customary designation for current, $\hat{I}$. The reason for this will be apparent in just a few pages (Section 10.5) and should cause little or no confusion with regard to the use of J (without a hat) or with regard to j, which can be used to represent the instantaneous branch current and $j = \sqrt{-1}$. The lowercase j to represent an instantaneous time domain branch current is rarely used and is unlikely to be confused with $j = \sqrt{-1}$.

■ **EXAMPLE 10.1**

Find all the branch currents and branch voltages in the ladder network of Fig. 10.1a.

Solution As in the case of the resistive ladder, the first step is to determine the impedance that is presented to the source. In order to do this, one must put the network into the phasor domain by noting that

$$\omega = 2\pi f = 2\pi \left(\frac{500}{2\pi} \right) = 500 \text{ rad/s}$$

$$X_{C1} = \frac{1}{\omega C_1} = \frac{1}{500(33.33 \times 10^{-6})} = 60 \ \Omega$$

$$X_{C2} = \frac{1}{\omega C_2} = \frac{1}{500(25 \times 10^{-6})} = 80 \ \Omega$$

$$X_L = \omega L = 500(0.080) = 40 \ \Omega$$

The phasor domain network with designations of branch current phasors is presented in Fig. 10.1b. Notice here that $\hat{I}_1 = -\hat{J}_1$. When the branch current phasors are obtained, the branch voltages may be obtained from the Ohm's law relationship $\hat{V} = Z\hat{I}$.

FIGURE 10.1   (a) A ladder network, (b) the same network in the phasor domain, and (c) the step-to-step evaluation of the impedance presented to the source

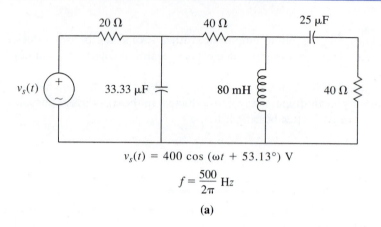

$$v_s(t) = 400 \cos(\omega t + 53.13°) \text{ V}$$

$$f = \frac{500}{2\pi} \text{ Hz}$$

(a)

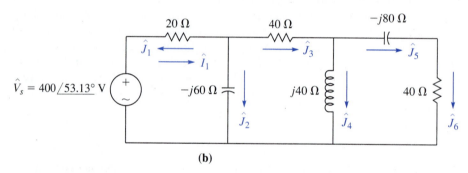

(b)

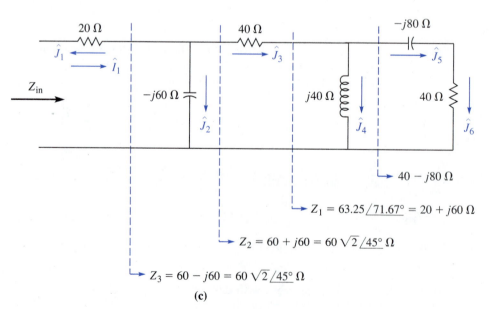

$$40 - j80 \ \Omega$$

$$Z_1 = 63.25 \underline{/71.67°} = 20 + j60 \ \Omega$$

$$Z_2 = 60 + j60 = 60\sqrt{2} \underline{/45°} \ \Omega$$

$$Z_3 = 60 - j60 = 60\sqrt{2} \underline{/45°} \ \Omega$$

(c)

Calculations for Z_{in} now follow, and these are summarized, step by step, in Fig. 10.1c.

$$Z_1 = \frac{(40 - j80)(j40)}{40 - j80 + j40} = \frac{3200 + j1600}{40 - j40}$$

$$= \frac{3577.71\,\underline{/26.57°}}{40\sqrt{2}\,\underline{/-45°}} = 63.25\,\underline{/71.57°} = 20 + j60\ \Omega$$

$$Z_2 = 40 + 20 + j60 = 60 + j60 = 60\sqrt{2}\,\underline{/45°}\ \Omega$$

$$Z_3 = \frac{(60 + j60)(-j60)}{60 + j60 - 60} = \frac{3600 - j3600}{60} = 60 - j60\ \Omega$$

$$Z_{in} = 20 + 60 - j60 = 80 - j60 = 100\,\underline{/-36.87°}\ \Omega$$

The branch currents may now be determined. First, use $\hat{I}_1 = -\hat{J}_1$:

$$\hat{I}_1 = \frac{\hat{V}_s}{Z_{in}} = \frac{400\,\underline{/53.13°}}{100\,\underline{/-36.87°}} = 4\,\underline{/90°} = 0 + j4\ A$$

$$\hat{J}_1 = -4\,\underline{/90°} = 4\,\underline{/-90°}\ A$$

Then, by a current division and KCL,

$$\hat{J}_2 = \frac{60 + j60}{60 + j60 - j60}\,\hat{I}_1 = \frac{60 + j60}{60}\,(4\,\underline{/90°})$$

$$= (1 + j)(j4) = -4 + j4 = 4\sqrt{2}\,\underline{/135°}\ A$$

$$\hat{J}_3 = \hat{I}_1 - \hat{J}_2 = 4\,\underline{/90°} - 4\sqrt{2}\,\underline{/135°} = 0 + j4 - (-4 + j4)$$

$$= 4 + j0 = 4 = 4\,\underline{/0°}\ A$$

Another current division yields

$$\hat{J}_4 = \frac{40 - j80}{40 - j80 + j40}\,\hat{J}_3 = \frac{40 - j80}{40 - j40}\,(4)$$

$$= \frac{357.77\,\underline{/-63.43°}}{40\sqrt{2}\,\underline{/-45°}} = 6.32\,\underline{/-18.43°} = 6 - j2\ A$$

and by KCL

$$\hat{J}_5 = \hat{J}_6 = \hat{J}_3 - \hat{J}_4 = 4 + j0 - (6 - j2) = -2 + j2 = 2\sqrt{2}\,\underline{/135°}\ A$$

The branch voltages are now determined:

$$\hat{V}_1 = Z_1\hat{J}_1 + \hat{V}_s = 20(4\,\underline{/-90°}) + 400\,\underline{/53.13°} = 80\,\underline{/-90°} + 400\,\underline{/53.13°}$$

$$= -j80 + 240 + j320 = 240 + j240 = 240\sqrt{2}\,\underline{/45°}\ V$$

$$\hat{V}_2 = Z_2\hat{J}_2 = (-j60)(4\sqrt{2}\,\underline{/135°}) = (60\,\underline{/-90°})(4\sqrt{2}\,\underline{/135°})$$

$$= 240\sqrt{2}\,\underline{/45°} = 240 + j240\ V$$

Reference to Fig. 10.1b shows that this checks because $\hat{V}_1$, which includes $\hat{V}_s$, must equal $\hat{V}_2$. Notice, however, that the voltage across the 20-Ω

resistor is

$$\hat{V}_{20} = R\hat{J}_1 = 20(4\underline{/-90°}) = 80\underline{/-90°} = 0 - j80 \text{ V}$$

Continuing, one obtains

$$\hat{V}_3 = Z_3\hat{J}_3 = 40(4 + j0) = 160 + j0 = 160\underline{/0°} \text{ V}$$

$$\hat{V}_4 = Z_4\hat{J}_4 = j40(6.32\underline{/-18.43°}) = (40\underline{/90°})(6.32\underline{/-18.43°})$$
$$= 252.98\underline{/71.57°} = 80 + j240 \text{ V}$$

Here, a KVL check can be made:

$$\hat{V}_2 = \hat{V}_3 + \hat{V}_4 = 160 + j0 + 80 + j240 = 240 + j240 = 240\sqrt{2}\underline{/45°} \text{ V}$$

which checks the previously calculated value of $\hat{V}_2$.
    Finally,

$$\hat{V}_5 = Z_5\hat{J}_5 = -j80(2\sqrt{2}\underline{/135°}) = (80\underline{/-90°})(2\sqrt{2}\underline{/135°})$$
$$= 160\sqrt{2}\underline{/45°} = 160 + j160 \text{ V}$$

$$\hat{V}_6 = Z_6\hat{J}_6 = 40(2\sqrt{2}\underline{/135°}) = 80\sqrt{2}\underline{/135°} = -80 + j80 \text{ V}$$

By KVL, these should add to $\hat{V}_4$:

$$\hat{V}_4 = \hat{V}_5 + \hat{V}_6 = 160 + j160 - 80 + j80 = 80 + j240 \text{ V}$$

The check is satisfactory.
    The results can be summarized in terms of branch current and branch voltage vectors, which are first presented here in polar form and in the phasor domain:

$$\hat{\mathbf{J}} = \begin{bmatrix} 4\underline{/-90°} \\ 4\sqrt{2}\underline{/135°} \\ 4\underline{/0°} \\ 6.32\underline{/-18.43°} \\ 2\sqrt{2}\underline{/135°} \\ 2\sqrt{2}\underline{/135°} \end{bmatrix} \text{A} \qquad \hat{\mathbf{V}} = \begin{bmatrix} 240\sqrt{2}\underline{/45°} \\ 240\sqrt{2}\underline{/45°} \\ 160\underline{/0°} \\ 252.98\underline{/71.57°} \\ 160\sqrt{2}\underline{/45°} \\ 80\sqrt{2}\underline{/135°} \end{bmatrix} \text{V}$$

In instantaneous form with maximum (not rms) amplitudes, the results are

$$\mathbf{J} = \begin{bmatrix} 4\cos(500t - 90°) \\ 4\sqrt{2}\cos(500t + 135°) \\ 4\cos(500t + 0°) \\ 6.32\cos(500t - 18.43°) \\ 2\sqrt{2}\cos(500t + 135°) \\ 2\sqrt{2}\cos(500t + 135°) \end{bmatrix} \text{A} \qquad \mathbf{V} = \begin{bmatrix} 240\sqrt{2}\cos(500t + 45°) \\ 240\sqrt{2}\cos(500t + 45°) \\ 160\cos(500t + 0°) \\ 252.98\cos(500t + 71.57°) \\ 160\sqrt{2}\cos(500t + 45°) \\ 80\sqrt{2}\cos(500t + 135°) \end{bmatrix} \text{V}$$

The relationships between branch currents and branch voltages in resistors, in ductors, and capacitors is of considerable interest. For the inductor,

$$Z_L = j\omega L = jX_L = X_L \underline{/90°}$$

and with $\hat{V} = V\underline{/\phi}$,

$$\hat{I} = \frac{V\underline{/\phi}}{X_L\underline{/90°}} = I\underline{/\phi - 90°}$$

This shows that for a pure inductor, the current *lags* the voltage by 90°. On the other hand, for the capacitor,

$$Z_C = -j\frac{1}{\omega C} = -jX_C = X_C\underline{/-90°}$$

and again with $\hat{V} = V\underline{/\phi}$,

$$\hat{I} = \frac{V\underline{/\phi}}{X_C\underline{/-90°}} = I\underline{/\theta + 90°}$$

and this shows that for a pure capacitor, the current *leads* the voltage by 90°. These two important facts are neatly summarized by the mnemonic

ELI the *ICE* man

which says that in an inductor (*L*), the voltage (*E* or *V*) comes before (or *leads*) the current (*I*) by 90°; and in a capacitor (*C*), the voltage (*E* or *V*) comes after (or *lags*) the current (*I*) by 90°.

Branches that contain a single resistor do not exhibit this characteristic. For the resistor, where $Z_R = R = R\underline{/0°}$,

$$\hat{I} = \frac{V\underline{/\phi}}{R\underline{/0°}} = I\underline{/\phi}$$

There is no lead or lag for the pure resistor, and the branch voltage and branch current are said to be in phase.

■ **EXAMPLE 10.2**

Repeat Example 10.1 by using the ladder method discussed in Section 2.9 and demonstrated in Example 2.5.

Solution There is a freedom of choice here. Let $\hat{V}_6 = 1 + j0 = 1\underline{/0°}$ V across the 40-Ω resistor, and determine all other current and voltage values as functions of this selected value of $\hat{V}_6$:

$$\hat{J}_5 = \hat{J}_6 = \frac{1 + j0}{40} = \left(\frac{1}{40} + j0\right) = \frac{1}{40}\underline{/0°} \text{ amperes}$$

$$\hat{V}_5 = -j80(\hat{J}_5) = -j80\left(\frac{1}{40}\underline{/0°}\right) = (2\underline{/-90°}) = (0 - j2) \text{ volts}$$

By KVL,

$$\hat{V}_4 = \hat{V}_5 + \hat{V}_6 = (0 - j2) + (1 + j0) = (1 - j2) = (\sqrt{5}\,\underline{/-63.43°})\ \text{volts}$$

and then

$$\hat{J}_4 = \frac{\hat{V}_4}{j40} = \frac{(\sqrt{5}\,\underline{/-63.43°})}{40\,\underline{/90°}} = (0.056\,\underline{/-153.43°})$$

$$= -\left(\frac{1}{20} + j\frac{1}{40}\right)\ \text{amperes}$$

By KCL

$$\hat{J}_3 = \hat{J}_4 + \hat{J}_5 = -\left(\frac{1}{20} + j\frac{1}{40}\right) + \left(\frac{1}{40} + j0\right)$$

$$= -\left(\frac{1}{40} + j\frac{1}{40}\right) = \left(\frac{1}{40}\sqrt{2}\,\underline{/-135°}\right)\ \text{amperes}$$

$$\hat{V}_3 = 40\hat{J}_3 = 40\left(\frac{1}{40}\sqrt{2}\,\underline{/-135°}\right) = (\sqrt{2}\,\underline{/-135°}) = -(1 + j)\hat{V}_6\ \text{volts}$$

so that by KVL,

$$\hat{V}_2 = \hat{V}_3 + \hat{V}_4 = -(1 + j) + (1 - j2) = (0 - j3) = (3\,\underline{/-90°})\ \text{volts}$$

The branch current $\hat{J}_2$ is

$$\hat{J}_2 = \frac{\hat{V}_2}{-j60} = \frac{(3\,\underline{/-90°})}{60\,\underline{/-90°}} = \left(\frac{1}{20}\,\underline{/0°}\right) = \left(\frac{1}{20} + j0\right)\ \text{amperes}$$

and by KCL,

$$\hat{I}_1 = -\hat{J}_1 = \hat{J}_2 + \hat{J}_3 = \left(\frac{1}{20} + j0\right) - \left(\frac{1}{40} + j\frac{1}{40}\right)$$

$$= \left(\frac{1}{40} - j\frac{1}{40}\right) = \left(\frac{\sqrt{2}}{40}\,\underline{/-45°}\right)\ \text{volts}$$

so that

$$\hat{V}_1 = 20\hat{I}_1 = 20\left(\frac{\sqrt{2}}{40}\,\underline{/-45°}\right) = \left(\frac{\sqrt{2}}{2}\,\underline{/-45°}\right) = \left(\frac{1}{2} - j\frac{1}{2}\right)\ \text{volts}$$

Finally, by KVL,

$$\hat{V}_s = \hat{V}_1 + \hat{V}_2 = \left(\frac{1}{2} - j\frac{1}{2}\right) + (0 - j3) = \left(\frac{1}{2} - j\frac{7}{2}\right)$$

$$= \left(\frac{\sqrt{50}}{2}\,\underline{/-81.87°}\right)\ \text{volts}$$

and because the actual value of $\hat{V}_s$ is $\hat{V}_s = 400\underline{/53.13°}$ V, all branch voltages ($\hat{V}_1$ through $\hat{V}_6$) and all branch currents ($\hat{J}_1$ through $\hat{J}_6$) must be modified by the ratio

$$\frac{400\underline{/53.13°}}{(\sqrt{50}/2)\underline{/-81.87°}} = 80\sqrt{2}\underline{/135°} = (-80 + j80)$$

For example, the actual value of $\hat{V}_6$ is

$$\hat{V}_6 = (-80 + j80)(1 + j0) = -80 + j80 = 80\sqrt{2}\underline{/-135°} \text{ volts}$$

and the actual value of $\hat{J}_3$ is

$$\hat{J}_3 = (-80 + j80)\left(-\frac{1}{40} + j\frac{1}{40}\right) = 4 + j0 = 4\underline{/0°} \text{ amperes,}$$

which confirms the values obtained in Example 10.1. ■

THE PHASOR DIAGRAM   SECTION 10.3

The branch currents and branch voltages summarized in phasor form as branch current and branch voltage vectors at the end of the previous section can be placed on the complex plane as shown in Fig. 10.2. This graphical representation is called a *phasor diagram*, and the entire picture can be considered as rotating counterclockwise at an angular velocity of 500 rad/s. Thus, each phasor is rotating at 500 rad/s, and all phasors are fixed with respect to one another. Because phasors behave like

Phasor diagram for the phasor domain network of Fig. 10.1b   FIGURE 10.2

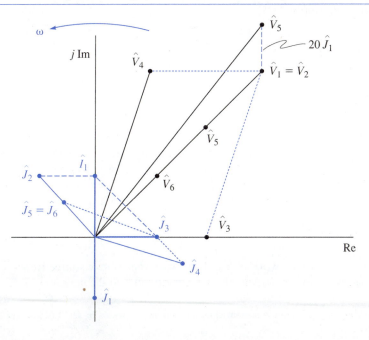

FIGURE 10.3   Network (Exercise 10.1)

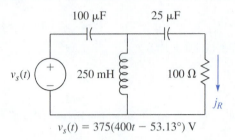

$$v_s(t) = 375(400t - 53.13°) \text{ V}$$

vectors, certain aspects of the analysis conducted in Section 10.2 can be confirmed by the phasor diagram.

In the network of Fig. 10.1b, branches 2 and 5 contain single capacitors. The *ELI* the *ICE*man rule says that $\hat{V}_2$ and $\hat{V}_5$ must lag $\hat{J}_2$ and $\hat{J}_5$ by 90°, and this may be confirmed in the phasor diagram. The diagram also shows that $\hat{J}_4$ lags $\hat{V}_4$ by 90°, which indicates that branch 4 must contain a single inductor (it does). The phasor diagram also indicates that all other branch currents and branch voltages are in phase, which is to be expected when purely resistive branches are considered. Here, the dashed line marked $20\hat{J}_1$ is the voltage across the 20-Ω resistor.

Observe here and in Fig. 10.1b that $\hat{V}_1 = \hat{V}_2$, which was determined to be the case in the actual analysis, and that branch voltage $\hat{V}_1$ is a KVL sum of $\hat{V}_s$ and $20\hat{J}_1$. KVL must also be satisfied in the mesh consisting of branches 2, 3, and 4, and this is indicated by the dotted lines showing that $\hat{V}_2$ is the sum of $\hat{V}_3$ and $\hat{V}_4$. The KCL relationships $\hat{I}_1 = \hat{J}_2 + \hat{J}_3$ and $\hat{J}_3 = \hat{J}_4 + \hat{J}_5$ are also confirmed in the diagram.

EXERCISE 10.1

Find the branch current $j_R(t)$ in Fig. 10.3.

Answer   3 cos 400t amperes.

SECTION 10.4   **NODE ANALYSIS**

10.4.1 An Example of Node Analysis

Complexity is the only difference between ac steady-state node, mesh, or loop analysis of RLC networks and resistive networks. Because alternating-current analysis is conducted in the phasor domain, the ac case requires a journey through the realm of complex number algebra, whereas the dc resistive case does not. In addition, ac analyses may include the effect of mutual inductance,[2] which does not appear in steady-state dc analysis.

■ **EXAMPLE 10.3**

Figure 10.4 is the ladder network of Fig. 10.1b with the voltage source transformed to a current source. It is in the phasor domain, and four nodes are

---

[2] The subject of mutual inductance is an important consideration and is treated in detail in Chapter 14.

The phasor domain ladder network of Fig. 10.1b with the voltage source transformed to a current source

FIGURE 10.4

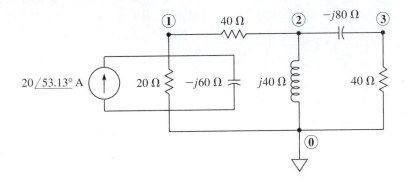

designated with numerals within circles. Use node 0 as the datum node, and use a node analysis to find $\hat{E}_1$.

Solution   The node equations are written by inspection in accordance with the procedure presented in Section 3.4.1. Recognize that the node analysis involves admittances, where

$$Y_R(j\omega) = G = \frac{1}{R} \qquad Y_L(j\omega) = \frac{-j}{X_L} \qquad Y_C(j\omega) = \frac{j}{X_C}$$

Figure 10.4, which is in the phasor domain, shows that the node equations are

$$\left(\frac{1}{20} + \frac{1}{40} + j\frac{1}{60}\right)\hat{E}_1 \qquad -\frac{1}{40}\hat{E}_2 \qquad\qquad = 20\underline{/53.13°}$$

$$-\frac{1}{40}\hat{E}_1 \qquad +\left(\frac{1}{40} - j\frac{1}{40} + j\frac{1}{80}\right)\hat{E}_2 \qquad -j\frac{1}{80}\hat{E}_3 \qquad = \qquad 0$$

$$-j\frac{1}{80}\hat{E}_2 \qquad +\left(\frac{1}{40} + j\frac{1}{80}\right)\hat{E}_3 = \qquad 0$$

or in matrix form,

$$\begin{bmatrix} \dfrac{9+j2}{120} & -\dfrac{1}{40} & 0 \\[3mm] -\dfrac{1}{40} & \dfrac{2-j}{80} & -j\dfrac{1}{80} \\[3mm] 0 & -j\dfrac{1}{80} & \dfrac{2+j}{80} \end{bmatrix} \begin{bmatrix} \hat{E}_1 \\[2mm] \hat{E}_2 \\[2mm] \hat{E}_3 \end{bmatrix} = \begin{bmatrix} 20\underline{/53.13°} \\[2mm] 0 \\[2mm] 0 \end{bmatrix}$$

The Cramer's rule solution for $\hat{E}_1$ can be set up after all elements in the foregoing matrix representation are multiplied by 240 and then put into polar form:

$$\begin{bmatrix} 18.44\underline{/12.53°} & -6\underline{/0°} & 0 \\ -6\underline{/0°} & 6.71\underline{/-26.57°} & -3\underline{/90°} \\ 0 & -3\underline{/90°} & 6.71\underline{/26.57°} \end{bmatrix}\begin{bmatrix} \hat{E}_1 \\ \hat{E}_2 \\ \hat{E}_3 \end{bmatrix} = \begin{bmatrix} 4800\underline{/53.13°} \\ 0 \\ 0 \end{bmatrix}$$

Thus,

$$\hat{E}_1 = \frac{\begin{vmatrix} 4800\underline{/53.13°} & -6\underline{/0°} & 0 \\ 0 & 6.71\underline{/-26.57°} & -3\underline{/90°} \\ 0 & -3\underline{/90°} & 6.71\underline{/26.57°} \end{vmatrix}}{\begin{vmatrix} 18.44\underline{/12.53°} & -6\underline{/0°} & 0 \\ -6\underline{/0°} & 6.71\underline{/-26.57°} & -3\underline{/90°} \\ 0 & -3\underline{/90°} & 6.71\underline{/26.57°} \end{vmatrix}}$$

$$= \frac{216,000\underline{/53.13°} - 43,200\underline{/-126.87°}}{829.76\underline{/12.53°} - 165.95\underline{/-167.47°} - 241.50\underline{/26.57°}}$$

$$= \frac{129,600 + j172,800 - (-25,920 - j34,560)}{810 + j180 - (-162 - j36) - (216 + j108)}$$

$$= \frac{155,520 + j207,360}{756 + j108}$$

$$= \frac{259,200\underline{/53.13°}}{763.68\underline{/8.13°}}$$

$$= 240\sqrt{2}\underline{/45°} \text{ V}$$

and this confirms the result in Section 10.2.

EXERCISE 10.2

Use node analysis to find the branch current $j_R(t)$ in Fig. 10.5.

Answer   3 cos 400t amperes.

FIGURE 10.5   Network (Exercise 10.2)

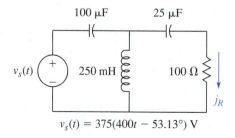

$$v_s(t) = 375(400t - 53.13°) \text{ V}$$

FIGURE 10.6

A network with a voltage-controlled current source

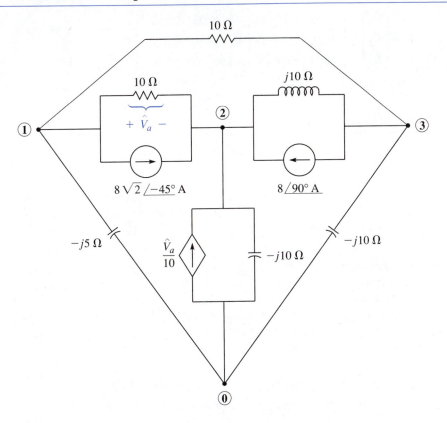

10.4.2 Node Analysis with Controlled Sources

■ **EXAMPLE 10.4**

A network in the phasor domain with a voltage-controlled voltage source and with nodes indicated by numerals within circles is shown in Fig. 10.6. Use node 0 as the datum node, and find all three node voltages.

Solution   If node 0 is selected as the datum node, three node equations may be carefully written, with $\hat{V}_a = \hat{E}_1 - \hat{E}_2$:

$$(0.1 + 0.1 + j0.2)\hat{E}_1 - 0.1\hat{E}_2 - 0.1\hat{E}_3 = -8\sqrt{2}\,\underline{/-45°}$$

$$-0.1\hat{E}_1 + (0.1 - j0.1 + j0.1)\hat{E}_2 - (-j0.1)\hat{E}_3$$
$$= 8\sqrt{2}\,\underline{/-45°} + 8\,\underline{/90°} + 0.1(\hat{E}_1 - \hat{E}_2)$$

$$-0.1\hat{E}_1 - (-j0.1)\hat{E}_2 + (0.1 - j0.1 + j0.1)\hat{E}_3 = -8\,\underline{/90°}$$

After adjustment and multiplication throughout by the factor 10, these may be put into matrix form:

$$\begin{bmatrix} (2\sqrt{2}\,\underline{/45°}) & -1 & -1 \\ -2 & 2 & 1\,\underline{/90°} \\ -1 & 1\,\underline{/90°} & 1 \end{bmatrix} \begin{bmatrix} \hat{E}_1 \\ \hat{E}_2 \\ \hat{E}_3 \end{bmatrix} = \begin{bmatrix} -80\sqrt{2}\,\underline{/-45°} \\ 80 \\ -80\,\underline{/90°} \end{bmatrix}$$

The node voltage vector, as the reader may wish to verify, can be found to be

$$
\hat{\mathbf{E}} = \begin{bmatrix} \hat{E}_1 \\ \hat{E}_2 \\ \hat{E}_3 \end{bmatrix} = \begin{bmatrix} -5.65 + j25.41 \\ 4.71 + j18.82 \\ 13.18 - j59.29 \end{bmatrix} = \begin{bmatrix} 26.03\underline{/102.53°} \\ 19.40\underline{/75.96°} \\ 60.74\underline{/-77.47°} \end{bmatrix} \text{V}
$$

■

SECTION 10.5   MESH ANALYSIS

10.5.1 An Example of Mesh Analysis

■ **EXAMPLE 10.5**

Figure 10.7 is the ladder network of Fig. 10.1b. It is in the phasor domain, and three clockwise mesh currents are indicated by the phasors $\hat{I}_1$, $\hat{I}_2$, and $\hat{I}_3$. What is the value of $\hat{I}_1$?

Solution The mesh equations can be written by inspection in accordance with the principles set forth in Section 3.4.1. The principal diagonal elements of the impedance coefficient matrix represent the self-impedances of each of the meshes, and all of these elements are positive. The off-diagonal elements of this matrix are the coupling impedances between meshes and are negative. The mesh equations are

$$
\begin{aligned}
(20 - j60)\hat{I}_1 \quad &- (-j60)\hat{I}_2 \quad &= 400\underline{/53.13°} \\
-(-j60)\hat{I}_1 + (40 - j20)\hat{I}_2 \quad &- (j40)\hat{I}_3 = \quad &0 \\
&-(j40)\hat{I}_2 + (40 - j40)\hat{I}_3 = \quad &0
\end{aligned}
$$

The matrix formulation of this mesh analysis problem is

$$
\begin{bmatrix} (20 - j60) & j60 & 0 \\ j60 & (40 - j20) & -j40 \\ 0 & -j40 & (40 - j40) \end{bmatrix} \begin{bmatrix} \hat{I}_1 \\ \hat{I}_2 \\ \hat{I}_3 \end{bmatrix} = \begin{bmatrix} 400\underline{/53.13°} \\ 0 \\ 0 \end{bmatrix}
$$

FIGURE 10.7   The phasor domain ladder network of Fig. 10.1b with three mesh currents

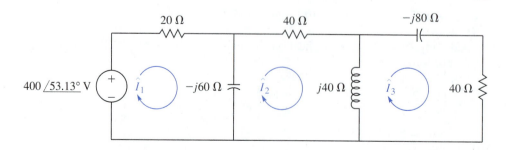

Notice that a magnitude symmetry in the coefficient matrix exists but that the two $j60$ entries are positive, which is due to the fact that the impedance of a pure capacitive element is negative.

The matrix formulation is easily adjusted to put the elements of the impedance matrix into polar form:

$$
\begin{bmatrix}
63.25\underline{/-71.57°} & 60\underline{/90°} & 0 \\
60\underline{/90°} & 44.72\underline{/-26.57°} & 40\underline{/-90°} \\
0 & 40\underline{/-90°} & 40\sqrt{2}\underline{/-45°}
\end{bmatrix}
\begin{bmatrix}
\hat{I}_1 \\
\hat{I}_2 \\
\hat{I}_3
\end{bmatrix}
=
\begin{bmatrix}
400\underline{/53.13°} \\
0 \\
0
\end{bmatrix}
$$

One may solve for $\hat{I}_1$ by Cramer's rule:

$$
\hat{I}_1 = \frac{
\begin{vmatrix}
400\underline{/53.13°} & 60\underline{/90°} & 0 \\
0 & 44.72\underline{/-26.57°} & 40\underline{/-90°} \\
0 & 40\underline{/-90°} & 40\sqrt{2}\underline{/-45°}
\end{vmatrix}
}{
\begin{vmatrix}
63.25\underline{/-71.57°} & 60\underline{/90°} & 0 \\
60\underline{/90°} & 44.72\underline{/-26.57°} & 40\underline{/-90°} \\
0 & 40\underline{/-90°} & 40\sqrt{2}\underline{/-45°}
\end{vmatrix}
}
$$

$$
= \frac{1{,}011{,}928.85\underline{/-18.43°} - 640{,}000\underline{/-126.87°}}{160{,}000\underline{/-143.13°} - 101{,}192.89\underline{/108.43°} - 203{,}646.75\underline{/135°}}
$$

$$
= \frac{960{,}000 - j320{,}000 - (-384{,}000 - j512{,}000)}{-128{,}000 - j96{,}000 - (-32{,}000 + j96{,}000) - (-144{,}000 + j144{,}000)}
$$

$$
= \frac{1344 + j192}{48 - j336} = \frac{1357.64\underline{/8.13°}}{339.41\underline{/-81.87°}} = 4\underline{/90°} = 0 + j4 \text{ A}
$$

and this confirms the result in Examples 10.1 and 10.2. ∎

EXERCISE 10.3

Use the method of mesh analysis to find the branch current $j_R(t)$ in Fig. 10.8.

Answer   3 cos 400t amperes.

Network (Exercise 10.3)

FIGURE 10.8

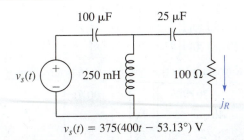

$v_s(t) = 375(400t - 53.13°)$ V

FIGURE 10.9

A network in the phasor domain with a voltage-controlled voltage source (VCVS) and with three mesh currents

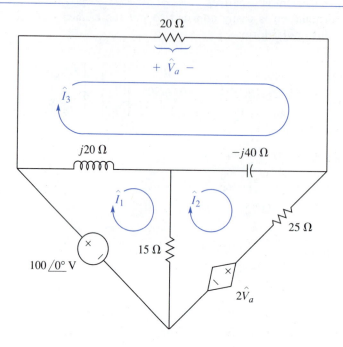

10.5.2 Mesh Analysis in the Presence of Controlled Sources

■ **EXAMPLE 10.6**

A controlled source is present in the network in Fig. 10.9. This figure is in the phasor domain, and three meshes with clockwise mesh currents $\hat{I}_1$, $\hat{I}_2$, and $\hat{I}_3$ can be noted. Notice that $\hat{V}_a = 20\hat{I}_3$, so that the magnitude of the controlled voltage source (VCVS) is $40\hat{I}_3$ volts. Find the value of the mesh current $\hat{I}_1$.

Solution The mesh equations are

$$
\begin{aligned}
(15 + j20)\hat{I}_1 &\quad - 15\hat{I}_2 &\quad - j20\hat{I}_3 &= 100\underline{/0°} \\
-15\hat{I}_1 &\quad + (40 - j40)\hat{I}_2 &\quad - (-j40)\hat{I}_3 &= -40\hat{I}_3 \\
-j20\hat{I}_1 &\quad - (-j40)\hat{I}_2 &\quad + (20 - j20)\hat{I}_3 &= 0
\end{aligned}
$$

and it is observed that the effect of the controlled source is on the right-hand side of the second equation. These equations may be put into matrix form,

$$
\begin{bmatrix}
(15 + j20) & -15 & -j20 \\
-15 & (40 - j40) & (40 + j40) \\
-j20 & j40 & (20 - j20)
\end{bmatrix}
\begin{bmatrix}
\hat{I}_1 \\
\hat{I}_2 \\
\hat{I}_3
\end{bmatrix}
=
\begin{bmatrix}
100\underline{/0°} \\
0 \\
0
\end{bmatrix}
$$

where it is seen that the controlled source upsets the symmetry. In polar form,

$$
\begin{bmatrix}
25\underline{/53.13°} & -15\underline{/0°} & -20\underline{/90°} \\
-15\underline{/0°} & 40\sqrt{2}\underline{/-45°} & 40\sqrt{2}\underline{/45°} \\
-20\underline{/90°} & 40\underline{/90°} & 20\sqrt{2}\underline{/-45°}
\end{bmatrix}
\begin{bmatrix}
\hat{I}_1 \\
\hat{I}_2 \\
\hat{I}_3
\end{bmatrix}
=
\begin{bmatrix}
100\underline{/0°} \\
0 \\
0
\end{bmatrix}
$$

The current $\hat{I}_1$ can be obtained through an application of Cramer's rule. Using N and D to designate the numerator and denominator determinants, respectively, gives

$$\hat{I}_1 = \frac{N}{D} = \frac{\begin{vmatrix} 100\underline{/0^\circ} & -15\underline{/90^\circ} & -20\underline{/90^\circ} \\ 0 & 40\sqrt{2}\underline{/-45^\circ} & 40\sqrt{2}\underline{/45^\circ} \\ 0 & 40\underline{/90^\circ} & 20\sqrt{2}\underline{/-45^\circ} \end{vmatrix}}{\begin{vmatrix} 25\underline{/53.13^\circ} & -15\underline{/0^\circ} & -20\underline{/90^\circ} \\ -15\underline{/0^\circ} & 40\sqrt{2}\underline{/-45^\circ} & 40\sqrt{2}\underline{/45^\circ} \\ -20\underline{/90^\circ} & 40\underline{/90^\circ} & 20\sqrt{2}\underline{/-45^\circ} \end{vmatrix}}$$

The numerator is evaluated as

$$\begin{aligned} N &= 160{,}000\underline{/-90^\circ} - 160{,}000\sqrt{2}\underline{/135^\circ} \\ &= -j160{,}000 - (-160{,}000 + j160{,}000) = 160{,}000 - j320{,}000 \\ &= 357{,}770.88\underline{/-63.43^\circ} \end{aligned}$$

and a similar procedure applies to the evaluation of the denominator:

$$\begin{aligned} D &= 40{,}000\underline{/-36.87^\circ} + 12{,}000\sqrt{2}\underline{/135^\circ} + 12{,}000\underline{/180^\circ} \\ &\quad - 16{,}000\sqrt{2}\underline{/135^\circ} - 40{,}000\sqrt{2}\underline{/-171.87^\circ} - 4500\sqrt{2}\underline{/-45^\circ} \\ &= (32{,}000 - j24{,}000) + (-12{,}000 + j12{,}000) - 12{,}000 \\ &\quad - (-16{,}000 + j16{,}000) - (-56{,}000 - j8000) - (4500 - j4500) \\ &= 75{,}500 - j15{,}500 = 77{,}074.64\underline{/-11.60^\circ} \end{aligned}$$

The value of $\hat{I}_1$ is

$$\hat{I}_1 = \frac{N}{D} = \frac{357{,}770.88\underline{/-63.43^\circ}}{77{,}074.64\underline{/-11.60^\circ}} = 4.64\underline{/-51.83^\circ} = 2.87 - j3.65 \text{ A} \quad \blacksquare$$

EXERCISE 10.4

Find the branch current, in instantaneous form, through the 50-Ω resistor in Fig. 10.10.

Answer $5 \cos(1000t + 126.87^\circ)$ amperes.

Ladder network (Exercise 10.4)

FIGURE 10.10

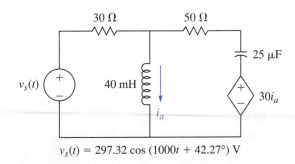

$v_s(t) = 297.32 \cos(1000t + 42.27^\circ)$ V

FIGURE 10.11   (a) The phasor domain ladder network of Fig. 10.1b, (b) an oriented graph of the
network, (c) one possible tree of the network, and (d) the network set up for loop
analysis

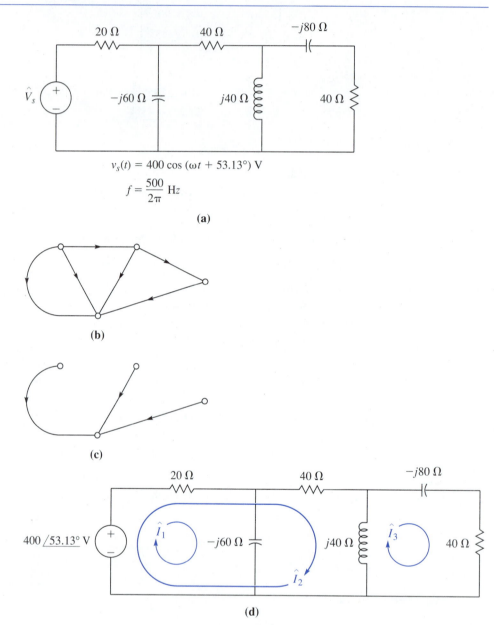

$$v_s(t) = 400 \cos(\omega t + 53.13°)\ \text{V}$$

$$f = \frac{500}{2\pi}\ \text{Hz}$$

(a)

(b)

(c)

(d)

SECTION 10.6   **LOOP ANALYSIS**

■ **EXAMPLE 10.7**

The phasor domain ladder network of Fig. 10.1b is repeated here as Fig.
10.11a, an oriented graph is shown in Fig. 10.11b, and one possible tree of
this graph is indicated in Fig. 10.11c. Use this tree to form three loops, and
use a loop analysis to find the current through the capacitor having the impedance of $-j80\ \Omega$.

Solution When the links in Fig. 10.11c are replaced one at a time, the network in Fig. 10.11d showing three loops results. The loop equations are written from an application of KVL to each loop:

$$20(\hat{I}_1 + \hat{I}_2) + (-j60)\hat{I}_1 - 400\underline{/53.13°} \qquad = 0$$

$$20(\hat{I}_1 + \hat{I}_2) + 40\hat{I}_2 + j40(\hat{I}_2 - \hat{I}_3) - 400\underline{/53.13°} = 0$$

$$j40(\hat{I}_3 - \hat{I}_2) + (-j80)\hat{I}_3 + 40\hat{I}_3 \qquad = 0$$

These may be rearranged to

$$(20 - j60)\hat{I}_1 \quad + 20\hat{I}_2 \qquad\qquad\qquad = 400\underline{/53.13°}$$

$$20\hat{I}_1 \quad + (60 + j40)\hat{I}_2 \quad - j40\hat{I}_3 \quad = 400\underline{/53.13°}$$

$$- j40\hat{I}_2 \quad + (40 - j40)\hat{I}_3 = \quad 0$$

and in matrix form, with the elements in polar form,

$$\begin{bmatrix} 63.25\underline{/-71.57°} & 20\underline{/0°} & 0 \\ 20\underline{/0°} & 72.11\underline{/33.69°} & 40\underline{/-90°} \\ 0 & 40\underline{/-90°} & 40\sqrt{2}\underline{/-45°} \end{bmatrix} \begin{bmatrix} \hat{I}_1 \\ \hat{I}_2 \\ \hat{I}_3 \end{bmatrix} = \begin{bmatrix} 400\underline{/53.13°} \\ 400\underline{/53.13°} \\ 0 \end{bmatrix}$$

Cramer's rule can be used to find $\hat{I}_3$:

$$\hat{I}_3 = \frac{N}{D} = \frac{\begin{vmatrix} 63.25\underline{/-71.57°} & 20\underline{/0°} & 400\underline{/53.13°} \\ 20\underline{/0°} & 72.11\underline{/33.69°} & 400\underline{/53.13°} \\ 0 & 40\underline{/-90°} & 0 \end{vmatrix}}{\begin{vmatrix} 63.25\underline{/-71.57°} & 20\underline{/0°} & 0 \\ 20\underline{/0°} & 72.11\underline{/33.69°} & 40\underline{/-90°} \\ 0 & 40\underline{/-90°} & 40\sqrt{2}\underline{/-45°} \end{vmatrix}}$$

The numerator of this may be evaluated as

$$N = 320,000\underline{/-36.87°} - 1,011,928.05\underline{/-108.43°}$$
$$= 256,000 - j192,000 - (-320,000 - j960,000)$$
$$= 576,000 + j768,000 = 960,000\underline{/53.13°}$$

and the denominator is

$$D = 257,992.20\underline{/-82.87°} - 101,192.89\underline{/108.43°} - 16,000\sqrt{2}\underline{/-45°}$$
$$= 32,000 - j256,000 - (-32,000 + j96,000) - (16,000 - j16,000)$$
$$= 48,000 - j336,000 = 339,411.25\underline{/-81.87°}$$

This makes

$$\hat{I}_3 = \frac{N}{D} = \frac{960,000\underline{/53.13°}}{339,411.25\underline{/-81.87°}} = 2\sqrt{2}\underline{/135°} = -2 + j2 \text{ A}$$

This is equal (as it should be) to the branch currents $\hat{J}_5$ and $\hat{J}_6$ in Examples 10.1 and 10.2. ■

FIGURE 10.12   Ladder network (Exercise 10.5)

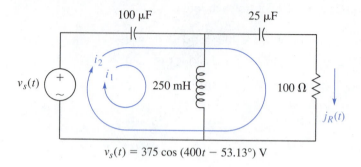

$$v_s(t) = 375 \cos (400t - 53.13°) \text{ V}$$

EXERCISE 10.5

Use the loops provided in Fig. 10.12 to find the branch current $j_R(t)$, using the method of loop analysis.

Answer   3 cos 400t amperes.

SECTION 10.7   AN APPLICATION OF THE SUPERPOSITION PRINCIPLE

■ **EXAMPLE 10.8**

Consider the network in Fig. 10.13a, where the angular frequency is 10,000 rad/s. With $X_L = \omega L = 10{,}000(0.020) = 200 \; \Omega$ and $X_C = 1/\omega C = 1/(10{,}000)(1 \times 10^{-6}) = 100 \; \Omega$, the phasor domain network is as shown in Fig. 10.13b, where an unknown current is represented by the phasor $\hat{I}$. Find this current by using the superposition principle.

Solution   Superposition says that the current $\hat{I}$ will be the sum of two currents $\hat{I}_1$ and $\hat{I}_2$, where $\hat{I}_1$ is the current obtained when the current source is acting alone (the voltage source is nulled by replacing it with a short circuit) and $\hat{I}_2$ is the current contribution due to the voltage source acting alone (with the current source nulled by replacing it with an open circuit). These current components of $\hat{I}$ are shown, respectively, in Figs. 10.8c and 10.8d.

In Fig. 10.8c, the impedance seen by the $8\underline{/0°}$-A source is not needed. The current $\hat{I}_a$ can be obtained from a current division involving a parallel combination of impedances:

$$Z_a = \frac{(100 - j100)(j200)}{100 - j100 + j200} = \frac{(100\sqrt{2}\underline{/-45°})(200\underline{/90°})}{100\sqrt{2}\underline{/45°}}$$

$$= 200\underline{/0°} = 200 \; \Omega$$

By current division

$$\hat{I}_a = \left(\frac{100}{100 + 300}\right)(8\underline{/0°}) = 2\underline{/0°} \text{ A}$$

FIGURE 10.13

(a) A network driven by two sources in which the value of $i(t)$ is sought, (b) the network in the phasor domain, (c) the network with the voltage source removed, and (d) the network with the current source removed

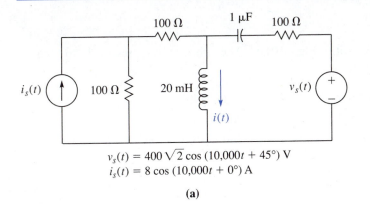

$$v_s(t) = 400 \sqrt{2} \cos (10{,}000t + 45°) \text{ V}$$
$$i_s(t) = 8 \cos (10{,}000t + 0°) \text{ A}$$

(a)

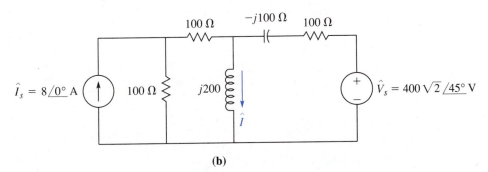

(b)

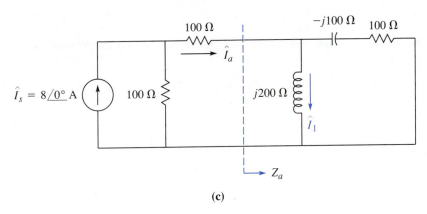

(c)

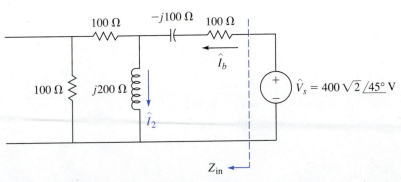

(d)

Then, another current division gives $\hat{I}_1$:

$$\hat{I}_1 = \left(\frac{100 - j100}{100 - j100 + j200}\right)\hat{I}_a = \left(\frac{100\sqrt{2}/-45°}{100\sqrt{2}/45°}\right)(2/0°)$$
$$= 2/-90° = 0 - j2 \text{ A}$$

Now, work with Fig. 10.8d. The impedance presented to the $400\sqrt{2}/45°$ V source is

$$Z_{in} = \frac{(200)(j200)}{200 + j200} + 100 - j100 = \frac{40{,}000/90°}{200\sqrt{2}/45°} + 100 - j100$$
$$= 100\sqrt{2}/45° + 100 - j100 = 100 + j100 + 100 - j100 = 200 \ \Omega$$

Hence,

$$\hat{I}_b = \frac{400\sqrt{2}/45°}{200} = 2\sqrt{2}/45° \text{ A}$$

and by a current division,

$$\hat{I}_2 = \left(\frac{200}{200 + j200}\right)\hat{I}_b = \left(\frac{200}{200\sqrt{2}/45°}\right)(2\sqrt{2}/45°) = 2/0° = 2 + j0 \text{ A}$$

The current $\hat{I}$ in Fig. 10.8b is

$$\hat{I} = \hat{I}_1 + \hat{I}_2 = -j2 + 2 = 2\sqrt{2}/-45° \text{ A} \qquad \blacksquare$$

EXERCISE 10.6

Use the method of superposition to find the voltage across the 40-mH inductor in Fig. 10.14.

Answer $20\sqrt{2} \cos(500t - 135°)$ volts.

FIGURE 10.14   Network (Exercise 10.6)

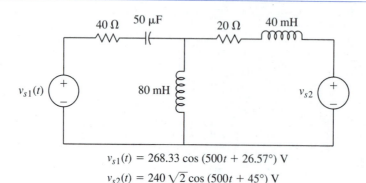

$$v_{s1}(t) = 268.33 \cos (500t + 26.57°) \text{ V}$$
$$v_{s2}(t) = 240\sqrt{2} \cos (500t + 45°) \text{ V}$$

AN APPLICATION OF THÉVENIN'S THEOREM

■ **EXAMPLE 10.9**

Find the branch current designated as $\hat{J}_6$ in Fig. 10.1b, repeated here as Fig. 10.15a, by an application of Thévenin's theorem.

Solution   In this event, the 40-Ω resistor is removed from the phasor domain network, as shown in Fig. 10.15b. The Thévenin equivalent voltage, also shown in Fig. 10.15b, may be determined from $\hat{V}_T = \hat{V}_{oc} = j40\hat{I}_2$ after $\hat{I}_2$ is obtained from a mesh analysis.

The mesh equations are

$$(20 - j60)\hat{I}_1 \quad + j60\hat{I}_2 \quad = 400\underline{/53.13°}$$
$$j60\hat{I}_1 \quad + (40 - j20)\hat{I}_2 = \quad 0$$

or with the impedances in polar form,

$$63.25\underline{/-71.57°}\hat{I}_1 \quad + 60\underline{/90°}\,\hat{I}_2 \quad = 400\underline{/53.13°}$$
$$60\underline{/90°}\,\hat{I}_1 \quad + 44.72\underline{/-26.57°}\hat{I}_2 = \quad 0$$

The mesh current $\hat{I}_2$ can be found by employing Cramer's rule:

$$\hat{I}_2 = \frac{\begin{vmatrix} 63.25\underline{/-71.57°} & 400\underline{/53.13°} \\ 60\underline{/90°} & 0 \end{vmatrix}}{\begin{vmatrix} 63.25\underline{/-71.57°} & 60\underline{/90°} \\ 60\underline{/90°} & 44.72\underline{/-26.57°} \end{vmatrix}}$$

$$= \frac{-24,000\underline{/143.13°}}{2828.43\underline{/-98.13°} - 3600\underline{/180°}} = \frac{-24,000\underline{/143.13°}}{-400 - j2800 + 3600}$$

$$= \frac{-24,000\underline{/143.13°}}{3200 - j2800} = \frac{-240\underline{/143.13°}}{42.52\underline{/-41.19°}} = -5.64\underline{/184.32°}$$

$$= 5.64\underline{/4.32°} \text{ A}$$

The value of $\hat{V}_T$ is

$$\hat{V}_T = j40\hat{I}_2 = (40\underline{/90°})(5.64\underline{/4.32°}) = 225.77\underline{/94.32°} \text{ V}$$

One way of finding Z_T is to obtain the impedance looking into the terminals after the source $\hat{V}_s$ is removed and replaced by a short circuit. This is shown in Fig. 10.15c, and it is seen that Z_1 is

$$Z_1 = \frac{(20)(-j60)}{20 - j60} = \frac{1200\underline{/-90°}}{63.25\underline{/-71.57°}} = 18.97\underline{/-18.43°} = 18 - j6 \ \Omega$$

FIGURE 10.15

(a) The ladder network of Fig. 10.1b, (b) the ladder network with the 40-Ω resistor removed, (c) steps employed in the computation of the Thévenin equivalent impedance, (d) the Thévenin equivalent network with the 40-Ω resistor connected across its terminals, and (e) the evaluation of the short circuit current

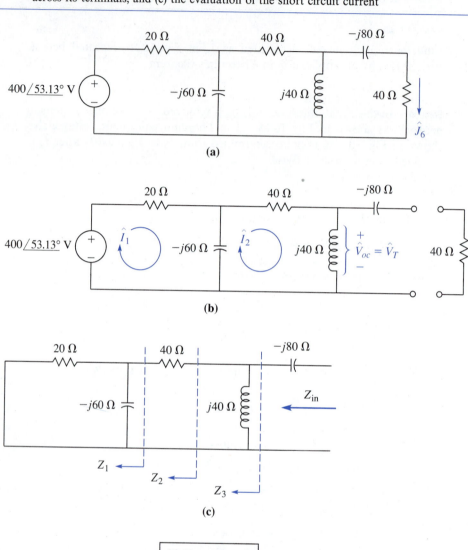

(a)

(b)

(c)

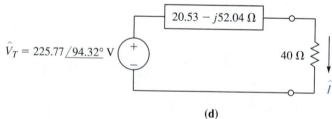

(d)

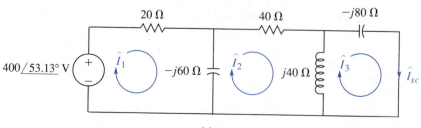

(e)

Then,

$$Z_2 = 40 + Z_1 = 40 + (18 - j6) = 58 - j6 \ \Omega$$

$$Z_3 = \frac{(58 - j6)(j40)}{58 - j6 + j40} = \frac{240 + j2320}{58 + j34} = \frac{2332.38 \underline{/84.09°}}{67.23 \underline{/30.38°}}$$

$$= 34.69 \underline{/53.71°} = 20.53 + j27.96 \ \Omega$$

The Thévenin equivalent impedance is

$$Z_T = Z_3 - j80 = 20.53 + j27.96 - j80 = 20.53 - j52.04 \ \Omega$$

The Thévenin equivalent impedance and the 40-Ω resistor are shown in Fig. 10.15d. This figure shows a single current $\hat{I}$, which has a value

$$\hat{I} = \frac{\hat{V}_T}{Z_T + 40} = \frac{225.77 \underline{/94.32°}}{20.53 - j52.04 + 40} = \frac{225.77 \underline{/94.32°}}{60.53 - j52.04}$$

$$= \frac{225.77 \underline{/94.32°}}{79.82 \underline{/-40.68°}} = 2\sqrt{2} \underline{/135°} \ A$$

which is equal to the value of $\hat{J}_6$ calculated in the ladder network analyses in Examples 10.1 and 10.2 in Section 10.2.

An alternative method for the determination of Z_T lies in the computation of $\hat{I}_{sc}$, which is mesh current $\hat{I}_3$ in Fig. 10.15e. Here, the mesh equations, with coefficients in polar form, are

$$63.25 \underline{/-71.57°} \hat{I}_1 \quad + 60 \underline{/90°} \hat{I}_2 \qquad\qquad = 400 \underline{/53.13°}$$

$$60 \underline{/90°} \hat{I}_1 \quad + 44.72 \underline{/-26.57°} \hat{I}_2 - 40 \underline{/90°} \hat{I}_3 = \quad 0$$

$$- 40 \underline{/90°} \hat{I}_2 \quad - 40 \underline{/90°} \hat{I}_3 = \quad 0$$

and a Cramer's rule solution for $\hat{I}_3$ is

$$\hat{I}_3 = \frac{\begin{vmatrix} 63.25 \underline{/-71.57°} & 60 \underline{/90°} & 400 \underline{/53.13°} \\ 60 \underline{/90°} & 44.72 \underline{/-26.57°} & 0 \\ 0 & -40 \underline{/90°} & 0 \end{vmatrix}}{\begin{vmatrix} 63.25 \underline{/-71.57°} & 60 \underline{/90°} & 0 \\ 60 \underline{/90°} & 44.72 \underline{/-26.57°} & -40 \underline{/90°} \\ 0 & -40 \underline{/90°} & -40 \underline{/90°} \end{vmatrix}}$$

$$= \frac{-960,000 \underline{/233.13°}}{-113,137.09 \underline{/-8.13°} - 101,192.89 \underline{/108.43°} + 144,000 \underline{/270°}}$$

$$= \frac{-960,000 \underline{/233.13°}}{-(112,000 - j16,000) - (-32,000 + j96,000) + (0 - j144,000)}$$

$$= \frac{-960,000 \underline{/233.13°}}{-80,000 - j224,000} = \frac{-960,000 \underline{/233.13°}}{237,857.10 \underline{/-109.65°}}$$

FIGURE 10.16   Network (Exercise 10.7)

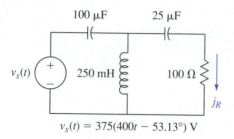

$$v_s(t) = 375(400t - 53.13°) \text{ V}$$

or

$$\hat{I}_3 = \hat{I}_{sc} = \hat{I}_N = -4.04\underline{/342.78°} = 4.04\underline{/162.78°} \text{ A}$$

The value of Z_T is

$$Z_T = \frac{\hat{V}_{oc}}{\hat{I}_{sc}} = \frac{225.77\underline{/94.32°}}{4.04\underline{/162.78°}} = 55.94\underline{/-68.47°} = 20.53 - j52.04 \ \Omega$$

which is the same result as obtained by considering the impedance looking into the terminals (Fig. 10.15c). ■

EXERCISE 10.7

Use Thévenin's theorem to find the branch current $j_R(t)$ in Fig. 10.16.

Answer   3 cos 400t amperes.

SECTION 10.9   ## THE TEE-PI (PI-TEE) TRANSFORMATION REVISITED

No matter what method is employed, the analysis of networks driven by ac sources usually involves more than just a simple exercise in the algebra of complex numbers. This is true in the evaluation of the inverse of the coefficient matrix or in the evaluation of determinants in applications where Cramer's rule is used. The computations can be expedited to a considerable extent through the use of the tee-to-pi or the pi-to-tee transformations. Of course, care must be exercised so that the branch currents and branch voltages of interest are not lost in the transformation process, and naturally, the analyst must decide whether the labor involved in executing the transformations offsets the eventual overall labor saved.

■ **EXAMPLE 10.10**

Once again, consider the phasor domain network in Fig. 10.1b, which is reproduced here as Fig. 10.17a. Use the tee-pi and pi-tee transformations to reduce the labor in finding the current through the 20-Ω ressistor, which was calculated in Sections 10.2 and 10.4 as $4\underline{/90°}$ A.

Solution   The three-mesh network in Fig. 10.17a can be reduced to the two-mesh network in Fig. 10.17b by transforming the pi network in the dashed

(a) The phasor domain network of Fig. 10.1b, (b) a possible simplification using pi-to-tee transformation, and (c) the network in the phasor domain with two meshes after the simplification is made

FIGURE 10.17

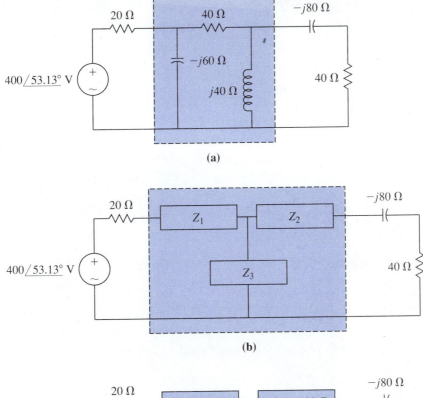

(a)

(b)

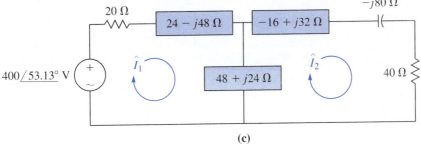

(c)

area of Fig. 10.17a to the tee network in the dashed area of Fig. 10.17b. A mesh analysis will then provide the desired result, and it is assumed that the calculations required in performing the pi-to-tee transformation do not offset the labor saved in the eventual analysis using two rather than three meshes.

The pi-to-tee transformation is made by using the summary in Fig. 2.27, with the R's replaced by Z's. Here,

$$Z_{12} = 40\ \Omega \qquad Z_{23} = j40\ \Omega \qquad Z_{31} = -j60\ \Omega$$

$$\sum Z = Z_{12} + Z_{23} + Z_{31} = 40 + j40 - j60 = 40 - j20$$

$$= 44.71\underline{/-26.57°}\ \Omega$$

Then,

$$Z_1 = \frac{Z_{12}Z_{31}}{\Sigma Z} = \frac{40(-j60)}{44.71 \, / -26.57°} = \frac{2400 \, / -90°}{44.71 \, / -26.57°}$$
$$= 53.67 \, / -63.43° = 24 - j48 \ \Omega$$

$$Z_2 = \frac{Z_{12}Z_{23}}{\Sigma Z} = \frac{40(j40)}{44.71 \, / -26.57°} = \frac{1600 \, / 90°}{44.71 \, / -36.57°}$$
$$= 35.78 \, / 116.57° = -16 + j32 \ \Omega$$

$$Z_3 = \frac{Z_{23}Z_{31}}{\Sigma Z} = \frac{(j40)(-j60)}{44.71 \, / -26.57°} = \frac{2400 \, / 0°}{44.71 \, / -26.57°}$$
$$= 53.67 \, / 26.57° = 48 + j24 \ \Omega$$

These impedances may be placed into Fig. 10.17b, and the network with the two meshes indicated is shown in Fig. 10.17c. The mesh equations are

$$(92 - j24)\hat{I}_1 - (48 + j24)\hat{I}_2 = 400 \, / 53.13°$$
$$(-48 + j24)\hat{I}_2 + (72 - j24)\hat{I}_2 = 0$$

or in polar form,

$$95.08 \, / -14.62° \, \hat{I}_1 - 53.67 \, / 26.57° \, \hat{I}_2 = 400 \, / 53.13°$$
$$-53.67 \, / 26.57° \, \hat{I}_1 + 75.89 \, / -18.43° \, \hat{I}_2 = 0$$

The Cramer's rule solution should yield $\hat{I}_1 = 4 \, / 90° = 0 + j4$ A:

$$\hat{I}_1 = \frac{\begin{vmatrix} 400 \, / 53.13° & -53.67 \, / 26.57° \\ 0 & 75.89 \, / -18.43° \end{vmatrix}}{\begin{vmatrix} 95.08 \, / -14.62° & -53.67 \, / 26.57° \\ -53.67 \, / 26.57° & 75.89 \, / -18.43° \end{vmatrix}}$$

$$= \frac{30{,}357.87 \, / 34.70°}{7215.98 \, / -33.06° - 2880 \, / 53.13°}$$

$$= \frac{33{,}357.87 \, / 34.70°}{6048 - j3936 - (1728 + j2304)}$$

$$= \frac{30{,}357.87 \, / 34.70°}{4320 - j6240} = \frac{30{,}357.87 \, / 34.70°}{7589.47 \, / -55.30°} = 4 \, / 90° = 0 + j4 \text{ A}$$

which is the sought-after result. ■

EXERCISE 10.8

In the phasor domain network of Fig. 10.18a, which contains seven meshes, only the current $\hat{I}$ is of interest. Draw a network containing only three meshes that retains the current $\hat{I}$ as one of the mesh currents.

Answer   See Fig. 10.18b.

(a) A seven-mesh network and (b) its reduction to a three-mesh network (Exercise 10.8)

FIGURE 10.18

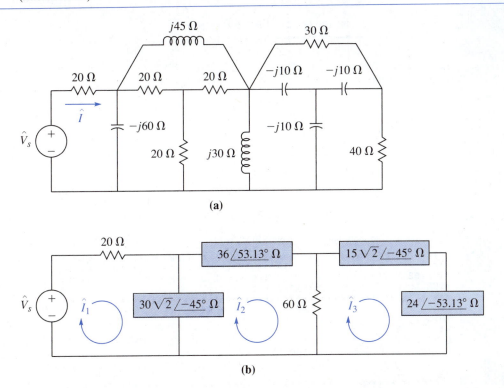

(a)

(b)

SPICE EXAMPLES

Reading: In addition to reading Sections C.1 through C.7, which were recommended for reading in previous chapters, the reader should read and understand Section C.8 before proceeding to the SPICE examples that follow.

EXAMPLE S10.1

This example will verify the result of Example 10.8. The network is shown in Fig. S10.1, and the network made ready for PSPICE analysis is shown in Fig. S10.2. The input file is displayed in Fig. S10.3, and a pertinent extract from the output file is provided in Fig. S10.4.

Figure S10.1   Network for SPICE Example S10.1

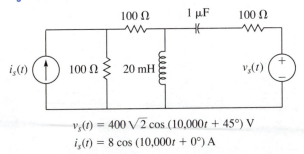

$v_s(t) = 400\sqrt{2}\cos(10{,}000t + 45°)$ V

$i_s(t) = 8\cos(10{,}000t + 0°)$ A

(continues)

Example S10.1 *(continued)*

Figure S10.2   Network made ready for SPICE analysis (Example S10.1)

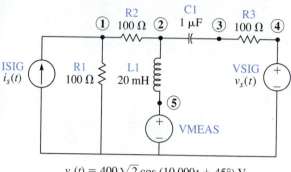

$$v_s(t) = 400\sqrt{2}\cos(10{,}000t + 45°)\ \text{V}$$
$$i_s(t) = 8\cos(10{,}000t + 0°)\ \text{A}$$

Figure S10.3   The input file for Example S10.1

```
SPICE PROBLEM CHAPTER 10 - NUMBER 1 - AC SUPERPOSITION EXAMPLE
***************************************************************
*THE OBJECTIVE IS TO CONFIRM THE VALUE OF THE CURRENT THROUGH
*THE INDUCTOR FOUND IN EXAMPLE 10.8.
***************************************************************
VSIG     4       0       AC      565.685         45
ISIG     0       1       AC      8
R1       1       0       100
R2       1       2       100
R3       3       4       100
C1       2       3       1U
L1       2       5       20M
***************************************************************
*HERE IS THE DEVICE TO MEASURE THE CURRENT
VMEAS    5       0       DC      0
***************************************************************
*HERE ARE THE CONTROL STATEMENTS
.AC      LIN     1       1591.55         1591.55
.PRINT   AC      IM(VMEAS)    IP(VMEAS)    IR(VMEAS)    II(VMEAS)
*DON'T FORGET THE END STATEMENT
.END
***************************************************************
```

Figure S10.4   A pertinent extract from the output file (Example S10.1)

```
SPICE PROBLEM CHAPTER 10 - NUMBER 1 - AC SUPERPOSITION EXAMPLE


****      AC ANALYSIS                          TEMPERATURE =    27.000 DEG C


*****************************************************************************


  FREQ          IM(VMEAS)    IP(VMEAS)    IR(VMEAS)    II(VMEAS)


  1.592E+03     2.828E+00   -4.500E+01    2.000E+00   -2.000E+00


              JOB CONCLUDED
```

EXAMPLE S10.2

This example will verify the result of Example 10.6. The network is shown in Fig. S10.5, and the network made ready for PSPICE analysis is shown in Fig. S10.6. The input file is displayed in Fig. S10.7, and a pertinent extract from the output file is provided in Fig. S10.8.

Figure S10.5   Network for SPICE Example S10.2

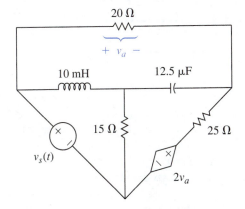

$$v_s(t) = 100 \cos (\omega t + 0°) \text{ V}$$
$$f = 318.13 \text{ Hz}$$

Figure S10.6   Network made ready for SPICE analysis (Example S10.2)

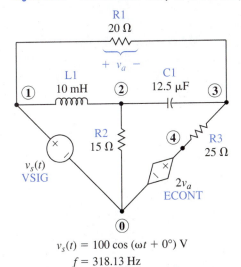

$$v_s(t) = 100 \cos (\omega t + 0°) \text{ V}$$
$$f = 318.13 \text{ Hz}$$

(continues)

Example S10.2 *(continued)*

Figure S10.7   The input file for Example S10.2

```
SPICE EXAMPLE - CHAPTER 10 - NUMBER 2 AC CONTROLLED SOURCE EXAMPLE
****************************************************************
*THE OBJECTIVE IS TO CONFIRM THE VALUE OF THE MESH CURRENT I1 IN
*IN EXAMPLE 10.6.
****************************************************************
VSIG     1       0       AC       100
ECONT    4       0       1        3        2
R1       1       3       20
R2       2       0       15
R3       3       4       25
C1       2       3       12.5U
L1       1       2       10M
****************************************************************
*BECAUSE OF THE POLARITY OF VSIG, THE PRINTED OUTPUT WILL BE THE
*NEGATIVE OF THE TRUE VALUE.
****************************************************************
*HERE ARE THE CONTROL STATEMENTS
.AC     LIN     1        318.13        318.13
.PRINT  AC      IR(VSIG)    II(VSIG)
*DON'T FORGET THE END STATEMENT
.END
****************************************************************
```

Figure S10.8   A pertinent extract from the output file (Example S10.2)

```
SPICE EXAMPLE - CHAPTER 10 - NUMBER 2 AC CONTROLLED SOURCE EXAMPLE


****        AC ANALYSIS                          TEMPERATURE =    27.000 DEG C



********************************************************************************



 FREQ          IR(VSIG)    II(VSIG)


  3.181E+02   -2.871E+00    3.651E+00


          JOB CONCLUDED
```

CHAPTER 10

SUMMARY

- Network analysis using sinusoidal signals (ac analysis) is performed in the phasor domain in exactly the same manner as nonsinusoidal analysis. The difficulty is greater, of course, because of the necessity to work with complex numbers.

- Ladder network analysis usually requires the evaluation of the driving-point impedance or admittance and then application of any or all of the following:
  - —The addition of impedances in series or the addition of admittances in parallel
  - —The combination of impedances in parallel or the combination of admittances in series
  - —The Kirchhoff current law
  - —The Kirchhoff voltage law
  - —The current division concept
  - —The voltage division concept
  - —The voltage-to-current source transformation
  - —The current-to-voltage source transformation

- Node analysis may be performed in exactly the same way that it is performed in purely resistive networks with dc signals. The same techniques apply when dummy nodes and controlled sources are present.

- Mesh analysis may be performed in exactly the same way that it is performed in purely resistive networks with dc signals. The same techniques apply when dummy meshes and controlled sources are present.

- Loop analysis is also performed in the same manner as in purely resistive networks.

- The superposition theorem and the network theorems of Thévenin and Norton are useful in performing sinusoidal steady-state analysis and are employed in the same manner as in the analysis of resistive networks with dc signals.

- When faced with a nonseries-parallel network and interest is focused on only part of the network, the use of both the tee-pi and the pi-tee transformations can save a significant amount of labor in the achievement of the result.

Additional Readings

Blackwell, W.A., and L.L. Grigsby. *Introductory Network Theory*. Boston: PWS Engineering, 1985, pp. 350–359.

Bobrow, L.S. *Elementary Linear Circuit Analysis*. 2d ed. New York: Holt, Rinehart and Winston, 1987, pp. 358–363, 375–385, 422–426.

Del Toro, V. *Engineering Circuits*. Englewood Cliffs, N.J.: Prentice-Hall, 1987, pp. 200, 204, 206, 208, 211, 217, 219–224, 230.

Dorf, R.C. *Introduction to Electric Circuits*. New York: Wiley, 1989, pp. 347–365, 390.

Hayt, W.H., Jr., and J.E. Kemmerly. *Engineering Circuit Analysis*. 4th ed. New York: McGraw-Hill, 1986, pp. 235–239, 264–272, 292, 470–477.

Irwin, J.D. *Basic Engineering Circuit Analysis*. 3d ed. New York: Macmillan, 1989, pp. 367, 384–394, 406–430, 453–458.

Johnson, D.E., J.L. Hilburn, and J.R. Johnson. *Basic Electric Circuit Analysis*. 4th ed. Englewood Cliffs, N.J.: Prentice-Hall, 1989, pp. 332–334, 341–359, 417–421.

Karni, S. *Applied Circuit Analysis*. New York: Wiley, 1988, pp. 272–293, 303, 344, 408–411.

Madhu, S. *Linear Circuit Analysis*. Englewood Cliffs, N.J.: Prentice-Hall, 1988, pp. 329–334, 354–359, 367–369, 375–379, 402–426, 437–449, 658.

Nilsson, J.W. *Electric Circuits*. 3d ed. Reading, Mass.: Addison-Wesley, 1990, pp. 336–354, 391.

Paul, C.R. *Analysis of Linear Circuits*. New York: McGraw-Hill, 1989, pp. 303–305.

PROBLEMS   CHAPTER 10

Section 10.2

10.1   Determine all branch currents and branch voltages in the ladder network shown in Fig. P10.1.

Figure P10.1

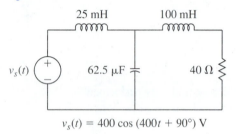

$$v_s(t) = 400 \cos (400t + 90°) \text{ V}$$

10.2   Determine all branch currents and branch voltages in the ladder network shown in Fig. P10.2.

Figure P10.2

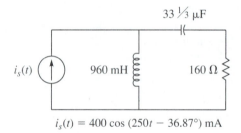

$$i_s(t) = 400 \cos (250t - 36.87°) \text{ mA}$$

10.3   Determine all branch currents and branch voltages in the ladder network shown in Fig. P10.3.

Figure P10.3

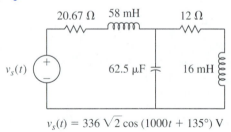

$$v_s(t) = 336 \sqrt{2} \cos (1000t + 135°) \text{ V}$$

10.4   Determine all branch currents and branch voltages in the ladder network shown in Fig. P10.4.

Figure P10.4

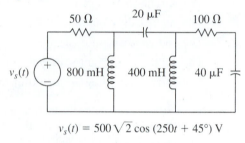

$$v_s(t) = 500 \sqrt{2} \cos (250t + 45°) \text{ V}$$

10.5   Determine all branch currents and branch voltages in the ladder network shown in Fig. P10.5.

Figure P10.5

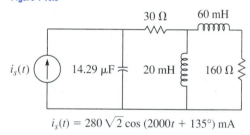

$$i_s(t) = 280 \sqrt{2} \cos (2000t + 135°) \text{ mA}$$

Section 10.3

10.6   Confirm the results obtained in Problem 10.1 by drawing a complete phasor diagram. Be sure to check all possible KVL and KCL relationships.

10.7   Confirm the results obtained in Problem 10.2 by drawing a complete phasor diagram. Be sure to check all possible KVL and KCL relationships.

10.8   Confirm the results obtained in Problem 10.3 by drawing a complete phasor diagram. Be sure to check all possible KVL and KCL relationships.

10.9   Confirm the results obtained in Problem 10.4 by drawing a complete phasor diagram. Be sure to check all possible KVL and KCL relationships.

10.10   Confirm the results obtained in Problem 10.5 by drawing a complete phasor diagram. Be sure to check all possible KVL and KCL relationships.

Section 10.4

10.11   Use node analysis to find the current through the 25-mH inductor and the voltage across the 62.5-μF capacitor in Fig. P10.1.

10.12 Use node analysis to find the current through the 960-mH inductor and the voltage across the 120 Ω resistor in Fig. P10.2.

10.13 Use node analysis to find the current through the 800-mH inductor and the voltage across the 40-μF capacitor in Fig. P10.4.

10.14 Use node analysis to find the current through the 50-mH inductor and the voltage across the 40-Ω resistor in Fig. P10.6.

Figure P10.6

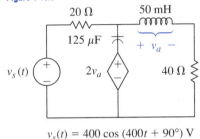

$$v_s(t) = 400 \cos (400t + 90°) \text{ V}$$

10.15 Use node analysis to find the current i in Fig. P10.7.

Figure P10.7

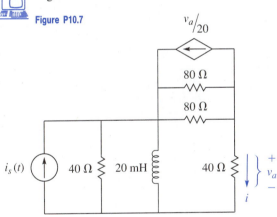

$$i_s(t) = 520 \cos (1000t + 22.62°) \text{ mA}$$

10.16 Use node analysis to find the current i in Fig. P10.8.

Figure P10.8

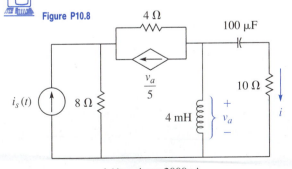

$$i_s(t) = 4 \cos 2000t \text{ A}$$

Section 10.5

10.17 Use mesh analysis to find the current through the 25-mH inductor and the voltage across the 62.5-μF capacitor in Fig. P10.1.

10.18 Use mesh analysis to find the current through the 960-mH inductor and the voltage across the 120 Ω resistor in Fig. P10.2.

10.19 Use mesh analysis to find the current through the 58-mH inductor and the voltage across the 62.5-μF capacitor in Fig. P10.3.

10.20 Use mesh analysis to find the current through the 800-mH inductor and the voltage across the 40-μF capacitor in Fig. P10.4.

10.21 Use mesh analysis to find the current through the 20-mH inductor and the voltage across the 30-Ω resistor in Fig. P10.5.

10.22 Use mesh analysis to find the current through the 50-mH inductor and the voltage across the 40-Ω resistor in Fig. P10.6.

10.23 Use mesh analysis to find the current i in Fig. P10.7.

10.24 Use mesh analysis to find the current i in Fig. P10.8.

Section 10.6

10.25 Use loop analysis to find the current through the 100-mH inductor and the voltage across the 62.5-μF capacitor in Fig. P10.1.

10.26 Use loop analysis to find the current through the 58-mH inductor and the voltage across the 62.5-μF capacitor in Fig. P10.3.

10.27 Use loop analysis to find the current through the 800-mH inductor and the voltage across the 40-μF capacitor in Fig. P10.4.

10.28 Use loop analysis to find the current through the 20-mH inductor and the voltage across the 30-Ω resistor in Fig. P10.5.

10.29 Use loop analysis to find the current through the 50-mH inductor and the voltage across the 40-Ω resistor in Fig. P10.6.

Section 10.7

10.30 Use superposition to find the current through the 25-mH inductor in Fig. P10.9.

Figure P10.9

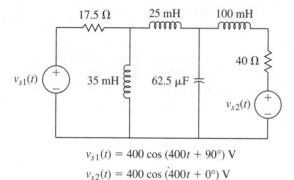

$$v_{s1}(t) = 400 \cos (400t + 90°) \text{ V}$$

$$v_{s2}(t) = 400 \cos (400t + 0°) \text{ V}$$

10.31 Use superposition to find the voltage across the 50-μF capacitor in Fig. P10.10.

Figure P10.10

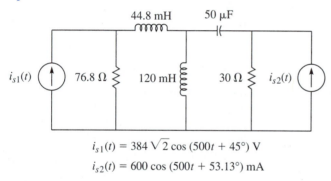

$$i_{s1}(t) = 384 \sqrt{2} \cos (500t + 45°) \text{ V}$$

$$i_{s2}(t) = 600 \cos (500t + 53.13°) \text{ mA}$$

10.32 Use superposition to find the current through the 58-mH inductor in Fig. P10.11.

Figure P10.11

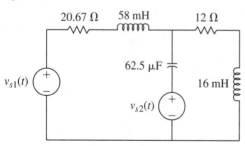

$$v_{s1}(t) = 336 \sqrt{2} \cos (1000t + 135°) \text{ V}$$

$$v_{s2}(t) = 260 \cos (1000t + 67.38°) \text{ V}$$

10.33 Use superposition to find the voltage across the 40-μF capacitor in Fig. P10.12.

Figure P10.12

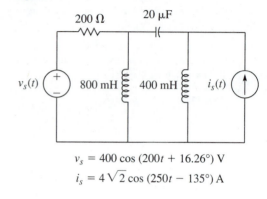

$$v_S = 400 \cos (200t + 16.26°) \text{ V}$$

$$i_s = 4 \sqrt{2} \cos (250t - 135°) \text{ A}$$

10.34 Use superposition to find the current through the 20-mH inductor in Fig. P10.13.

Figure P10.13

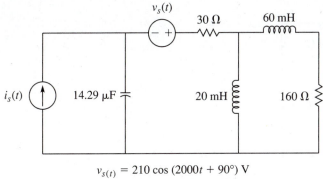

$$v_{s(t)} = 210 \cos (2000t + 90°) \text{ V}$$
$$i_{s(t)} = 16 \cos (2000t + 0°) \text{ mA}$$

Section 10.8

10.35 Use Thévenin's theorem to find the current through the 40-Ω resistor in Fig. P10.1.

10.36 Use Thévenin's theorem to find the current through the 16-mH inductor in Fig. P10.3.

10.37 Use Thévenin's theorem to find the current through the 40-μF capacitor in Fig. P10.4.

10.38 Use Thévenin's theorem to find the current through the 160-Ω resistor in Fig. P10.5.

10.39 Use Thévenin's theorem to find the current through the 10-Ω resistor in Fig. P10.8.

Section 10.9

10.40 Determine the impedance looking into terminals a–b in Fig. P10.14.

Figure P10.14

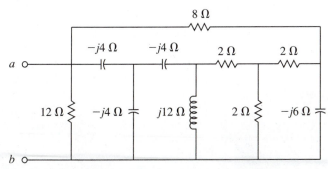

10.41   Determine the impedance looking into terminals
       a–b in Fig. P10.15.

Figure P10.15

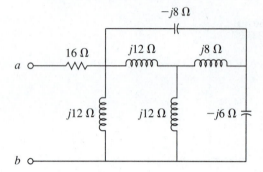

10.42   Determine the impedance looking into terminals
       a–b in Fig. P10.16.

Figure P10.16

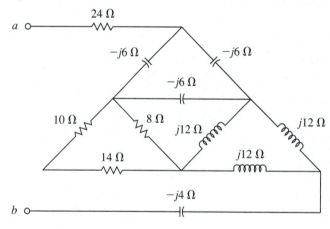

AC POWER

OBJECTIVES

The objectives of this chapter are to:

- Consider the average power in individual network elements and the general network impedance.

- Introduce the subject of complex power.

- Discuss the power factor and its correction.

- Determine the conditions for the transfer of maximum power to a load.

- Reconsider Tellegen's theorem.

INTRODUCTION   SECTION 11.1

When one or more sources of electric energy are applied to an electric network, energy flows; and the rate at which energy is generated by the sources or absorbed by the network elements is the power. The instantaneous power is always equal to the product of the instantaneous voltage and current:

$$p = vi$$

A knowledge of the maximum value of the instantaneous power can be extremely useful.

This chapter is concerned with the instantaneous and average ac power in individual network elements and in the general case when many elements are assembled in a network. This leads to a discussion of complex power, power factor, and power factor correction. Maximum power transfer is reintroduced for the ac network, and Tellegen's theorem is revisited.

Care must be exercised in the designation of the amplitude in the instantaneous form of the voltage and current sinusoids, because power calculations utilize effective or rms values. Thus, the V_m in

$$v = V_m \cos(\omega t + \alpha)$$

and

$$\hat{V} = V_m \underline{/\alpha}$$

represent maximum values for the voltage in the instantaneous and phasor forms. As shown in Chapter 5, these are related to the rms values, which are designated without a subscript via

$$V_m = \sqrt{2}\, V$$

so that $v = \sqrt{2}\, V \cos(\omega t + \alpha)$ and

$$\hat{V} = \sqrt{2}\, V \underline{/\alpha}$$

The same convention applies to the current

$$I_m = \sqrt{2}\, I$$

where I is the rms value and

$$i = I_m \cos(\omega t + \beta) = \sqrt{2}\, I \cos(\omega t + \beta)$$

or

$$\hat{I} = I_m \underline{/\beta} = \sqrt{2} I \underline{/\beta}$$

In this chapter, a V or an I without a subscript will indicate an effective or rms value.

AVERAGE POWER IN NETWORK ELEMENTS

11.2.1 The Resistor

If a voltage

$$v = V_m \cos \omega t \text{ volts}$$

is applied across a resistor, the current through the resistor will be

$$i = I_m \cos \omega t \text{ amperes}$$

and the instantaneous power will be

$$p = vi = V_m I_m \cos^2 \omega t \text{ watts}$$

By the trigonometric identity,

$$\cos^2 \omega t = \frac{1}{2}(1 + \cos 2\omega t)$$

the instantaneous power becomes

$$p = \frac{1}{2} V_m I_m (1 + \cos 2\omega t) \text{ watts} \tag{11.1}$$

Equation (11.1) indicates that the instantaneous power drawn and dissipated by a resistor is always positive (never less than zero) and that the power flow is at an apparent frequency that is twice the applied frequency.

The average power over one cycle can be obtained from an integration of eq. (11.1):

$$P_{av} = \frac{1}{2\pi} \int_0^{2\pi} \frac{1}{2} V_m I_m (1 + \cos 2\omega t) \, d(\omega t)$$

$$= \frac{1}{4\pi} V_m I_m \left[\int_0^{2\pi} d(\omega t) + \int_0^{2\pi} \cos 2\omega t \, d(\omega t) \right]$$

The second term within the brackets is the integration of a cosine function over one cycle, and this is numerically equal to zero regardless of the frequency. Thus,

$$P_{av} = \frac{1}{4\pi} V_m I_m (\omega t) \Big|_0^{2\pi} = \frac{1}{2} V_m I_m = \frac{V_m}{\sqrt{2}} \frac{I_m}{\sqrt{2}} = VI$$

where V and I are rms or effective values of voltage and current. Observe once again that great care has been taken to indicate maximum or peak amplitudes of the current or voltage signal by the subscript m. The corresponding rms or effective values are designated by V and I without a subscript.

Recall that power is the rate of energy transport. Because the flow of energy from a source to a resistor is never negative, and because the resistor irreversibly transforms the energy delivered to it as heat, the power flow to a resistor is commonly referred to as *real power*.

11.2.2 The Inductor

For the inductor, where the current lags the voltage by 90°, if

$$v = V_m \cos \omega t$$

then

$$i = I_m \cos(\omega t - 90°)$$

and the instantaneous power will be

$$p = vi = V_m I_m \cos \omega t \cos(\omega t - 90°)$$

But

$$\cos(\omega t - 90°) = \sin \omega t$$

so that

$$p = vi = V_m I_m \cos \omega t \sin \omega t$$

and by the trigonometric identity

$$\sin 2\omega t = 2 \cos \omega t \sin \omega t$$

it is observed that the instantaneous power becomes

$$p = \frac{1}{2} V_m I_m \sin 2\omega t \text{ watts} \tag{11.2}$$

Equation (11.2) indicates that the power drawn by the inductor can be positive or negative, the power flow is at an apparent frequency that is twice the applied frequency, and the average power over one cycle is equal to zero:

$$p_{av} = \frac{1}{2\pi} \int_0^{2\pi} \frac{V_m I_m}{2} \sin 2\omega t \, d(\omega t) = \frac{V_m I_m}{4\pi} \int_0^{2\pi} \sin 2\omega t \, d(\omega t) = 0$$

because the integral of the sine function over one cycle, regardless of the frequency, is equal to zero.

Physically, this means that a source will deliver energy to the inductor as long as p is positive, and the energy will be contained in the magnetic field surrounding the inductor. When p is negative, the magnetic field collapses and the energy will be returned to the source. Because the inductor is referred to as a reactive element, the rate of this energy flow is called *reactive power*.

11.2.3 The Capacitor

For the capacitor, with

$$v = V_m \cos \omega t$$

the current leads the applied voltage by 90°. Here,

$$i = I_m \cos(\omega t + 90°)$$

and the instantaneous power is

$$p = vi = V_m I_m \cos \omega t \cos(\omega t + 90°)$$

But because

$$\cos(\omega t + 90°) = -\sin \omega t$$

the instantaneous power can be represented by

$$p = vi = -V_m I_m \cos \omega t \sin \omega t$$

and by the trigonometric identity employed in the inductor discussion,

$$p = -\frac{1}{2} V_m I_m \sin 2\omega t \text{ watts} \tag{11.3}$$

Here, too, the power ranges over positive and negative values, the power flow is at an apparent frequency that is twice the applied frequency, and as in the inductor, the average power is equal to zero.

In this case, when p is positive, the energy delivered by the source to the capacitor is stored in the form of an electric field set up between the plates of the capacitor. When p is negative, the energy is returned to the source as the electric field collapses. The rate of this energy flow is also referred to as *reactive power*.

THE AVERAGE POWER FOR THE GENERAL IMPEDANCE SECTION 11.3

In the general case, where a network possesses a driving-point impedance

$$Z = |Z|\underline{/\theta}$$

a voltage source

$$v = V_m \cos(\omega t + \alpha)$$

written in phasor form as

$$\hat{V} = V_m \underline{/\alpha}$$

produces a current phasor

$$\hat{I} = \frac{\hat{V}}{Z} = \frac{V_m\underline{/\alpha°}}{|Z|\underline{/\theta}} = I_m\underline{/\alpha - \theta} = I_m\underline{/\beta}$$

where it is noted that the impedance angle θ equals $\alpha - \beta$.

The instantaneous power will be

$$p = vi = V_m I_m \cos(\omega t + \alpha)\cos(\omega t + \beta)$$

or because $\alpha = \theta + \beta$,

$$p = vi = V_m I_m \cos(\omega t + \theta + \beta)\cos(\omega t + \beta) \tag{11.4}$$

This time, the trigonometric identity

$$\cos A \cos B = \frac{1}{2}[\cos(A + B) + \cos(A - B)]$$

is used, with $A = \omega t + \theta + \beta$ and $B = \omega t + \beta$, to yield

$$p = vi = \frac{1}{2} V_m I_m[\cos(2\omega t + 2\beta + \theta) + \cos\theta]$$

The average power over one cycle is

$$p_{av} = \frac{1}{2\pi}\int_0^{2\pi}\frac{1}{2}V_m I_m[\cos(2\omega t + 2\beta + \theta)\,d(\omega t) + \cos\theta\,dt(\omega t)]$$

$$= \frac{1}{4\pi}V_m I_m\left[\int_0^{2\pi}\cos(2\omega t + 2\beta + \theta)\,d(\omega t) + \int_0^{2\pi}\cos\theta\,d(\omega t)\right]$$

The first integral within the brackets is equal to zero because it represents the integral of a cosine over one cycle. Thus,

$$P_{av} = \frac{1}{4\pi} V_m I_m (\cos \theta)(\omega t)\Big|_0^{2\pi} = \frac{1}{2} V_m I_m \cos \theta = \frac{V_m}{\sqrt{2}} \frac{I_m}{\sqrt{2}} \cos \theta$$

or

$$P = p_{av} = VI \cos \theta \tag{11.5}$$

where V and I are rms quantities where P is called the real power and where $\cos \theta$ is called the *power factor*, PF:

$$\text{PF} \equiv \cos \theta \tag{11.6}$$

Observe that the power factor is the cosine of the network's impedance angle; and the power factor may range from 1, when the network is purely resistive ($\theta = 0°$) to 0, when the network is purely reactive ($\theta = \pm 90°$). This, of course, means that a *reactive* component of power must be provided by the source and that for *ac*, the power factor can be considered as the ratio of the average power to the *apparent power* delivered to the network elements:

$$\text{PF} = \frac{p_{av}}{VI}$$

SECTION 11.4   COMPLEX POWER

The instantaneous power flowing to a network with impedance $Z = |Z|\underline{/\theta}$ is given by eq. (11.4):

$$p = vi = V_m I_m \cos(\omega t + \theta + \beta) \cos(\omega t + \beta) \tag{11.4}$$

If the reference is set at the time when the current is at a positive maximum, this may be written as

$$p = vi = V_m I_m \cos(\omega t + \theta) \cos \omega t \tag{11.7}$$

Two trigonometric identities are needed in the development that follows. One of them has been employed to develop the average power,

$$\cos A \cos B = \frac{1}{2} \cos(A - B) + \frac{1}{2} \cos(A + B) \tag{a}$$

and the other is

$$\cos(A + B) = \cos A \cos B - \sin A \sin B \tag{b}$$

If identity (a) with $A = \omega t + \theta$ and $B = \omega t$ is used in eq. (11.7), the result is

$$p = \frac{V_m I_m}{2} [\cos \theta + \cos(2\omega t + \theta)]$$

The $\cos\theta$ within the brackets leads to the real power, which was developed in Section 11.3,

$$P = VI\cos\theta \tag{11.6}$$

where V and I are the rms values of the voltage and current. The second term within the brackets may be evaluated by resorting to identity (b), with $A = 2\omega t$ and $B = \theta$, so that

$$p = \frac{V_m I_m}{2}\cos\theta + \frac{V_m I_m}{2}(\cos\theta\cos 2\omega t - \sin\theta\sin 2\omega t)$$

and as a result,

$$p = VI\cos\theta + VI\cos\theta\cos 2\omega t - VI\sin\theta\sin 2\omega t \tag{11.8}$$

Here, the second two terms represent instantaneous power, which produces no effect on the average power because the integral of both $\cos 2\omega t$ and $\sin 2\omega t$ over one cycle is zero. Yet it is important to note that the coefficient of the sine term represents the *reactive power*,

$$Q = VI\sin\theta \tag{11.9}$$

where again V and I represent rms values and where the unit of reactive power is the var, which stands for *volt-amperes-reactive*.

Because energy can be neither created nor destroyed, the sum of P and Q, with proper account taken for the *quadrature* of Q, leads to a consideration of the apparent power:

$$S = P + jQ = VI\cos\theta + jVI\sin\theta$$

or

$$S = P + jQ = |S|\underline{/\theta} \tag{11.10}$$

Here, one may note that

$$P = \text{Re}(S) = \text{Re}(VI\underline{/\theta}) = VI\cos\theta \tag{11.11a}$$

$$Q = \text{Im}(S) = \text{Im}(VI\underline{/\theta}) = VI\sin\theta \tag{11.11b}$$

Moreover,

$$P = I^2 R = \frac{V^2}{R} \tag{11.12a}$$

$$Q = I^2 X = \frac{V^2}{X} \tag{11.12b}$$

and the magnitude of S is related to the magnitude of Z:

$$|S| = I^2|Z| = \frac{V^2}{|Z|} \tag{11.12c}$$

A simple relationship for the complex power in terms of the voltage and current phasors exists. For an impedance $Z = |Z|\underline{/\theta}$, let $\hat{V} = V\underline{/\alpha}$ and $\hat{I} = I\underline{/\beta}$, where V and

I are the rms values of the voltage and current. Then, because

$$P = VI \cos \theta$$

and

$$\cos \theta = \text{Re}(e^{j\theta})$$

P can be expressed as

$$P = \text{Re}(VIe^{j\theta})$$

But $\theta = \alpha - \beta$, so that

$$P = \text{Re}(VIe^{j(\alpha-\beta)}) = \text{Re}(Ve^{j\theta}Ie^{-j\beta})$$

Recall that the conjugate is indicated by an asterisk, and observe that $Ie^{-j\beta} = (Ie^{j\beta})^* = \hat{I}^*$. Thus,

$$P = \text{Re}(\hat{V}\hat{I}^*)$$

and because $P = \text{Re}(S)$, it is seen that the complex power may be represented by

$$S = \hat{V}\hat{I}^* \tag{11.13}$$

Again, note that θ is the power factor angle, which is equal to the impedance angle, and that the foregoing discussion illustrates the phenomenon of complex power. Observe that $\hat{S}$ is a complex quantity, and strictly speaking, one cannot describe the complex power by providing a single number in volt-amperes. One can, however, give the magnitude of the complex number by a single number.

■ EXAMPLE 11.1

For the phasor domain network of Fig. 11.1, find the real power, the apparent power, the magnitude of the apparent power, and the power factor.

Solution Three methods are available. The first involves the determination of Z:

$$Z = \frac{(40 - j30)(j50)}{40 - j30 + j50} = \frac{1500 + j2000}{40 + j20} = \frac{2500\underline{/53.13^\circ}}{44.72\underline{/26.57^\circ}}$$
$$= 55.90\underline{/26.57^\circ} = 50 + j25 \ \Omega$$

FIGURE 11.1 A network in the phasor domain

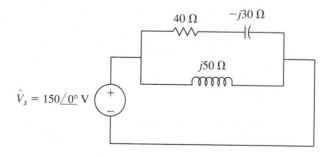

The current drawn by the network will be

$$\hat{I} = \frac{\hat{V}}{Z} = \frac{150\underline{/0°}}{55.90\underline{/26.57°}} = 2.68\underline{/-26.57°} \text{ A}$$

and the magnitude of the current phasor is 2.68 A.
   The real power is

$$P = I^2R = (2.68)^2(50) = 360 \ W$$

The reactive power is

$$Q = I^2X = (2.68)^2(25) = 180 \text{ VAR}$$

and the apparent power is

$$S = P + jQ = 360 + j180 = 402.49\underline{/26.57°} \text{ VA}$$

This makes the magnitude of the apparent power 402.49 VA and the power factor

$$PF = \cos 26.57° = 0.894$$

Notice that the power factor angle is equal to the impedance angle.
   The second method exploits the fact that the total power delivered by the source must be equal to the sum of the power drawn by the network elements. Let I_1 and I_2 be the magnitudes of the current phasors through $Z_1 = 40 + j30 = 50\underline{/36.87°} \ \Omega$ and $Z_2 = j50 = 50\underline{/90°} \ \Omega$, respectively. Then,

$$I_1 = \frac{150}{50} = 3 \text{ A} \quad \text{and} \quad I_2 = \frac{150}{50} = 3 \text{ A}$$

There is one component of real power,

$$P_R = I_1^2R = (3)^2(40) = 360 \text{ W}$$

There are two components of reactive power. For X_1, which is capacitive,

$$Q_1 = I_1^2X_1 = (3)^2(-30) = -270 \text{ VAR}$$

and for X_2, which is inductive,

$$Q_2 = I_2^2X_2 = (3)^2(50) = 450 \text{ VAR}$$

Thus,

$$P = 360 \ W$$

$$Q = Q_1 + Q_2 = -270 + 450 = 180 \text{ VAR}$$

which checks the previously calculated values.

The second method demonstrates how the capacitive reactance is handled because it is, in reality, a negative component of the total reactive power. Both methods are, of course, equivalent.

The third method works directly with the phasors $\hat{V}$ and $\hat{I}$. Here,

$$\hat{V} = 150\underline{/0°} \text{ V}$$

and $\hat{I}$ has been determined as

$$\hat{I} = 2.68\underline{/-26.57°} \text{ A}$$

This makes

$$S = \hat{V}\hat{I}^* = (150\underline{/0°})(2.68\underline{/26.57°}) = 402.49\underline{/26.57°}$$

Here, it is noted that the magnitude of the apparent power is

$$|S| = 402.49 \text{ VA}$$

the power factor is

$$\text{PF} = \cos 26.57° = 0.894$$

the real power is

$$P = 402.49 \cos 26.57° = 360 \text{ W}$$

and the reactive power is

$$Q = 402.49 \sin 26.57° = 180 \text{ VAR}$$ ■

EXERCISE 11.1

Determine the real power, the reactive power, the magnitude of the apparent power, and the power factor in the phasor domain network of Fig. 11.2.

Answer $P = 243.6 \text{ W}, Q = -25.2 \text{ VAR}, |S| = 244.9 \text{ VA}, \text{ and PF} = 0.995.$

FIGURE 11.2   A phasor domain network (Exercise 11.1)

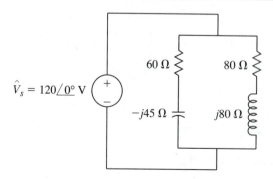

POWER FACTOR CORRECTION

A low power factor indicates a large power factor angle and a magnitude of apparent power delivered that is much greater than the real power drawn. This is depicted in Fig. 11.3; and in this situation, if S is the power delivered by a public utility and P is the power paid for by the consumer, a power factor correction charge may be incurred. Utilities will not deliver large quantities of volt-amperes to customers who pay for only a small quantity of watts without some form of penalty. This is an example of the *pay now or pay later* trick, where *pay now* refers to a small expenditure to correct the power factor.

For a highly inductive load, the power factor may be corrected by deliberately adding capacitive reactive power to reduce the power factor angle and increase the power factor. Notice that in this case, the power factor angle is positive and capacitive reactance is added. If the power factor angle is negative, the load is capacitive, and inductive reactance is added. Figure 11.3b describes a situation where the overall load is inductive with a positive power factor angle θ. Some authors refer to this case as one of a *lagging power factor*, because with an inductive load, the current *lags* the voltage. Indeed, this also leads to a consideration of a *leading power factor*, where the current *leads* the voltage. This leading and lagging terminology will not be used hereafter in this book.

The real power is designated as P, and the reactive power is

$$Q = P \tan \theta \tag{11.14}$$

If it is desired to correct the power factor to PF_d (d for *desired*) while the real power remains unchanged, then

$$\theta_d = \arccos PF_d \tag{11.15}$$

and the reactive power that yields the desired power factor will be

$$Q_d = P \tan \theta_d \tag{11.16}$$

(a) Real and apparent power for a load that is highly inductive and (b) the power components used to correct the power factor

FIGURE 11.3

The triangles in (b) are often referred to as power triangles.

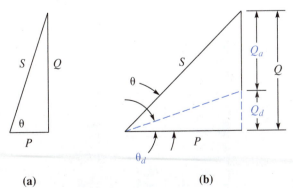

(a)   (b)

This would make the magnitude of the correction, Q_a (a for *added*)

$$Q_a = Q_d - Q \qquad\qquad (11.17)$$

The component to be installed to correct the power factor is derived from the appropriate reactance. For either a capacitive or an inductive correction,

$$X_a = \frac{V^2}{|Q_a|}$$

If the correction is capacitive,

$$C_a = \frac{1}{X_a \omega}$$

and if the correction is inductive,

$$L_a = \frac{X_a}{\omega}$$

These components are usually installed *across* or in parallel with the load.

■ EXAMPLE 11.2

The power distribution system in Fig. 11.4 operates at 60 Hz and consists of three loads. Find the total real and reactive power, the magnitude of the apparent power, and the correction necessary to make the power factor 0.950. Then, determine what single reactive component must be obtained in order to achieve the correction.

FIGURE 11.4   The power system for Example 11.2

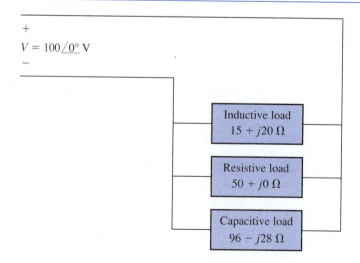

$+$

$V = 100\underline{/0°}$ V

$-$

Inductive load
$15 + j20\ \Omega$

Resistive load
$50 + j0\ \Omega$

Capacitive load
$96 - j28\ \Omega$

Solution   The three loads are designated by number,

$$Z_1 = 15 + j20 = 25\underline{/53.13°}\ \Omega$$

$$Z_2 = 50 + j0 = 50\underline{/0°}\ \Omega$$

$$Z_3 = 96 - j28 = 100\underline{/-16.26°}\ \Omega$$

so that

$$|Z_1| = 25\ \Omega \qquad |Z_2| = 50\ \Omega \qquad |Z_3| = 100\ \Omega$$

The rms currents through each of the loads will have magnitudes

$$I_1 = \frac{100}{25} = 4\ \text{A} \qquad I_2 = \frac{100}{50} = 2\ \text{A} \qquad I_3 = \frac{100}{100} = 1\ \text{A}$$

Because the power drawn must equal the power delivered,

$$P = P_1 + P_2 + P_3 = I_1^2 R_1 + I_2^2 R_2 + I_3^2 R_3$$
$$= (4)^2(15) + (2)^2(50) + (1)^2(96) = 240 + 200 + 96 = 536\ \text{W}$$

and this is the total real power.

The reactive power is determined in the same manner but with caution exercised in dealing with the capacitive load, where a minus sign is required:

$$Q = Q_1 + Q_2 + Q_3 = I^2 X_1 + I_2^2 X_2 + I_2^2 X_3$$
$$= (4)^2(20) + (2)^2(0) + (1)^2(-28) = 320 + 0 - 28 = 292\ \text{VAR}$$

and this is the total reactive power.

The apparent power is

$$S = 536 + j292 = 610.38\underline{/28.58°}\ \text{VA}$$

which makes the magnitude of the apparent power 610.38 VA and the power factor

$$\text{PF} = \cos 28.58° = 0.878$$

To determine the correction necessary to make the power factor 0.950, set the desired power factor to $\text{PF}_d = 0.950$ and use eqs. (11.15) through (11.17):

$$\theta_d = \arccos \text{PF}_d = \arccos 0.950 = 18.19°$$

$$\tan \theta_d = \tan 18.19° = 0.329$$

$$Q_d = P \tan \theta_d = 536(0.329) = 176.17\ \text{VAR}$$

$$Q_a = Q_d - Q = 176.17 - 292 = -115.83\ \text{VAR}$$

The minus sign indicates a capacitive correction of magnitude 115.83 VAR.

The actual capacitance required is determined from the capacitive reactance:

$$X_a = \frac{V^2}{Q_a} = \frac{100^2}{115.83} = 86.34\ \Omega$$

FIGURE 11.5

(a) A convenient tabular form for making complex power calculations and (b) the associated power triangles for all loads, including the capacitive load, for the correction of the power factor

Item	Load 1	Load 2	Load 3	Σ
$Z\,(\Omega)$	$15 + j20$	$50 + j0$	$96 - j28$	
$|Z|\,(\Omega)$	25	50	100	
$I = \dfrac{V}{|Z|}\,(A)$	4	2	1	
$R\,(\Omega)$	15	50	96	
$X\,(\Omega)$	20	0	-28	
$P = I^2R\,(W)$	240	200	96	536
$Q = I^2X\,(VAR)$	320	0	-28	292

(a)

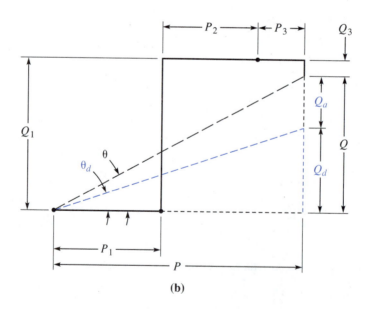

(b)

and because $X_{corr} = 1/\omega C$, the capacitance required will be

$$C = \frac{1}{\omega X_{corr}} = \frac{1}{2\pi(60)(86.34)} = 30.72\ \mu F$$

Figure 11.5 illustrates an expeditious calculation procedure in tabular form and an overall power triangle showing the pertinent quantities.

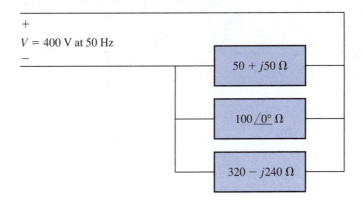

+

$V = 400$ V at 50 Hz

−

$50 + j50 \; \Omega$

$100 \underline{/0°} \; \Omega$

$320 - j240 \; \Omega$

EXERCISE 11.2

The power distribution system in Fig. 11.6 operates at 50 Hz and consists of three loads. Find the total real and reactive power, the magnitude of the apparent power, and the correction necessary to make the power factor 0.960. Then, determine what single reactive component must be obtained in order to achieve the correction.

Answer $P = 3520$ W, $Q = 1360$ VAR, $|S| = 3373.6$ VA, PF $= 0.933$, $X_a = 333.3$ var (capacitive), and $C = 6.63 \; \mu$F.

MAXIMUM POWER TRANSFER

SECTION 11.6

It was shown in Section 4.6 that for a resistive network with dc excitation, maximum power could be transferred from a network to a load if the load resistor were matched to the Thévenin equivalent resistance of the network. This is but a special case, and it is the intent here to develop a more general theory.

The network in Fig. 11.7 is in the phasor domain and shows a load impedance $Z_o = R_o + jX_o$ connected to the Thévenin equivalent of a network having a Thévenin

A network represented by its Thévenin equivalent and a load connected across its terminals

FIGURE 11.7

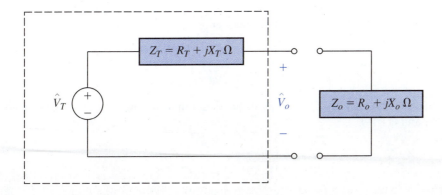

equivalent voltage $\hat{V}_T$ and a Thévenin equivalent impedance $Z_T = R_T + jX_T$. Let the load and Thévenin impedance be designated, respectively, by

$$Z_o = |Z_o|\underline{/\theta_o} = |Z_o|e^{j\theta_o} \tag{11.18}$$

$$Z_T = |Z_o|\underline{/\theta_T} = |Z_T|e^{j\theta_o} \tag{11.19}$$

where θ_o and θ_T may take on positive or negative values. If the Thévenin equivalent voltage is designated as

$$\hat{V}_T = V_T\underline{/\alpha} = V_T e^{j\alpha}$$

then as indicated in Section 11.4, the real power to the load can be determined from the real part of the product of the load voltage and the complex conjugate of the load current,

$$P_o = \mathrm{Re}(\hat{V}_o \hat{I}*)$$

where the load voltage can be found by considering a voltage division,

$$\hat{V}_o = \frac{Z_o \hat{V}_T}{Z_o + Z_T}$$

and where

$$\hat{I}* = \left(\frac{\hat{V}_T}{Z_o + Z_T}\right)^* = \frac{\hat{V}_T^*}{Z_o^* + Z_T^*}$$

Thus,

$$P_o = \mathrm{Re}\left[\left(\frac{Z_o \hat{V}_T}{Z_o + Z_T}\right)\left(\frac{\hat{V}_T^*}{Z_o^* + Z_T^*}\right)\right]$$

If $\hat{V}_T = V_T\underline{/\alpha}$, then

$$\hat{V}_T \hat{V}_T^* = V_T^2$$

so that

$$Z_T Z_T^* = |Z_T|^2$$

The power in the load can be represented as

$$P = \mathrm{Re}\left(\frac{V_T^2 |Z_o| e^{j\theta_o}}{|Z_T|^2 + |Z_o|^2 + Z_T Z_o^* + Z_T^* Z_o}\right)$$

Look at the term $Z_T Z_o^* + Z_T^* Z_o$ in the denominator and expand it by using eqs. (11.18) and (11.19):

$$Z_T Z_o^* + Z_T^* Z_o = |Z_T||Z_o|\underline{/\theta_T - \theta_o} + |Z_T||Z_o|\underline{/\theta_o - \theta_T}$$
$$= |Z_T||Z_o|[e^{j(\theta_o - \theta_T)} + e^{-j(\theta_o - \theta_T)}]$$

The bracketed term is $2\cos(\theta_o - \theta_T)$, and therefore,

$$P_o = \text{Re}\left[\frac{V_T^2 |Z_o| e^{j\theta_o}}{|Z_T|^2 + |Z_o|^2 + 2|Z_T||Z_o|\cos(\theta_o - \theta_T)}\right]$$

or

$$P_0 = \frac{V_T^2 |Z_o| \cos\theta_o}{|Z_T^2| + |Z_o|^2 + 2|Z_T||Z_o|\cos(\theta_o - \theta_T)} \qquad (11.20)$$

For a given network operating at a prescribed voltage, with $Z_T = |Z_T|\underline{/\theta_T}$, the real power drawn by any load of impedance $Z_o = |Z_o|\underline{/\theta_o}$ is a function of just two variables, $|Z_o|$ and θ_o. If the power is to be a maximum, the network analyst must select $|Z_o|$ and θ_o, and there are three alternatives:

1. Both $|Z_o|$ and θ_o are at the designer's discretion, and both are allowed to vary in any manner to achieve the desired result.
2. The angle θ_o is fixed but the magnitude $|Z_o|$ is allowed to vary. For example, the analyst may select and fix $\theta_o = 0°$. This requires that the load be resistive (Z is entirely real).
3. The magnitude of the load impedance $|Z_o|$ can be fixed but the impedance angle θ_o is allowed to vary.

The three cases all apply to eq. (11.20), and the criteria are derived by the standard method of determining a maximum or minimum: finding the point where the derivative or derivatives of eq. (11.20) vanish. The mathematics involved, while not difficult, is somewhat tedious and space consuming, and only the results are described in what follows.

If $|Z_o|$ and θ_o are allowed to vary in any manner whatsoever, eq. (11.20) is used to determine the conditions under which $\partial P_o/\partial|Z_o| = \partial P_o/\partial\theta_o = 0$. In this case, it is found that $|Z_o| = |Z_T|$ and $\theta_o = -\theta_T$, or

$$Z_o = Z_T^* \qquad (11.21)$$

If θ_o is fixed and only $|Z_o|$ can vary, then eq. (11.20) is used to find $\partial P_o/\partial|Z_o| = 0$, with the result that

$$|Z_o| = |Z_T| \qquad (11.22)$$

which indicates that for maximum power to the load, the magnitude of the load impedance must equal the magnitude of the network Thévenin equivalent impedance.

Finally, if θ_o can be allowed to vary and the magnitude of Z_o is fixed, then eq. (11.20) is used to find $\partial P_o/\partial\theta_o = 0$. In this case, the maximum power transfer to the load is seen to occur when

$$\theta_o = \arcsin\left(-\frac{2|Z_o||Z_T|\sin\theta_T}{|Z_o|^2 + |Z_T|^2}\right) \qquad (11.23)$$

Before proceeding to an example, consider the case depicted in Fig. 11.7, where both R_o and X_o are allowed to vary. Admittedly, what will now transpire is not as elegant as the optimizing process that is applied to eq. (11.20), but perhaps it will prove helpful.

The magnitude of the current in the network in Fig. 11.7 is

$$I = \frac{V_T}{\sqrt{(R_T + R_o)^2 + (X_T + X_o)^2}}$$

and the real power drawn from the line will be $P_o = I^2 R_o$, or

$$P_o = \frac{V_T^2 R_o}{(R_T + R_o)^2 + (X_T + X_o)^2}$$

The desire to achieve maximum P_o begins to be fullfilled when $X_T = -X_o$, so that the denominator is as small as possible for any R_T and R_o. Thus, with $X_T = -X_o$,

$$P_o = \frac{V_T^2 R_o}{(R_T + R_o)^2}$$

Now, take the derivative $\partial P_o / \partial R_o$ and determine where it vanishes:

$$\frac{\partial P_o}{\partial R_o} = \frac{\partial}{\partial R_o}\left[\frac{V_T^2 R_o}{(R_T + R_o)^2}\right] = \frac{\partial}{\partial R_o}\left[V_T^2 R_o(R_T + R_o)^{-2}\right] = 0$$

or

$$V_T^2(R_T + R_o)^{-2} - 2V_T^2 R_o(R_T + R_o)^{-3} = 0$$

The V_T^2 terms may be canceled, and after multiplication by $(R_T + R_o)^3$,

$$R_T + R_o = 2R_o \quad \text{or} \quad R_T = R_o$$

Thus, when both R_o and X_o can be varied, maximum power will be transferred to the load when the load impedance is the complex conjugate of the Thévenin equivalent impedance. This, of course, is the statement of eq. (11.21).

■ **EXAMPLE 11.3**

A load impedance is to be placed across terminals $a-b$ in the network shown in Fig. 11.8a. What is the value of Z_o if its elements are unrestricted, if it is a single resistor, or if the magnitude of Z_o must be 20 Ω but its angle is adjustable? The angular frequency is $\omega = 50,000$ rad/s.

Solution   The first step is to obtain the Thévenin equivalent for the network terminated at points a and b. The phasor domain network is displayed in Fig. 11.8b, where

$$X_L = \omega L = 50,000(0.0008) = 40 \ \Omega$$

and the two X_c's are

$$X_{C1} = \frac{1}{\omega C_1} = \frac{1}{50,000(1 \times 10^{-6})} = 20 \ \Omega$$

$$X_{C2} = \frac{1}{\omega C_2} = \frac{1}{50,000(8.33 \times 10^{-7})} = 24 \ \Omega$$

(a) A network for which the load Z_o is to be selected for maximum power transfer, (b) the phasor domain network, (c) the network showing the steps involved in the determination of the Thévenin equivalent impedance, and (d) the Thévenin equivalent network with the load impedance Z_o connected

FIGURE 11.8

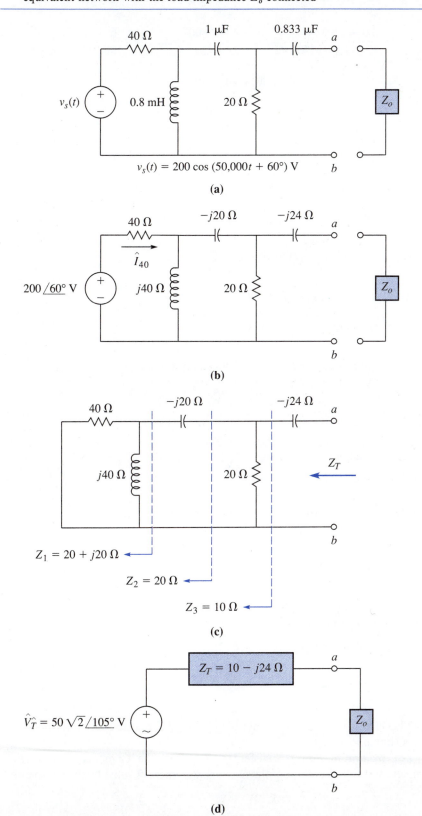

(a)

$v_s(t) = 200 \cos (50{,}000t + 60°) \text{ V}$

(b)

(c)

$Z_1 = 20 + j20 \ \Omega$

$Z_2 = 20 \ \Omega$

$Z_3 = 10 \ \Omega$

$\hat{V}_T = 50\sqrt{2}\,\underline{/105°}\text{ V}$

$Z_T = 10 - j24 \ \Omega$

(d)

The Thévenin equivalent voltage is the voltage across the 20-Ω resistor in Fig. 11.8b. The impedance presented to the $200\underline{/60°}$ V source is

$$Z_{in} = \frac{(20 - j20)(j40)}{20 - j20 + j40} + 40 = \frac{800 + j800}{20 + j20} + 40$$

$$= \frac{800\sqrt{2}\underline{/45°}}{20\sqrt{2}\underline{/45°}} + 40 = 40 + 40 = 80 \ \Omega$$

This yields a current through the 40-Ω resistor of

$$\hat{I}_{40} = \frac{200\underline{/60°}}{80} = \frac{5}{2}\underline{/60°} \ A$$

and a voltage

$$\hat{V}_{40} = 40\left(\frac{5}{2}\underline{/60°}\right) = 100\underline{/60°} \ V$$

By KVL, the voltage across the j40-Ω inductor is

$$\hat{V}_{j40} = 200\underline{/60°} - 100\underline{/60°} = 100\underline{/60°} \ V$$

This is the voltage across the (20 − j20)-Ω combination. Voltage division can then be employed to find $\hat{V}_T$:

$$\hat{V}_T = \left(\frac{20}{20 - j20}\right)\hat{V}_{j40} = \frac{20}{20\sqrt{2}\underline{/-45°}} \ 100\underline{/60°}$$

$$= \frac{2000\underline{/60°}}{20\sqrt{2}\underline{/-45°}} = 50\sqrt{2}\underline{/105°} \ V$$

The Thévenin equivalent impedance may be determined by considering the impedance looking into terminals a–b after the voltage source is replaced by a short circuit, as indicated in Fig. 11.8c. Here,

$$Z_1 = \frac{40(j40)}{40 + j40} = \frac{1600\underline{/90°}}{40\sqrt{2}\underline{/45°}} = 20\sqrt{2}\underline{/45°} = 20 + j20 \ \Omega$$

$$Z_2 = 20 + j20 - j20 = 20 \ \Omega$$

$$Z_3 = \frac{20(20)}{20 + 20} = \frac{400}{40} = 10 \ \Omega$$

$$Z_T = 10 - j24 \ \Omega$$

The Thévenin equivalent impedance with the load connected across terminals a–b is shown in Fig. 11.8d.

For maximum power transfer to Z_o when the elements of Z_o are completely at the discretion of the network designer, Z_o must be the complex conjugate of Z_T:

$$Z_o = Z_T^* = 10 + j24 \ \Omega$$

If Z_o is to be a single resistor R_o, then the magnitude of $Z_o = R_o$ must be equal to the magnitude of Z_T. Here,

$$Z_T = 10 - j24 = 26\underline{/-67.38°}$$

so that

$$R_o = |Z_o| = 26 \ \Omega$$

If the magnitude of Z_o must be 20 Ω but the angle is adjustable, the required angle is calculated from eq. (11.23):

$$\theta_o = \arcsin\left(-\frac{2|Z_o||Z_T|}{|Z_o|^2 + |Z_T|^2} \sin \theta_T\right)$$

$$= \arcsin\left[-\frac{2(20)(26)}{(20)^2 + (26)^2} \sin(-67.38°)\right] = \arcsin 0.892 = 63.15°$$

This makes Z_o

$$Z_o = 20\underline{/63.15°} = 9.03 + j17.84 \ \Omega$$

The power in each case is quickly determined. When $Z_o = 10 + j24 \ \Omega$, the total impedance is

$$Z = Z_T + Z_o = 10 - j24 + 10 + j24 = 20 \ \Omega$$

and

$$I = \frac{V_T}{Z} = \frac{50\sqrt{2}}{20} = \frac{5\sqrt{2}}{2} \ A$$

Thus,

$$P_o = I^2R_o = \left(\frac{5\sqrt{2}}{2}\right)^2 (10) = 125 \ W$$

When $Z_o = 26 \ \Omega$, the total impedance is

$$Z = Z_T + Z_o = 10 - j24 + 26 = 36 - j24 = 43.27\underline{/-33.69°} \ \Omega$$

and

$$I = \frac{V_T}{|Z|} = \frac{50\sqrt{2}}{43.27} = 1.63 \ A$$

which yields

$$P_o = I^2R_o = (1.63)^2(26) = 69.43 \ W$$

FIGURE 11.9   Network (Exercise 11.3)

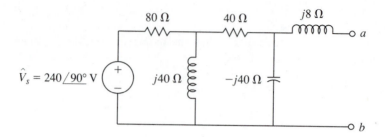

Finally, when $Z_o = 9.03 + j17.84 \ \Omega$, the total impedance is

$$Z = Z_T + Z_o = 10 - j24 + 9.03 + j17.84 = 19.03 - j6.16$$
$$= 20.00 \underline{/-17.94°} \ \Omega$$

Here,

$$I = \frac{V_T}{|Z|} = \frac{50\sqrt{2}}{20} = \frac{5}{2}\sqrt{2} \ \text{A}$$

$$P_o = I^2 R_o = \left(\frac{5\sqrt{2}}{2}\right)^2 (9.03) = 112.88 \ \text{W}$$

EXERCISE 11.3

In the phasor domain network shown in Fig. 11.9, what load should be placed across terminals a and b to ensure that maximum power is transferred to the load? Determine the value of this power.

Answer $Z_o = 28 - j28 \ \Omega$ and $P_o = 51.43 \ \text{W}$.

SECTION 11.7   **TELLEGEN'S THEOREM REVISITED**

Tellegen's theorem was introduced in Section 4.4. It is summarized by eq. (4.7).

$$\sum_{k=1}^{b} v_k j_k = 0 \tag{4.9}$$

and it may be extended to consider alternating current using phasor notation,

$$\sum_{k=1}^{b} \hat{V}_k \hat{J}_k = 0 \tag{11.24}$$

This says that the products of all branch voltages and branch currents must total zero. This may also be represented by the vector scalar product $\mathbf{V} \cdot \mathbf{I}$

$$\hat{\mathbf{V}} \cdot \hat{\mathbf{I}} = \hat{\mathbf{V}}^T \hat{\mathbf{J}} = \hat{\mathbf{J}}^T \hat{\mathbf{V}} = 0$$

where $\hat{\mathbf{V}}$ and $\hat{\mathbf{J}}$ are branch voltage and branch current vectors, respectively.

■ **EXAMPLE 11.4**

In Section 10.2, the network shown in Fig. 10.1a was considered, and, for its six branches, the branch current and branch voltage vectors in phasor form were determined as

$$\hat{\mathbf{J}} = \begin{bmatrix} 4\underline{/-90°} \\ 4\sqrt{2}\underline{/135°} \\ 4\underline{/0°} \\ 6.32\underline{/-18.43°} \\ 2\sqrt{2}\underline{/135°} \\ 2\sqrt{2}\underline{/135°} \end{bmatrix} \text{V} \quad \text{and} \quad \hat{\mathbf{V}} = \begin{bmatrix} 240\sqrt{2}\underline{/45°} \\ 240\sqrt{2}\underline{/45°} \\ 160\underline{/0°} \\ 252.98\underline{/71.57°} \\ 160\sqrt{2}\underline{/45°} \\ 80\sqrt{2}\underline{/135°} \end{bmatrix} \text{A}$$

Verify that Tellegen's theorem holds.

Solution   Although the vectors $\hat{\mathbf{V}}$ and $\hat{\mathbf{J}}$ can definitely be certified to be correct by a variety at methods, an additional warm feeling will occur if Tellegen's theorem holds. All that is necessary is to show that $\hat{\mathbf{V}}^T\hat{\mathbf{J}} = \hat{\mathbf{J}}^T\hat{\mathbf{V}} = 0$. Consider $\hat{\mathbf{V}} \cdot \hat{\mathbf{J}}$ and obtain

$$\begin{aligned}
\hat{\mathbf{V}}^T\hat{\mathbf{J}} &= (4\underline{/-90°})(240\sqrt{2}\underline{/45°}) + (4\sqrt{2}\underline{/135°})(240\sqrt{2}\underline{/45°}) \\
&\quad + (4\underline{/0°})(160\underline{/0°}) + (6.32\underline{/-18.43°})(252.98\underline{/71.57°}) \\
&\quad + (2\sqrt{2}\underline{/135°})(160\sqrt{2}\underline{/45°}) + (2\sqrt{2}\underline{/135°})(80\sqrt{2}\underline{/135°}) \\
&= 960\sqrt{2}\underline{/-45°} + 1920\underline{/180°} + 640\underline{/0°} \\
&\quad + 1600\underline{/53.13°} + 640\underline{/180°} + 320\underline{/270°} \\
&= 960 - j960 - 1920 + j0 + 640 + j0 \\
&\quad + 960 + j1280 - 640 + j0 + 0 - j320 \\
&= 0
\end{aligned}$$

This confirms that $\hat{\mathbf{V}}^T\hat{\mathbf{J}} = 0$.   ■

CHAPTER 11

SUMMARY

- Effective or *rms* values of voltage and current are used in ac power calculations.

- Resistors dissipate real power; the power transferred to and from inductors and capacitors is called reactive power.

- The real power dissipated at a particular sinusoidal frequency by a general impedance $Z = |Z|\underline{/\theta}$ is

$$P = VI \cos \theta \text{ W}$$

where $\cos \theta$ is called the power factor,

$$PF = \cos \theta$$

- Reactive power flows to, from, and between inductive and capacitive reactances and is given by

$$Q = VI \sin \theta \text{ VAR}$$

- Complex power is represented by an apparent power,

$$S = P + jQ = |S|\underline{/\theta} \text{ VA}$$

and in an impedance $Z = |Z|\underline{/\theta}$, the apparent power is also given by

$$S = \hat{V}\hat{I}^*$$

- In a power system exhibiting a low power factor, the power factor may be corrected.
  —If the power factor is inductive, the power factor may be corrected by adding capacitive reactance.
  —If the power factor is capacitive, the power factor may be corrected by adding inductive reactance.

- For maximum power transfer to a load impedance,

$$Z_o = |Z_o|\underline{/\theta_o}$$

Both $|Z_o|$ and θ_o must be selected in accordance with certain well-defined criteria after the formation of a Thévenin equivalent network having Thévenin equivalent impedance $Z_T = R_T + jX_T$.
  —If both $|Z_o|$ and θ_o are at the designer's discretion, and if both are allowed to vary in any manner, select

$$Z_o = Z_T^*$$

  —If the angle θ_o is fixed but the magnitude $|Z_o|$ is allowed to vary, select

$$|Z_o| = |Z_T|$$

  —If the magnitude of the load impedance $|Z_o|$ is fixed but the impedance angle θ_o is allowed to vary, select

$$\theta_o = \arcsin\left(-\frac{2|Z_o||Z_T| \sin \theta_T}{|Z_o|^2 + |Z_T|^2} \right)$$

Additional Readings

Blackwell, W.A., and L.L. Grigsby. *Introductory Network Theory.* Boston; PWS Engineering, 1985, pp. 337–345, 372–375.

Bobrow, L.S. *Elementary Linear Circuit Analysis.* 2d ed. New York: Holt, Rinehart and Winston, 1987, pp. 399–410.

Del Toro, V. *Engineering Circuits.* Englewood Cliffs, N.J.: Prentice-Hall, 1987, pp. 190–192.

Dorf, R.C. *Introduction to Electric Circuits.* New York: Wiley, 1989, pp. 380–391, 394–400.

Hayt, W.H., Jr., and J.E. Kemmerly. *Engineering Circuit Analysis*. 4th ed. New York: McGraw-Hill, 1986, pp. 284–293, 299–304.

Irwin, J.D. *Basic Engineering Circuit Analysis*. 3d ed. New York: Macmillan, 1989, pp. 444–458, 462–474, 532, 571.

Johnson, D.E., J.L. Hilburn, and J.R. Johnson. *Basic Electric Circuit Analysis*. 4th ed. Englewood Cliffs, N.J.: Prentice-Hall, 1989, pp. 371–381, 384–392.

Karni, S. *Applied Circuit Analysis*. New York: Wiley, 1988, pp. 324–333, 337–347.

Madhu, S. *Linear Circuit Analysis*. Englewood Cliffs, N.J.: Prentice-Hall, 1988, pp. 326–328, 381–391, 437–443.

Nilsson, J.W. *Electric Circuits*. 3d ed. Reading, Mass.: Addison-Wesley, 1990, pp. 367–373, 377–387, 391–397.

Paul, C.R. *Analysis of Linear Circuits*. New York: McGraw-Hill, 1989, pp. 284–294, 305–315.

CHAPTER 11	PROBLEMS

Note: In all problems, rms values of voltage and current are given, unless stated otherwise.

Section 11.3

11.1 Determine the average power and the power factor for an *RLC* series network operating at 400 Hz with $R = 100\ \Omega$, $L = 39.79$ mH, and $C = 15.92\ \mu F$ if the network is subjected to a sinusoidal voltage with a maximum amplitude of 180 V.

11.2 Determine the average power and the power factor for an *RLC* parallel network operating at 250 Hz with $R = 50\ \Omega$, $L = 50.93$ mH, and $C = 21.22\ \mu F$ if the network is subjected to a sinusoidal current with a maximum amplitude of 16 A.

11.3 Two impedances, $Z_1 = 40\sqrt{2}\underline{/-45^\circ}\ \Omega$ and $Z_2 = 80 - j60\ \Omega$, are connected in parallel, and the combination is connected in series with a 24-Ω resistor. If the combination is connected across a 120-V source, how much power is drawn and what is the power factor?

11.4 Two impedances, $Z_1 = 25\underline{/53.13^\circ}\ \Omega$ and $Z_2 = 39\underline{/67.38^\circ}\ \Omega$, are connected in parallel, and the combination is connected in series with $Z_3 = 17\underline{/28.07^\circ}\ \Omega$. If the combination is connected across a 20-A current source, how much power is drawn and what is the power factor?

11.5 An *RLC* series network is subjected to a voltage source, $V = V_m \cos \omega t$ volts, where $V_m = 200$ V and $\omega = 400$ rad/s. If $R = 40\ \Omega$, $L = 387.5$ mH, and $P = 640$ W, and if an impedance meter measures the impedance angle as 36.87°, what is the value of C?

Section 11.4

11.6 Determine the total power dissipated by all of the resistors in the phasor domain network of Fig. P11.1.

Figure P11.1

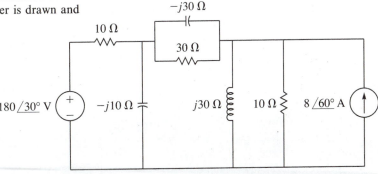

11.7 Find the complex power delivered to the phasor domain network in Fig. P11.2.

Figure P11.2

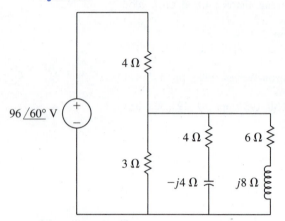

11.8 What is the reactive power in the $j30 \, \Omega$ inductor and the $-j60 \, \Omega$ capacitor in the phasor domain network in Fig. P11.3?

Figure P11.3

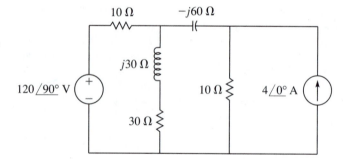

11.9 How much power is dissipated by the 30-Ω resistor in the phasor domain network of Fig. P11.3?

11.10 What is the complex power delivered to the phasor domain network of Fig. P11.4?

Figure P11.4

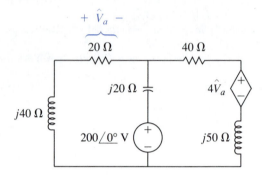

11.11 Determine the reactive power for each inductor and capacitor in the phasor domain network of Fig. P11.4.

11.12 What is the complex power delivered to the phasor domain network of Fig. P11.5?

Figure P11.5

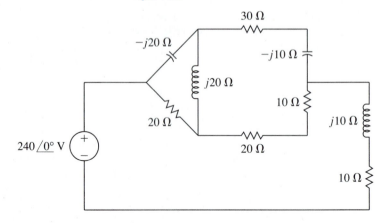

11.13 What is the complex power delivered to the network of Fig. P11.6?

Figure P11.6

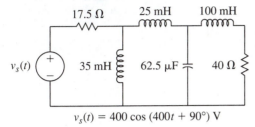

$$v_s(t) = 400 \cos (400t + 90°) \text{ V}$$

11.14 What is the complex power delivered to the network of Fig. P11.7?

Figure P11.7

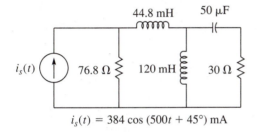

$$i_s(t) = 384 \cos (500t + 45°) \text{ mA}$$

11.15 What is the complex power delivered to the network of Fig. P11.8?

Figure P11.8

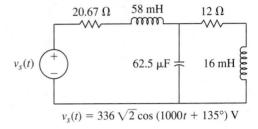

$$v_s(t) = 336 \sqrt{2} \cos (1000t + 135°) \text{ V}$$

11.16 What is the complex power delivered to the network of Fig. P11.9?

Figure P11.9

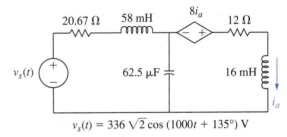

$$v_s(t) = 336 \sqrt{2} \cos (1000t + 135°) \text{ V}$$

11.17 What is the complex power delivered to the network of Fig. P11.10?

Figure P11.10

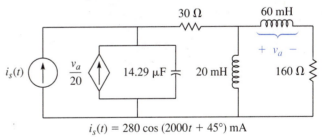

$$i_s(t) = 280 \cos (2000t + 45°) \text{ mA}$$

Section 11.5

11.18 For the power system shown in Fig. P11.11, determine the following:

a. The real power

b. The reactive power

c. The magnitude of the apparent power

d. The power factor

e. The correction necessary to make the power factor 0.950

f. The component necessary to achieve the correction in part e

Figure P11.11

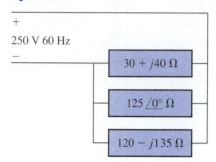

+

250 V 60 Hz

−

$30 + j40\ \Omega$

$125\ \underline{/0°}\ \Omega$

$120 - j135\ \Omega$

11.19 For the power system shown in Fig. P11.12 determine the following:

a. The real power

b. The reactive power

c. The magnitude of the apparent power

d. The power factor

e. The correction necessary to make the power factor 0.935

f. The component necessary to achieve the correction in part e

Figure P11.12

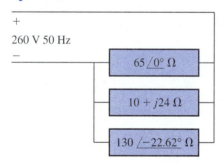

+

260 V 50 Hz

−

$65\ \underline{/0°}\ \Omega$

$10 + j24\ \Omega$

$130\ \underline{/-22.62°}\ \Omega$

11.20 For the power system shown in Fig. P11.13, determine the following:

a. The real power

b. The reactive power

c. The magnitude of the apparent power

d. The power factor

e. The correction necessary to make the power factor 0.920

f. The component necessary to achieve the correction in part e

Figure P11.13

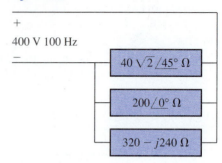

+

400 V 100 Hz

−

$40\ \sqrt{2}\ \underline{/45°}\ \Omega$

$200\ \underline{/0°}\ \Omega$

$320 - j240\ \Omega$

11.21 In the network of Fig. P11.14, what elements should be placed across terminals a–b to make the power factor unity?

Figure P11.14

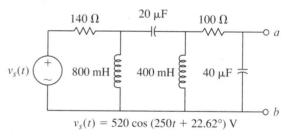

$v_s(t) = 520 \cos(250t + 22.62°)\ V$

11.22 In the network of Fig. P11.15, what elements should be placed across terminals a–b to make the power factor unity?

Figure P11.15

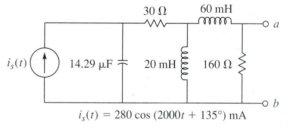

$i_s(t) = 280 \cos(2000t + 135°)\ mA$

11.23 In the phasor domain network of Fig. P11.16, what phasor domain elements should be placed across terminals *a–b* to make the power factor unity? Here $f = 100$ Hz.

Figure P11.16

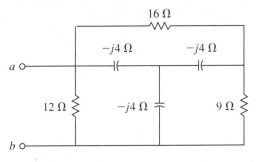

11.24 In the phasor domain network of Fig. P11.18, what phasor domain elements should be placed across terminals *a–b* to make the power factor unity? Here $f = 1000$ Hz.

Figure P11.17

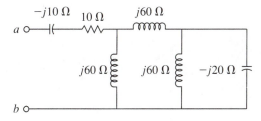

11.25 In the phasor domain network of Fig. 11.18, what phasor domain elements should be placed across the terminals *a–b* to make the power factor unity? Here $f = 1200$ Hz.

Figure P11.18

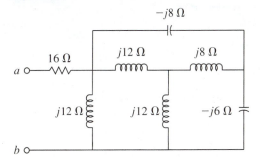

Section 11.6

11.26 If the designer has control over the specification of *R* and *X* as a load, what load should be placed across terminals *a–b* in the phasor domain network of Fig. P11.19 in order to make the power transferred to the load a maximum?

Figure P11.19

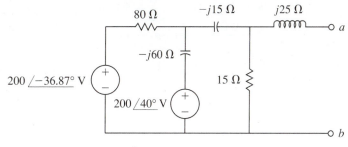

11.27 If the load impedance is to be completely resistive, what load should be placed across terminals a–b in the phasor domain network of Fig. P11.20 in order to make the power transferred to the load a maximum?

Figure P11.20

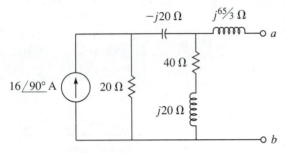

11.28 If the magnitude of the load impedance is to be 40 Ω, what resistance and reactance should be placed across terminals a–b in the phasor domain network of Fig. P11.21 in order to make the power transferred to the load a maximum?

Figure P11.21

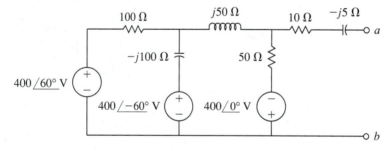

THREE-PHASE POWER SYSTEMS

OBJECTIVES

The objectives of this chapter are to:

- Consider the single-phase, three-wire system used in residential applications.

- Develop the equations that relate three-phase voltages, currents, and power.

- Discuss the wye and delta connections between three-phase generators and load.

- Provide examples of power computation for four types of generator-load connections.

- Consider the case of an unbalanced system.

- Discuss the measurement of three-phase power.

INTRODUCTION   SECTION 12.1

Alternating current, because of the ease with which it can be transformed from low to high (or, indeed, high to low) voltage, is used for the generation and transmission of electric power at high efficiency and reasonable cost. The frequency of 60 Hz in North America and 50 Hz elsewhere was selected as a compromise between the economy of generator use and the fact the inductive reactance of the transmission line tends to increase as the frequency increases. Efficient use of the windings of the synchronous generator is the reason for the use of three-phase systems, and these systems have the advantage of providing a steady rather than a pulsating output.

In the electrical systems discussed thus far, a load has been a network or element to which two connections have been made. Such a network is termed a *single-phase* network, and it is connected to a single-phase source via a single-phase *transmission line.*

Of more practical importance in the arena of power generation and distribution is the *polyphase* power system and the polyphase network. Here, *poly* means more than one, and the three- and four-wire, *three-phase* system is treated here in some detail because the use of the three-phase power system is so prevalent in actual practice.

A *balanced three-phase system* is one in which three voltages are generated. All are equal in magnitude, and each is separated in *phase angle* by 120°. The balanced voltages are connected across equal total impedances (the sum of the load and transmission line impedances). In this case, *balanced* currents flow. Balanced three-phase systems can be analyzed expeditiously by working with any one of the phases and treating it as a single-phase network.

Three-phase systems are widely used because three-phase windings, equally spaced around the generator shaft, make more efficient use of the generator's magnetic material and magnetic paths. Thus, three-phase machines are easier to start and the power flow tends to be more steady.

SECTION 12.2   THE SINGLE-PHASE, THREE-WIRE SYSTEM

Figure 12.1a shows a single-phase, three-wire source, where the three wires terminate at points that bear the labels a, n, and b. The use of the terminology *single-phase* is well-conceived; both equal-phase voltages with magnitude V (either rms or peak-to-peak) have the same phase angle, ϕ, Thus, in Fig. 12.1b,

$$\hat{V}_{an} = V\underline{/\phi} \tag{12.1a}$$

$$\hat{V}_{nb} = V\underline{/\phi} \tag{12.1b}$$

FIGURE 12.1   (a) A generating system (the box) with two voltage sources and three wires coming out of the box and (b) the voltage sources and the three wires, along with two impedances, of the system

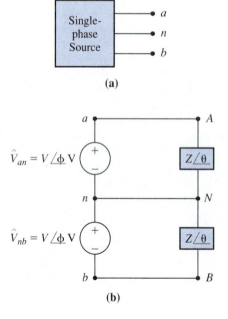

(a)

(b)

A single-phase, three-wire system with two identical loads

FIGURE 12.2

The nomenclature is shown for the line and phase currents.

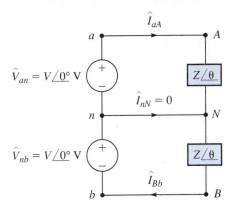

and by KVL,

$$\hat{V}_{ab} = \hat{V}_{an} + \hat{V}_{nb} = V\underline{/\phi} + V\underline{/\phi} = 2V\underline{/\phi} \tag{12.2}$$

Observe that the double-subscript notation indicates that the pair an and nb mean that $\hat{V}_{an}$ and $\hat{V}_{nb}$ are, respectively, the voltages of point a with respect to point n and the voltage of point n with respect to point b. In Fig. 12.1, the point n is referred to as the *neutral* point, or merely the *neutral*, and this double-subscript notation will be used throughout this chapter.

If a pair of identical loads $Z\underline{/\theta}$ are connected between points A, N, and B, as in Fig. 12.2, where $\hat{V}_{an} = \hat{V}_{nb} = \hat{V}\underline{/0°}$, then because of the short circuit between n and N (points n and N are at the same potential),

$$\hat{I}_{aA} = \hat{I}_{AN} = \frac{\hat{V}_{an}}{|Z|\underline{/\theta}} = \frac{V\underline{/0°}}{|Z|\underline{/\theta}} = I\underline{/-\theta} \tag{12.3a}$$

$$\hat{I}_{Bb} = \hat{I}_{NB} = \frac{\hat{V}_{nb}}{|Z|\underline{/\theta}} = \frac{V\underline{/0°}}{|Z|\underline{/\theta}} = I\underline{/-\theta} \tag{12.3b}$$

Then by KCL at point N,

$$\hat{I}_{nN} = \hat{I}_{NA} + \hat{I}_{NB} = \hat{I}_{Aa} + \hat{I}_{Bb} = -\hat{I}_{aA} + \hat{I}_{Bb} = 0 \tag{12.4}$$

which shows that when identical loads are employed, no current flows in the neutral wire. Indeed, in Fig. 12.3, where Z_s, Z_l, and Z_o represent source, transmission line, and load impedances, the system is balanced if $Z_{s,a} = Z_{s,b}$, $Z_{l,a} = Z_{l,b}$, and $Z_{o,a} = Z_{o,b}$. In this case, no current flows in the neutral wire, even though it may possess an impedance $Z_{l,n}$.

Analysis of a balanced, single-phase, three-wire system is quite straightforward. But, unfortunately, in household (residential) systems, loads are rarely balanced; and the analyst is confronted with a problem in node, mesh, or loop analysis. Example 12.1 shows how such an analysis is handled.

FIGURE 12.3   The system of Fig. 12.2 with source, line, and load impedances

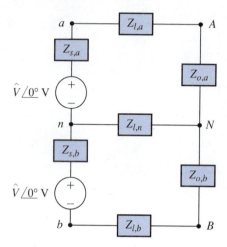

■ EXAMPLE 12.1

Consider the single-phase, three-wire system shown in Fig. 12.4a. The reactive load between A and B might represent an induction motor driving an air-conditioning machine, and the loads between A and N and B and N might represent lighting loads. What is the power drawn by the system?

Solution   The system is redrawn in Fig. 12.4c with three loop currents, which have been obtained from the tree shown in Fig. 12.4b. Three independent loop equations may therefore be written:

$$
\begin{aligned}
60\hat{I}_1 \;\;\; -\; 2\hat{I}_2 \;\;\;\;\;\;\;\;\;\;\; +\; \hat{I}_3 &= 115\underline{/0^\circ}\\
-2\hat{I}_1 +\; 100\hat{I}_2 \;\;\;\;\;\;\;\; +\; \hat{I}_3 &= 115\underline{/0^\circ}\\
\hat{I}_1 \;\;\;\; +\; \hat{I}_2 + (65 + j65)\hat{I}_3 &= 230\underline{/0^\circ}
\end{aligned}
$$

These may be solved in any number of ways, and the reader may verify that the solution vector in both rectangular and polar form is

$$
\hat{\mathbf{I}} =
\begin{bmatrix}
1.93\underline{/0.87^\circ}\\
1.17\underline{/0.98^\circ}\\
2.47\underline{/-45^\circ}
\end{bmatrix}
=
\begin{bmatrix}
1.93 + j0.03\\
1.17 + j0.02\\
1.75 - j1.75
\end{bmatrix} \text{A}
$$

The line current I_{aA} leaves generator a:

$$
\begin{aligned}
\hat{I}_{aA} = \hat{I}_{na} = \hat{I}_1 + \hat{I}_3 &= (1.93 + j0.03) + (1.75 - j1.75)\\
&= 3.68 - j1.72 = 4.06\underline{/-25.05^\circ} \text{ A}
\end{aligned}
$$

The line current $\hat{I}_{Bb}$ flows through generator b:

$$
\begin{aligned}
\hat{I}_{Bb} = \hat{I}_{bn} = \hat{I}_2 + \hat{I}_3 &= (1.17 + j0.02) + (1.75 - j1.75)\\
&= 2.92 - j1.73 = 3.39\underline{/-30.65^\circ} \text{ A}
\end{aligned}
$$

FIGURE 12.4

(a) An unbalanced, single-phase, three-wire system, (b) a tree formed from an oriented graph of the system and (c) the system showing three loop currents, which have been obtained from the tree

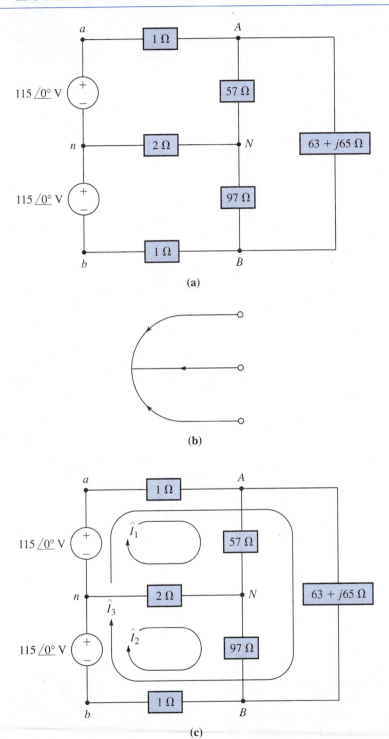

The line current in the neutral wire is

$$\hat{I}_{Nn} = \hat{I}_1 - \hat{I}_2 = (1.93 + j0.03) - (1.17 + j0.02)$$
$$= 0.76 + j0.01 = 0.76\,\underline{/0.75°}\ \text{A}$$

The power drawn by each load is

$$P_{AN} = (I_1)^2(57) = (1.93)^2(57) = 212.3\ \text{W}$$

$$P_{NB} = (I_2)^2(97) = (1.17)^2(97) = 132.8\ \text{W}$$

$$P_{AB} = (I_3)^2(63) = (2.47)^2(63) = 384.4\ \text{W}$$

The total power supplied to all three loads is

$$P_0 = 212.3 + 132.8 + 384.4 = 729.5\ \text{W}$$

The line losses are

$$P_{aA} = (I_{aA})^2(1) = (4.06)^2(1) = 16.5\ \text{W}$$

$$P_{Bb} = (I_{Bb})^2(1) = (3.39)^2(1) = 11.5\ \text{W}$$

$$P_{Nn} = (I_{Nn})^2(2) = (0.76)^2(2) = 1.1\ \text{W}$$

so that the total of all line losses is

$$P_l = 16.5 + 11.5 + 1.1 = 29.1\ \text{W}$$

The two generators must supply

$$P_s = P_o + P_\ell = 729.5 + 29.1 = 758.6\ \text{W}$$

and this may be confirmed by looking at the generators individually. For generator a, $\hat{I}_{na} = 4.06\,\underline{/-25.05°}$ A, so that the power factor is $\cos\theta = \cos(-25.05°) = 0.906$. This makes the power delivered by generator a equal to

$$P_a = VI\cos\theta = 115(4.06)(0.906) = 423.1\ \text{W}$$

For generator b, $\hat{I}_{bn} = 3.39\,\underline{/-30.65°}$ A, and here, $\cos\theta = \cos(-30.65°) = 0.860$, so that

$$P_b = VI\cos\theta = 115(3.39)(0.860) = 335.5\ \text{W}$$

The total power supplied is

$$P_s = P_a + P_b = 423.1 + 335.5 = 758.6\ \text{W}$$

and this confirms the previous value determined from the power absorbed by the lines and the loads.

FIGURE 12.5

A single-phase, three-wire power system (Exercise 12.1)

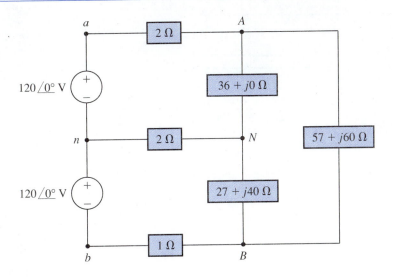

EXERCISE 12.1

What is the total power drawn by the loads in the single-phase, three-wire system shown in Fig. 12.5?

Answer   912.7 W

THREE-PHASE VOLTAGE AND POWER

Figure 12.6 shows three generators that provide identical rms voltages and frequency to a power transmission system. It is helpful to imagine that the windings of each of the three different generators are wound equally spaced around the same shaft so that the phase angle between the output of each generator is fixed at exactly 120°. It may then be pointed out that the winding design of the common synchronous generator leads to the same effect. The voltages at the terminals of each of the three *phases* on this single system can be designated by a double subscript, and they may be written in instantaneous form, using V without a subscript as the rms magnitude, as

$$v_{an}(t) = V \cos \omega t \tag{12.5a}$$

$$v_{bn}(t) = V \cos(\omega t - 120°) \tag{12.5b}$$

$$v_{cn}(t) = V \cos(\omega t - 240°)$$

or

$$v_{cn}(t) = V \cos(\omega t + 120°) \tag{12.5c}$$

FIGURE 12.6   A balanced three-phase voltage source

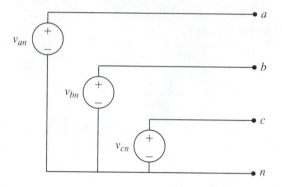

In phasor form, these become

$$\hat{V}_{an} = V\underline{/0°} \tag{12.6a}$$

$$\hat{V}_{bn} = V\underline{/-120°} \tag{12.6b}$$

$$\hat{V}_{cn} = V\underline{/120°} \tag{12.6c}$$

Here, too, n is used to designate the neutral point; and because all of the *phase volt-ages* possess equal amplitudes, they are said to be balanced. However, because the angles are in the sequence $0°$, $-120°$, $-240°$, they are said to be in an *abc* or *positive* sequence, because $\hat{V}_{an}$ leads $\hat{V}_{bn}$, which, in turn, leads $\hat{V}_{cn}$.

FIGURE 12.7   Phasor diagrams for the three-phase system in (a) the *abc* or positive sequence and (b) the *acb* or negative sequence

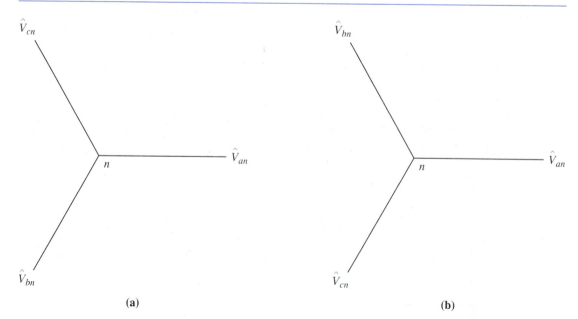

(a)   (b)

If the balanced voltages are such that

$$\hat{V}_{an} = V\underline{/0°} \tag{12.7a}$$

$$\hat{V}_{bn} = V\underline{/120°} \tag{12.7b}$$

$$\hat{V}_{cn} = V\underline{/240°} \quad \text{or} \quad \hat{V}_{cn} = V\underline{/-120°} \tag{12.7c}$$

they are said to be in an *acb* or *negative* sequence, because $\hat{V}_{cn}$ leads $\hat{V}_{bn}$, which leads $\hat{V}_{an}$. Both sequences are indicated in Fig. 12.7, where both phasor diagrams are assumed to be rotating in the counterclockwise direction at angular velocity ω.

When the three generators, whose voltages are given by eqs. (12.5), are delivering power to identical loads $|Z|\underline{/\theta}$ connected across their terminals, then for identical transmission line impedances, the entire configuration is balanced and is referred to as *a balanced three-phase network* or *system*. In this event, the phase currents are

$$i_{an}(t) = I\cos(\omega t - \theta) \tag{12.8a}$$

$$i_{bn}(t) = I\cos(\omega t - 120° - \theta) \tag{12.8b}$$

$$i_{cn}(t) = I\cos(\omega t + 120° - \theta) \tag{12.8c}$$

where I designates the magnitude of the rms current.

The *total* instantaneous power delivered by the three generators will be

$$p = p_a + p_b + p_c \tag{12.9}$$

where it is emphasized (see Sections 11.2 and 11.3) that the total instantaneous power derives from a consideration of the peak values of voltages and current. Thus, use of eqs. (12.5) and (12.8), but with a subscript m used to indicate a peak or maximum value, gives

$$p_a = v_{an}i_{an} = V_m I_m \cos\omega t \cos(\omega t - \theta)$$

$$p_b = v_{bn}i_{bn} = V_m I_m \cos(\omega t - 120°)\cos(\omega t - 120° - \theta)$$

$$p_c = v_{cn}i_{cn} = V_m I_m \cos(\omega t + 120°)\cos(\omega t + 120° - \theta)$$

The foregoing expressions for the power in each phase can be simplified by invoking the trigonometric identity

$$\cos A \cos B = \frac{1}{2}[\cos(A + B) + \cos(A - B)]$$

so that for phase a, with $A = \omega t$ and $B = \omega t - \theta$,

$$p_a = \frac{1}{2}V_m I_m[\cos(2\omega t - \theta) + \cos\theta] \tag{12.10a}$$

For phase b, with $A = \omega t - 120°$ and $B = \omega t - 120° - \theta$,

$$p_b = \frac{1}{2}V_m I_m[\cos(2\omega t - 240° - \theta) + \cos\theta] \tag{12.10b}$$

And for phase c, with $A = \omega t + 120°$ and $B = \omega t + 120° - \theta$,

$$p_c = \frac{1}{2} V_m I_m [\cos(2\omega t + 240° - \theta) + \cos \theta] \qquad (12.10c)$$

Equations (12.10) may be added in accordance with eq. (12.9) to yield a sum of two terms:

$$p = \frac{3}{2} V_m I_m \cos \theta + \frac{1}{2} V_m I_m [\cos(2\omega t - \theta) + \cos(2\omega t - 240° - \theta)$$
$$+ \cos(2\omega t + 240° - \theta)] \qquad (12.11)$$

The bracketed term in eq. (12.11) is equal to zero. To show this, imagine that this term can be represented as three phasors rotating at an angular velocity of $\alpha = 2\omega t - \theta$. The three phasors

$$\hat{V}_1 = \frac{1}{2} V_m I_m \underline{/0°} = \frac{1}{2} V_m I_m (1 + j0)$$

$$\hat{V}_2 = \frac{1}{2} V_m I_m \underline{/-240°} = \frac{1}{2} V_m I_m \left(-\frac{1}{2} + j\frac{\sqrt{3}}{2} \right)$$

$$\hat{V}_3 = \frac{1}{2} V_m I_m \underline{/240°} = \frac{1}{2} V_m I_m \left(-\frac{1}{2} - j\frac{\sqrt{3}}{2} \right)$$

will possess a sum

$$\frac{1}{2} V_m I_m \left[(1 + j0) + \left(-\frac{1}{2} + \frac{\sqrt{3}}{2} \right) + \left(-\frac{1}{2} - \frac{\sqrt{3}}{2} \right) \right] = 0$$

Thus, the total instantaneous power is

$$p = 3VI \cos \theta \qquad (12.12)$$

where V and I are rms quantities and where $\cos \theta$ is the power factor, which will be identical for each phase because the system is a balanced system. Equation (12.12) shows that the instantaneous power is constant, in contrast to the fluctuating nature of single-phase power, which, as demonstrated in Section 11.2, is dissipated or stored at a frequency that is twice the frequency of the impressed voltage.

The sum of a balanced set of voltages at any instant of time is zero:

$$v_{an} + v_{bn} + v_{cn} = 0 \qquad (12.13a)$$

or

$$\hat{V}_{an} + \hat{V}_{bn} + \hat{V}_{cn} = 0 \qquad (12.13b)$$

This is clearly observed by referring to either of the phase sequences indicated in Fig. 12.7. For Fig. 12.7a,

$$\hat{V}_{an} = V(1 + j0) \qquad \hat{V}_{bn} = \frac{1}{2} V(-1 - j\sqrt{3}) \qquad \hat{V}_{cn} = \frac{1}{2} V(-1 + j\sqrt{3})$$

Thus,

$$\hat{V}_{an} + \hat{V}_{bn} + \hat{V}_{cn} = \frac{1}{2}\left[(2 + j0) + (-1 - j\sqrt{3}) + (-1 + j\sqrt{3})\right] = 0$$

and a similar development can be constructed by using Fig. 12.7b.

WYE AND DELTA CONNECTIONS

SECTION 12.4

Four wires leave the closed system shown in Fig. 12.8a. They are labeled *a*, *b*, *c*, and *n*. This may be considered as a three-phase system having three generators arranged so that their phase voltages are in either positive or negative sequence. The over-all generator may be connected in a *wye* configuration, with a neutral wire, as shown in Fig. 12.8b, or in a *delta* configuration, without a neutral wire, as in Fig. 12.8c. The resemblance of these connections to the familiar *tee* and *pi* networks is unmistakable.

(a) A four-wire system that is assumed to be a generator, (b) the wye connection, and (c) the delta connection

FIGURE 12.8

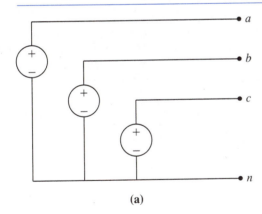

(a)

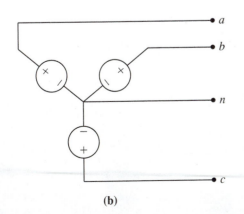

(b)

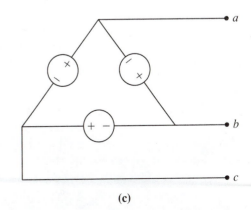

(c)

FIGURE 12.9   (a) A balanced, three-phase, wye generator, (b) the phasor diagram for the development of one of the line voltages, (c) the phasor diagram showing both phase and line voltages for the positive sequence, and (d) the phasor diagram showing both phase and line voltages for the negative sequence

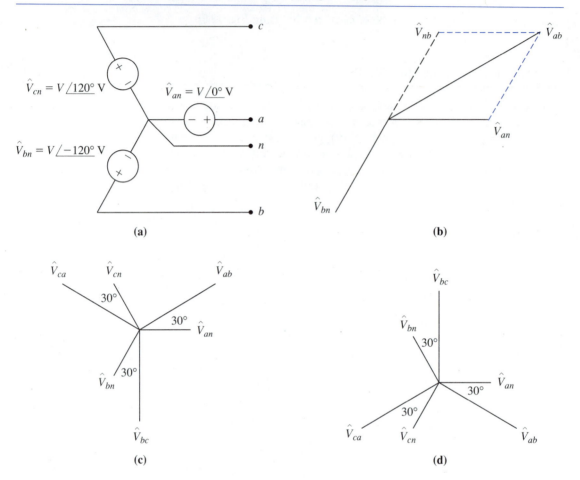

12.4.1 The Wye- (Y-) Connected Generator

Consider the balanced-wye arrangement shown in Fig. 12.9a, and assume that the system is delivering power to a balanced load connected in either wye or delta. Here, the phase voltages are $\hat{V}_{an}$, $\hat{V}_{bn}$, and $\hat{V}_{cn}$, and they are connected in the positive sequence. The line-to-line voltages (hereinafter referred to merely as the *line voltages*) can be obtained from a consideration of KVL:

$$\hat{V}_{ab} = \hat{V}_{an} - \hat{V}_{bn} = V\underline{/0°} - V\underline{/-120°} = V\left[(1+j0) - \left(-\frac{1}{2} - j\frac{\sqrt{3}}{2}\right)\right]$$

$$= V\left(\frac{3}{2} + j\frac{\sqrt{3}}{2}\right)$$

or

$$\hat{V}_{ab} = \sqrt{3}\,V\underline{/30°}$$

This can be substantiated by referring to the phasor diagrams in Figs. 12.9b and 12.9c. A similar development will yield

$$\hat{V}_{bc} = \sqrt{3}\,V\underline{/-90°} \qquad \text{and} \qquad \hat{V}_{ca} = \sqrt{3}\,V\underline{/150°}$$

All of the phase and line voltages for the positive-sequence, wye-connected generator are indicated in Fig. 12.9c, and a similar summary (after an identical development) is provided for the negative-sequence, wye-connected generator in Fig. 12.9d.

Regardless of whether the phase sequence is positive or negative, for the wye connection, the magnitude of the line and the phase voltages are related by

$$V_\ell = \sqrt{3}\,V_p \qquad\qquad\qquad\qquad\qquad \textbf{(12.14)}$$

where the subscripts ℓ and p are intended to refer, respectively, to line and phase quantities.

12.4.2 The Delta- (Δ-) Connected Generator

It is observed in Fig. 12.10a that the delta connection, merely by its nature, is a connection in which the line voltages are equal to the phase voltages. Thus,

$$V_\ell = V_p \qquad\qquad\qquad\qquad\qquad\qquad \textbf{(12.15)}$$

The phasor diagram for these voltages is shown in Fig. 12.10b for the positive sequence and in Fig. 12.10c for the negative sequence.

(a) The delta-connected generator, (b) the phasor diagram for the line and phase voltages in the positive phase sequence and (c) the phasor diagram for the line and phase voltages in the negative phase sequence

FIGURE 12.10

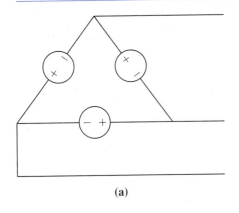

(a)

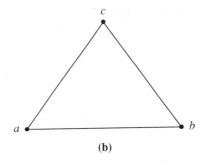

(b)

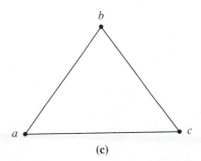

(c)

Four possible balanced generator-load arrangements: (a) wye-wye, (b) wye-delta,
(c) delta-wye, and (d) delta-delta

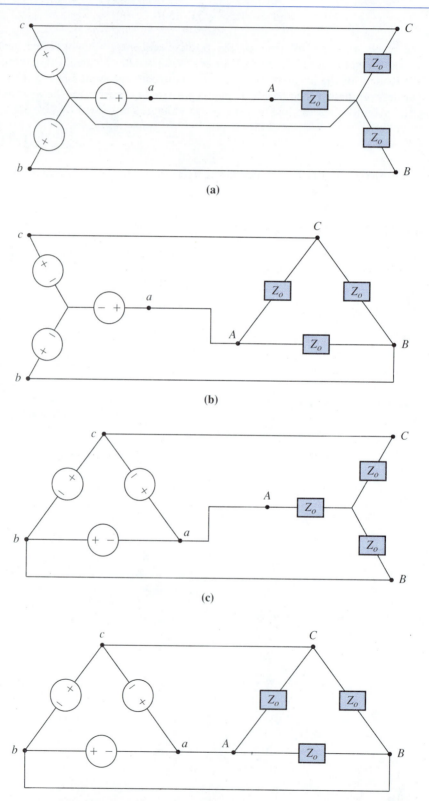

(a)

(b)

(c)

(d)

LOAD CONNECTIONS

The system loads may also be connected in wye or delta, and as indicated in Fig. 12.11, there are four possible arrangements of the generator-load entity. In all cases, the impedances in each load phase are equal, so that the system is balanced. Notice that lowercase letters are used to indicate the terminals of the generator and that uppercase letters are employed to indicate the corresponding terminals at the load. This nomenclature permits the designation of line quantities with the subscripts aA, bB, and cC.

The phase currents in either the wye-connected or the delta-connected load are set by the phase voltage and the impedance per phase, and they will lead or lag the phase voltage by the impedance angle. But for the wye generator or load, the nature of the configuration indicates that the line currents are equal to the phase currents. Thus,

$$I_\ell = I_p \tag{12.16}$$

But in the delta configuration, KCL requires that at point a in Fig. 12.12, which is for the positive sequence,

$$\hat{I}_{aA} = \hat{I}_{AB} + \hat{I}_{AC} = I_p \underline{/0^\circ} + I_p \underline{/-60^\circ} = I_p\left[(1 + j0) + \left(\frac{1}{2} - j\frac{\sqrt{3}}{2}\right)\right]$$

$$= \sqrt{3}\,I_p\left(\frac{\sqrt{3}}{2} - j\frac{1}{2}\right) = \sqrt{3}\,I_p \underline{/-30^\circ}\ \text{A}$$

All of the phase and line currents for the delta-connected load can be established by an identical procedure and are shown for the positive sequence in Fig. 12.12b; a similar picture for the negative sequence is shown in Fig. 12.12c. In all cases pertaining to the delta configuration,

$$I_\ell = \sqrt{3}\,I_p \tag{12.17}$$

THE POWER IN THE LOAD

Equations (12.14) through (12.17) indicate that there are a pair of simple relationships for the power based on either phase or line values of voltage or current. Certainly, in the balanced system, the total power delivered to the load will be three times the power delivered to each of the three phases. Thus, by eq. (12.12),

$$P = 3V_pI_p \cos\theta \tag{12.12}$$

For the wye connection, eqs. (12.14) and (12.16) can be used in eq. (12.12):

$$P = 3V_pI_p \cos\theta = 3\left(\frac{V_\ell}{\sqrt{3}}\right)I_\ell \cos\theta = \sqrt{3}\,V_\ell I_\ell \cos\theta$$

For the delta connection, an identical procedure using eqs. (12.15) and (12.17) yields

$$P = 3V_pI_p \cos\theta = 3V_\ell\left(\frac{I_\ell}{\sqrt{3}}\right)\cos\theta = \sqrt{3}\,V_\ell I_\ell \cos\theta$$

FIGURE 12.12

(a) The phase currents in a delta-connected load, where voltage is supplied in the positive sequence, (b) the phasor diagram for both line and phase currents in the positive sequence, and (c) the phasor diagram for both line and phase currents in the negative sequence

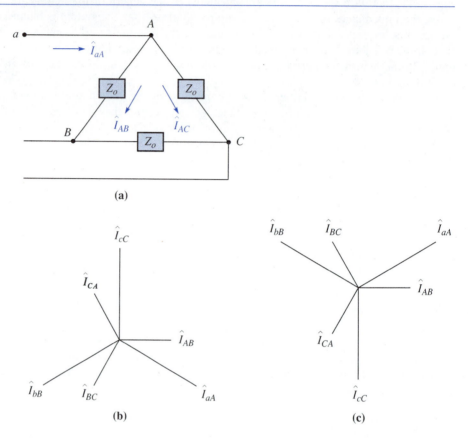

(a)

(b)

(c)

Thus, regardless of the type of connection, on the basis of line quantities,

$$P = \sqrt{3}\, V_\ell I_\ell \cos \theta \tag{12.18}$$

Table 12.1 shows the relationships between line and phase quantities in the wye and delta arrangements.

TABLE 12.1

Summary of line and phase relationships for wye and delta connections

Wye	Delta
$V_\ell = \sqrt{3}\, V_p$	$V_\ell = V_p$
$I_\ell = I_p$	$I_\ell = \sqrt{3}\, I_p$

THREE-PHASE POWER COMPUTATIONS

All four of the generator-load arrangements shown in Fig. 12.11 will now be considered in detailed examples.

12.7.1 The Wye-to-Wye Configuration

■ **EXAMPLE 12.2**

The configuration shown in Fig. 12.13 is a balanced three-phase system. The generator supplies a rated rms line voltage of 208 V at 60 Hz in the positive-phase sequence. All generator and line impedances are negligible, and each phase of the load possesses an equal impedance of $Z_o = 5\underline{/45°} = 3.54 + j3.54 \ \Omega$. Determine all line and phase voltages, line and load phase currents, and the power delivered to the load.

Solution The neutral wire in this balanced system does not carry any current, and the magnitude of the phase voltages in the wye-connected source is, by eq. (12.14),

$$V_p = \frac{V_\ell}{\sqrt{3}} = \frac{208}{\sqrt{3}} = 120.09 \ \text{V}$$

If the phase voltage $\hat{V}_{an}$ is chosen as the reference,

$$\hat{V}_{an} = 120.09 \underline{/0°} \ \text{V}$$

and in accordance with Fig. 12.7 for the positive sequence,

$$\hat{V}_{bn} = 120.09 \underline{/-120°} \ \text{V} \qquad \text{and} \qquad \hat{V}_{cn} = 120.09 \underline{/120°} \ \text{V}$$

These are also the phase voltages at the load:

$$\hat{V}_{AN} = 120.09 \underline{/0°} \ \text{V} \qquad \hat{V}_{BN} = 120.09 \underline{/-120°} \ \text{V}$$
$$\hat{V}_{CN} = 120.09 \underline{/120°} \ \text{V}$$

Wye generator and balanced-wye load (a balanced wye-wye system) for Example 12.2

FIGURE 12.13

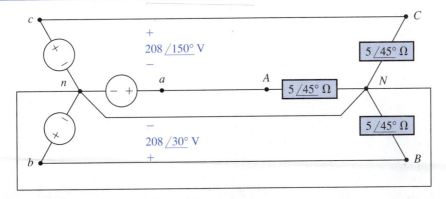

Figure 12.9c shows that for the positive-phase sequence, the line voltages must lead the phase voltages in the wye-connected generator by 30°, so that

$$\hat{V}_{ab} = \hat{V}_{AB} = 208\underline{/30°}\text{ V} \qquad \hat{V}_{bc} = \hat{V}_{BC} = 208\underline{/-90°}\text{ V}$$
$$\hat{V}_{ca} = \hat{V}_{CA} = 208\underline{/150°}\text{ V}$$

For this wye-wye arrangement, eq. (12.16) shows that the phase currents are equal to the line currents. Thus,

$$\hat{I}_{aA} = \hat{I}_{AN} = \frac{V_{AN}}{Z_{AN}} = \frac{120.09\underline{/0°}}{5\underline{/45°}} = 24.02\underline{/-45°}\text{ A}$$

and because the load is balanced,

$$\hat{I}_{bB} = \hat{I}_{BN} = 24.02\underline{/-165°}\text{ A} \qquad \hat{I}_{cC} = \hat{I}_{CN} = 24.02\underline{/75°}\text{ A}$$

All of the foregoing phasors are displayed in the phasor diagram in Fig. 12.14.

Equation (12.18) gives the power. With the power factor angle equal to the impedance angle in this balanced system, $\cos\theta = \cos 45° = 0.707$, and

$$P = \sqrt{3}\ V_\ell I_\ell \cos\theta = \sqrt{3}(208)(24.02)(0.707) = 6118.5\text{ W}$$

This can be confirmed by checking the power drawn by the total resistive load,

$$P = 3I_p^2 R_p = 3(24.02)^2(3.54) = 6118.5\text{ W}$$

It is of interest to note that the reactive power is

$$Q = 3I_p^2 X_p = 3(24.02)^2(3.54) = 6118.5\text{ VAR.}$$

FIGURE 12.14   Phasor diagram for Example 12.2

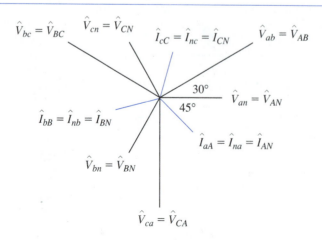

FIGURE 12.15

Three-phase power system (Exercise 12.2)

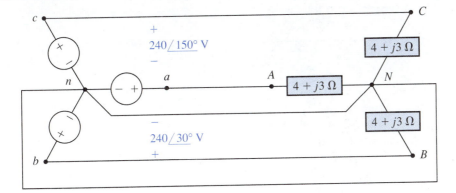

and that the system transmits an apparent power

$$S = 6118.5 + j6118.5 = 8652.8\underline{/45°}\ \text{VA}$$

with the magnitude of the apparent power being 8652.8 VA.

EXERCISE 12.2

The configuration shown in Fig. 12.15 is a balanced three-phase system. The generator supplies rated rms line voltages of 240 V at 60 Hz in the positive-phase sequence. All generator and line impedances are negligible, and each phase of the load possesses an equal impedance of $Z_o = 5\underline{/36.87°} = 4 + j3\ \Omega$. Determine all line and phase voltages, line and load phase currents and the power delivered to the load.

Answer

$$\hat{V}_{an} = \hat{V}_{AN} = 138.56\underline{/0°}\ \text{V} \qquad \hat{V}_{ca} = \hat{V}_{CA} = 240\underline{/150°}\ \text{V}$$
$$\hat{V}_{bn} = \hat{V}_{BN} = 138.56\underline{/-120°}\ \text{V} \qquad \hat{I}_{aA} = \hat{I}_{AN} = 27.71\underline{/-36.87°}\ \text{A}$$
$$\hat{V}_{cn} = \hat{V}_{CN} = 138.56\underline{/120°}\ \text{V} \qquad \hat{I}_{bB} = \hat{I}_{BN} = 24.71\underline{/-156.87°}\ \text{A}$$
$$\hat{V}_{ab} = \hat{V}_{AB} = 240\underline{/30°}\ \text{V} \qquad \hat{I}_{cC} = \hat{I}_{CN} = 27.71\underline{/83.13°}\ \text{A}$$
$$\hat{V}_{bc} = \hat{V}_{BC} = 240\underline{/-90°}\ \text{V} \qquad P = 9215.7\ \text{W}$$

12.7.2 The Wye-to-Delta Configuration

EXAMPLE 12.3

For the balanced wye-to-delta system indicated in Fig. 12.16, the generator is the same generator used in Example 12.2 (208 V, 60 Hz, and positive-phase sequence). The delta-connected load has three identical impedances $Z_{AB} = Z_{BC} = Z_{CA} = 15\underline{/45°} = 10.61 + j10.61\ \Omega$. Determine all phase and load voltages, load and line currents, and the power transmitted.

Solution   Here, too,

$$V_p = \frac{V_\ell}{\sqrt{3}} = 120.09\ \text{V}$$

FIGURE 12.16   Wye generator and balanced-delta load (a balanced wye-delta system) for Example 12.3

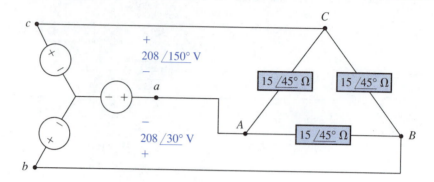

and with $\hat{V}_{an}$ selected as the reference,

$$\hat{V}_{an} = 120.09 \underline{/0°} \text{ V} \qquad \hat{V}_{bn} = 120.09 \underline{/-120°} \text{ V} \qquad \hat{V}_{cn} = 120.09 \underline{/120°} \text{ V}$$

The phase voltages in the load, by eq. (12.15), are the same as the line voltages. Because of the positive generator sequence,

$$\hat{V}_{ab} = \hat{V}_{AB} = 208 \underline{/30°} \text{ V} \qquad \hat{V}_{bc} = \hat{V}_{BC} = 208 \underline{/-90°} \text{ V}$$
$$\hat{V}_{ca} = V_{CA} = 208 \underline{/150°} \text{ V}$$

Then,

$$\hat{I}_{AB} = \frac{\hat{V}_{AB}}{Z_{AB}} = \frac{208 \underline{/30°}}{15 \underline{/45°}} = 13.87 \underline{/-15°} \text{ A}$$
$$\hat{I}_{BC} = 13.87 \underline{/-135°} \text{ A} \qquad \hat{I}_{CA} = 13.87 \underline{/105°} \text{ A}$$

The line currents have a magnitude, as shown by eq. (12.17), of

$$I_\ell = \sqrt{3} I_p = \sqrt{3}(13.87) = 24.02 \text{ A}$$

and in accordance with Fig. 12.12b,

$$\hat{I}_{aA} = 24.02 \underline{/-45°} \text{ A} \qquad \hat{I}_{bB} = 24.02 \underline{/-165°} \text{ A} \qquad \hat{I}_{cC} = 24.02 \underline{/75°} \text{ A}$$

A phasor diagram with all of the foregoing quantities displayed is shown in Fig. 12.17.

The power in this case (again with $PF = \cos\theta = 0.707$) will be

$$P = \sqrt{3} V_\ell I_\ell \cos\theta = \sqrt{3}(208)(24.02)(0.707) = 6118.5 \text{ W}$$

and this can be confirmed:

$$P = 3I_p^2 R_p = 3(13.87)^2(10.61) = 6118.5 \text{ W}$$

FIGURE 12.17

Phasor diagram for Example 12.3

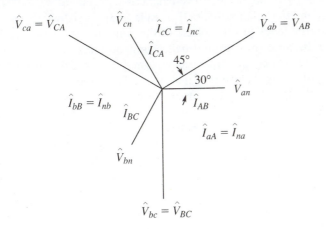

EXERCISE 12.3

The configuration shown in Fig. 12.18 is a balanced three-phase system. The generator supplies rated rms line voltages of 240 V at 60 Hz in the positive-phase sequence. All generator and line impedances are negligible and each phase of the load possesses an equal impedance of $Z_o = 10\underline{/53.13°} = 6 + j8 \, \Omega$. Determine all line and phase voltages, line and load phase currents, and the power delivered to the load.

Answer

$$\hat{V}_{an} = 138.56\underline{/0°} \text{ V} \qquad \hat{I}_{AB} = 24\underline{/-23.13°} \text{ A}$$

$$\hat{V}_{bn} = 138.56\underline{/-120°} \text{ V} \qquad \hat{I}_{BC} = 24\underline{/-143.13°} \text{ A}$$

$$\hat{V}_{cn} = 138.56\underline{/120°} \text{ V} \qquad \hat{I}_{CA} = 24\underline{/96.87°} \text{ A}$$

$$\hat{V}_{ab} = \hat{V}_{AB} = 240\underline{/30°} \text{ V} \qquad \hat{I}_{aA} = \hat{I}_{na} = 41.57\underline{/-53.13°} \text{ A}$$

$$\hat{V}_{bc} = \hat{V}_{BC} = 240\underline{/-90°} \text{ V} \qquad \hat{I}_{bB} = \hat{I}_{nb} = 41.57\underline{/-173.13°} \text{ A}$$

$$\hat{V}_{ca} = \hat{V}_{CA} = 240\underline{/150°} \text{ V} \qquad \hat{I}_{cC} = \hat{I}_{nc} = 41.57\underline{/66.87°} \text{ A}$$

$$P = 10,368 \text{ W}$$

FIGURE 12.18

Three-phase power system (Exercise 12.3)

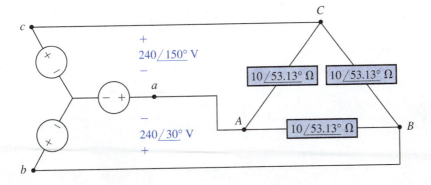

12.7.3 Digression—The Single-Phase Equivalent

Examples 12.2 and 12.3 show that for identical balanced-wye generators, the balanced-wye (Example 12.2) and the balanced-delta (Example 12.3) loads draw the same power. In these examples, the loads in each phase of the delta are equal to exactly 3 times those in the wye. Because the wye and delta configurations are merely arrangements of tee and pi networks, the balanced wye may be obtained from the balanced delta by merely dividing each impedance by 3. Thus, the $Z = 15\underline{/45°}$-Ω delta impedance becomes the $Z = 5\underline{/45°}$-Ω tee impedance.

The question of using a single-phase, wye-to-wye equivalent for balanced three-phase system calculations has merit and is advocated by some. The only real caution is in the fact that there will be a neutral wire that will carry the single-phase equivalent line current. This should not be misleading, because when all of the phases are reassembled, the neutral wire will contain three current components, which will sum to zero.

12.7.4 The Delta-to-Wye-Configuration

■ **EXAMPLE 12.4**

The arrangement in Fig. 12.19 shows a delta-connected generator with rated rms line voltages of 208V operating in the positive sequence at 60 Hz. The balanced-wye load contains equal impedances, $Z_{AN} = Z_{BN} = Z_{CN} = 5\underline{/45°} = 3.54 + j3.54\ \Omega$. The line and phase voltages, the line and load currents, and the transmitted power are to be determined.

Solution   This time, let $\hat{V}_{AB}$ be the reference, so that

$$\hat{V}_{AB} = \hat{V}_{ab} = 208\underline{/0°}\ \text{V} \qquad \hat{V}_{BC} = \hat{V}_{bc} = 208\underline{/-120°}\ \text{V}$$
$$\hat{V}_{CA} = \hat{V}_{ca} = 208\underline{/120°}\ \text{V}$$

The phase voltages at the load are

$$V_p = \frac{V_\ell}{\sqrt{3}} = \frac{208}{\sqrt{3}} = 120.09\ \text{V}$$

FIGURE 12.19   Delta source and balanced-wye load arrangement for Example 12.4

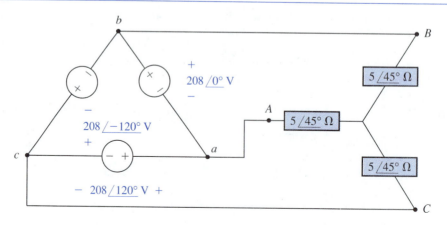

FIGURE 12.20

Phasor diagram for Example 12.4

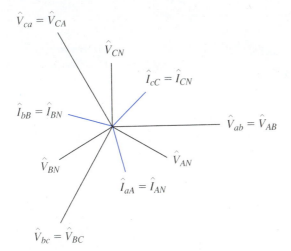

and thus,

$$\hat{V}_{AN} = 120.09\underline{/-30°}\ \text{V} \qquad \hat{V}_{BN} = 120.09\underline{/-150°}\ \text{V}$$
$$\hat{V}_{CN} = 120.09\underline{/90°}\ \text{V}$$

The load currents, which are the line currents, are

$$\hat{I}_{AN} = \hat{I}_{aA} = \frac{\hat{V}_{AN}}{Z_{AN}} = \frac{120.09\underline{/-30°}}{5\underline{/45°}} = 24.02\underline{/-75°}\ \text{A}$$

$$\hat{I}_{BN} = \hat{I}_{bB} = 24.02\underline{/165°}\ \text{A} \qquad \hat{I}_{CN} = \hat{I}_{cC} = 24.02\underline{/45°}\ \text{A}$$

These quantities are displayed in the phasor diagram in Fig. 12.20.
The power ($\cos\theta = 0.707$) will be

$$P = \sqrt{3}\,V_\ell I_\ell \cos\theta = \sqrt{3}(208)(24.02)(0.707) = 6118.5\ \text{W}$$

and this may be confirmed by

$$P = 3I_p^2 R_p = 3(24.02)^2(3.54) = 6118.5\ \text{W}$$

EXERCISE 12.4

The configuration shown in Fig. 12.21 is a balanced three-phase system. The generator supplies rated rms line voltages of 240 V at 60 Hz in the positive-phase sequence. All generator and line impedances are negligible, and each phase of the load possesses an equal impedance of $Z_o = 13\underline{/22.62°} = 12 + j5\ \Omega$. Determine all line and phase voltages, line and load phase currents, and the power delivered to the load.

FIGURE 12.21   Three-phase power system (Exercise 12.4)

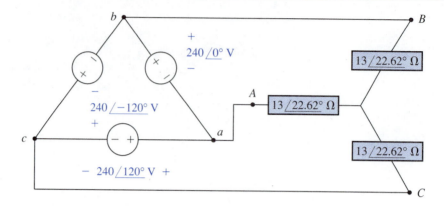

Answer

$$\hat{V}_{AN} = 138.56\underline{/-30°}\ \text{V} \qquad \hat{V}_{ca} = \hat{V}_{CA} = 240\underline{/120°}\ \text{V}$$

$$\hat{V}_{BN} = 138.56\underline{/-150°}\ \text{V} \qquad \hat{I}_{aA} = \hat{I}_{AN} = 10.66\underline{/-52.62°}\ \text{A}$$

$$\hat{V}_{CN} = 138.56\underline{/90°}\ \text{V} \qquad \hat{I}_{bB} = \hat{I}_{BN} = 10.66\underline{/-172.62°}\ \text{A}$$

$$\hat{V}_{ab} = \hat{V}_{AB} = 240\underline{/0°}\ \text{V} \qquad \hat{I}_{cC} = \hat{I}_{CN} = 10.66\underline{/67.38°}\ \text{A}$$

$$\hat{V}_{bc} = V_{BC} = 240\underline{/-120°}\ \text{V} \qquad P = 4089.9\ \text{W}$$

12.7.5 The Delta-to-Delta Configuration

■ **EXAMPLE   12.5**

The delta generator in Fig. 12.22 operates in the positive sequence at 60 Hz with rms line voltages of 208 V. The delta loads are balanced, with $Z_{AB} = Z_{BC} = Z_{CA} = 15\underline{/45°} = 10.61 + j10.61\ \Omega$; and the task is to determine all line and phase voltages, line and load phase currents, and the power flow.

FIGURE 12.22   Delta source and balanced-delta load arrangement for Example 12.5

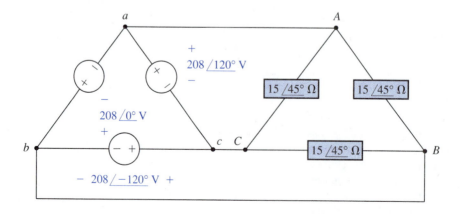

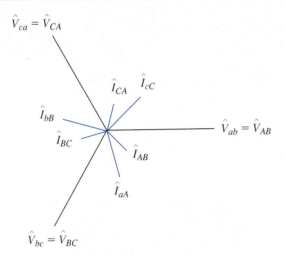

Solution   For this delta-delta configuration, the line voltages are the equal to the phase voltages. With $\hat{V}_{AB}$ as the reference,

$$\hat{V}_{AB} = \hat{V}_{ab} = 208\,\underline{/0°}\ \text{V} \qquad \hat{V}_{BC} = \hat{V}_{bc} = 208\,\underline{/-120°}\ \text{V}$$
$$\hat{V}_{CA} = \hat{V}_{ca} = 208\,\underline{/120°}\ \text{V}$$

The phase currents are

$$\hat{I}_{AB} = \frac{\hat{V}_{AB}}{Z_{AB}} = \frac{208\,\underline{/0°}}{15\,\underline{/45°}} = 13.87\,\underline{/-45°}\ \text{A}$$
$$\hat{I}_{BC} = 13.87\,\underline{/-165°}\ \text{A} \qquad \hat{I}_{CA} = 13.87\,\underline{/75°}\ \text{A}$$

Then, in accordance with eq. (12.17),

$$I_\ell = \sqrt{3}\,I_p = \sqrt{3}(13.87) = 24.02\ \text{A}$$

and the line currents are

$$\hat{I}_{aA} = 24.02\,\underline{/-75°}\ \text{A} \qquad \hat{I}_{bB} = 24.02\,\underline{/165°}\ \text{A} \qquad \hat{I}_{cC} = 24.02\,\underline{/45°}\ \text{A}$$

All of these quantities are displayed in the phasor diagram in Fig. 12.23.
The power flow, with $\cos\theta = 0.707$, is

$$P = \sqrt{3}\ V_\ell I_\ell \cos\theta = \sqrt{3}(208)(24.02)(0.707) = 6118.5\ \text{W}$$

and indeed,

$$P = 3I_p^2 R_p = 3(13.x7)^2(10.64) = 6118.5\ \text{W}$$

FIGURE 12.24   Three-phase power system (Exercise 12.5)

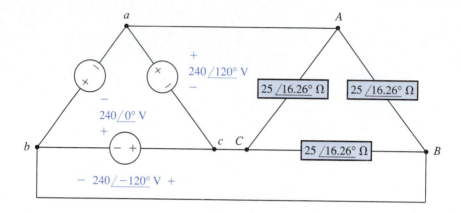

EXERCISE 12.5

The configuration shown in Fig. 12.24 is a balanced three-phase system. The generator supplies rated rms line voltages of 240 V at 60 Hz in the positive-phase sequence. All generator and line impedances are negligible, and each phase of the load possesses an equal impedance of $Z_o = 25\underline{/16.26°} = 24 + j7\ \Omega$. Determine all line and phase voltages, line and load phase currents, and the power delivered to the load.

Answer

$$\hat{V}_{ab} = V_{AB} = 240\underline{/0°}\ \text{V} \qquad \hat{I}_{CA} = 9.60\underline{/103.74°}\ \text{A}$$

$$\hat{V}_{bc} = \hat{V}_{BC} = 240\underline{/-120°}\ \text{V} \qquad \hat{I}_{aA} = 16.63\underline{/-46.26°}\ \text{A}$$

$$\hat{V}_{ca} = V_{CA} = 240\underline{/120°}\ \text{V} \qquad \hat{I}_{bB} = 16.63\underline{/-166.26°}\ \text{A}$$

$$\hat{I}_{AB} = 9.60\underline{/-16.26°}\ \text{A} \qquad \hat{I}_{cC} = 16.63\underline{/73.74°}\ \text{A}$$

$$\hat{I}_{BC} = 9.60\underline{/-136.26°}\ \text{A} \qquad P = 6635.5\ \text{W}$$

SECTION 12.8   AN EXAMPLE OF AN UNBALANCED SYSTEM

If all of the load impedances, whether in wye or delta, are not equal, then the system is unbalanced. This, of course, is a frequent state of affairs; and the one way to analyze such a system is by node, mesh, or loop analysis.

■ EXAMPLE 12.6

An unbalanced system, with a balanced-wye generator, with rms phase voltages of 200 V operating at 60 Hz in the positive sequence, and with a delta-connected load, is shown in Fig. 12.25a. Determine the total power transmitted to the load.

(a) An unbalanced three-phase system with a wye generator and a delta-connected load and (b) the system with mesh currents inserted

FIGURE 12.25

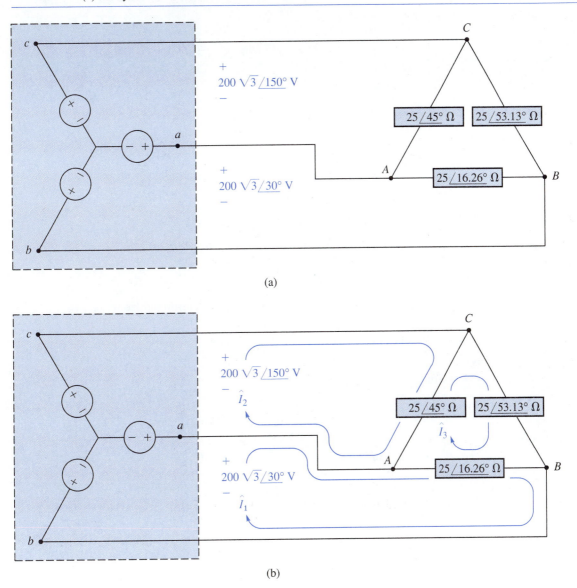

(a)

(b)

Solution   A mesh analysis will be performed, and this requires that the network be redrawn showing the mesh currents, as in Fig. 12.25b. Observe that all of the load impedances have a magnitude of 25 Ω, but all possess a different impedance angle. The mesh equations for the network in Fig. 12.25b are based on the positive sequence, with line voltages having a magnitude of $200\sqrt{3}$ V:

$$25\underline{/16.26^\circ}\,\hat{I}_1 \qquad\qquad -25\underline{/16.26^\circ}\,\hat{I}_3 = 200\sqrt{3}\underline{/30^\circ}$$

$$25\underline{/45^\circ}\,\hat{I}_2 \qquad -25\underline{/45^\circ}\,\hat{I}_3 = 200\sqrt{3}\underline{/150^\circ}$$

$$-25\underline{/16.26^\circ}\,\hat{I}_1 - 25\underline{/45^\circ}\,\hat{I}_2 + 72.17\underline{/38.25^\circ}\,\hat{I}_3 = 0$$

These may be put into a matrix representation with the elements of the coefficient matrix in rectangular form:

$$
\begin{bmatrix}
(24 + j7) & 0 & -(24 + j7) \\
0 & (17.68 + j17.68) & -(17.68 + j17.68) \\
-(24 + j7) & -(17.68 + j17.68) & (56.68 + j44.68)
\end{bmatrix}
\begin{bmatrix}
\hat{I}_1 \\
\hat{I}_2 \\
\hat{I}_3
\end{bmatrix}
$$

$$
= \begin{bmatrix}
200\sqrt{3}\underline{/30°} \\
200\sqrt{3}\underline{/150°} \\
0
\end{bmatrix}
$$

The reader may verify that this set of equations produces a solution vector

$$
\hat{I} = \begin{bmatrix} \hat{I}_1 \\ \hat{I}_2 \\ \hat{I}_3 \end{bmatrix} = \begin{bmatrix} 24.55 + j11.60 \\ 7.50 + j21.70 \\ 11.09 + j8.31 \end{bmatrix} = \begin{bmatrix} 27.15\underline{/25.30°} \\ 22.96\underline{/70.94°} \\ 13.86\underline{/36.87°} \end{bmatrix} \text{ A}
$$

The power delivered to the load can be determined from

$$ P = I_{CA}^2(17.68) + I_{CB}^2(15.00) + I_{AB}^2(24.00) $$

Here,

$$
\hat{I}_{CA} = \hat{I}_2 - \hat{I}_3 = (7.50 + j21.70) - (11.09 + j8.31)
$$
$$
= -3.59 + j13.39 = 13.86\underline{/105.01°} \text{ A}
$$
$$
\hat{I}_{CB} = 13.86\underline{/36.87°} \text{ A}
$$
$$
\hat{I}_{AB} = \hat{I}_1 - \hat{I}_3 = (24.55 + j11.60) - (11.09 + j8.31)
$$
$$
= 13.46 + j3.29 = 13.86\underline{/13.74°} \text{ A}
$$

Thus

$$ P = (13.86)^2(17.68) + (13.86)^2(15.00) + (13.86)^2(24.00) = 10.888.2 \text{ W} \quad \blacksquare $$

EXERCISE 12.6

An unbalanced system with a balanced-wye generator and an unbalanced-wye load is shown in Fig. 12.26. The generator line voltages have an rms magnitude of 240 V at 60 Hz and are in the positive-phase sequence. All generator and line impedances are negligible. Determine the power delivered to the load.

Answer   3520.6 W

FIGURE 12.26

Three-phase power system (Exercise 12.6)

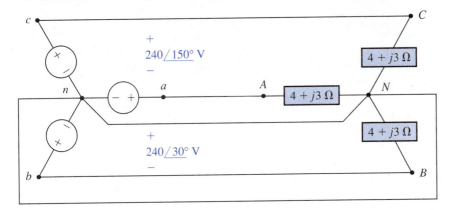

THE MEASUREMENT OF THREE-PHASE POWER

Figure 12.27 shows a three-phase system with an unbalanced-delta load which is connected to a balanced-wye generator. Two wattmeters are connected to the system, one between lines aA and cC and the other between lines bB and cC. Here, eq. (5.18) in Section 5.7 points out that the average power read by the two wattmeters will be

$$p_{av} = \frac{1}{T}\int_0^T v_{ac}i_{aA}\,dt + \frac{1}{T}\int_0^T v_{bc}i_{bB}\,dt$$

where T is related to the frequency at which the system operates, $T = 1/f$. Use of KCL at points A and B allows the average power to be written as

$$p_{av} = \frac{1}{T}\int_0^T v_{ac}(i_{AB} + i_{AC})\,dt + \frac{1}{T}\int_0^T v_{bc}(i_{BA} + i_{BC})\,dt$$

Three-phase system with an unbalanced delta-connected load and two wattmeters

FIGURE 12.27

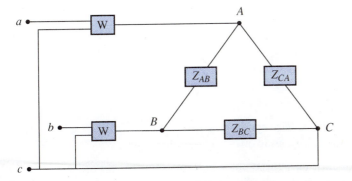

Because the line and phase voltages are equal, this can be rearranged to

$$p_{av} = \frac{1}{T} \int_0^T \left[v_{AC}i_{AC} + v_{BC}i_{BC} + (v_{AC}i_{AB} + v_{BC}i_{BA}) \right] dt$$

Consider the term within the parentheses. If the subscripts on i_{AB} and v_{BC} are reversed in the second term, the entire bracketed term becomes

$$v_{AC}i_{AB} + v_{BC}i_{BA} = (v_{AC} + v_{CB})i_{AB}$$

Then, with $v_{AC} + v_{CB} = v_{AB}$,

$$(v_{AC} + v_{CB})i_{AB} = v_{AB}i_{AB}$$

Thus, the average power may be written as

$$p_{av} = \frac{1}{T} \int_0^T (v_{AC}i_{AC} + v_{BC}i_{BC} + v_{AB}i_{AB}) \, dt \tag{12.19}$$

The voltage-current products here represent the instantaneous power in the three impedances in the delta load, and it is observed that the two wattmeters read the total power. And because only line voltages and currents are involved, eq. (12.19) can be applied to wye-connected loads as well.

CHAPTER 12

SUMMARY

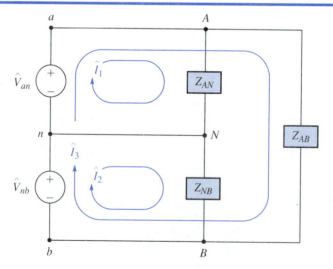

- In a single-phase, three-wire system, the loads are rarely balanced, and the system may be analyzed by node, mesh or loop analysis.

- The phase sequence in a three-phase system refers to the order in which the voltages reach their maximum value.

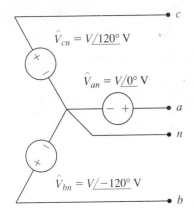

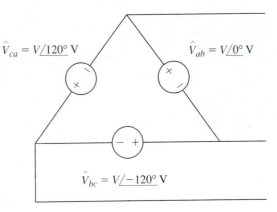

- Three-phase generators and loads may be connected in the wye or delta configurations.

- The relationship between line and phase voltages and currents is as follows:

Wye	Delta
$V_\ell = \sqrt{3}\,V_p$	$V_\ell = V_p$
$I_\ell = I_p$	$I_\ell = \sqrt{3}\,I_p$

- The total power delivered to the load in a balanced three-phase system is given by

$$P = 3V_p I_p \cos\theta$$

or

$$P = \sqrt{3}\,V_\ell I_\ell \cos\theta.$$

- Unbalanced three-phase systems may be expeditiously analyzed by node, mesh or loop analysis.

- The power in a three-phase system may be measured by two wattmeters.

Additional Readings

Blackwell, W.A., and L.L. Grigsby. *Introductory Network Theory*. Boston: PWS Engineering, 1985, pp. 389–405.

Bobrow, L.S. *Elementary Linear Circuit Analysis*. 2d ed. New York: Holt, Rinehart and Winston, 1987, pp. 410–421, 426–432.

Del Toro, V. *Engineering Circuits*. Englewood Cliffs, N.J.: Prentice-Hall, 1987, pp. 233–242.

Hayt, W.H., Jr., and J.E. Kemmerly. *Engineering Circuit Analysis*. 4th ed. New York: McGraw-Hill, 1986, pp. 311–325.

Irwin, J.D. *Basic Engineering Circuiit Analysis*. 3d ed. New York: Macmillan, 1989, pp. 495–534.

Johnson, D.E., J.L. Hilburn, and J.R. Johnson. *Basic Electric Circuit Analysis*. 4th ed. Englewood Cliffs, N.J.: Prentice-Hall, 1989, pp. 401–417, 424–428.

Karni, S. *Applied Circuit Analysis*. New York: Wiley, 1988, pp. 374–407.

Madhu, S. *Linear Circuit Analysis*. Englewood Cliffs, N.J.: Prentice-Hall, 1988, pp. 405–520.

Nilsson, J.W. *Electric Circuits*. 3rd ed. Reading, Mass.: Addison-Wesley, 1990, pp. 403–437.

Paul, C.R. *Analysis of Linear Circuits*. New York: McGraw-Hill, 1989, pp. 305–315.

PROBLEMS   **CHAPTER 12**

Section 12.2

12.1 In Fig. P12.1, $\hat{V}_{AN} = \hat{V}_{NB} = 120\underline{/0°}$ V. If $Z_1 = Z_2 = Z_3 = 40\,\Omega$, find the total power delivered to the system of loads.

Figure P12.1

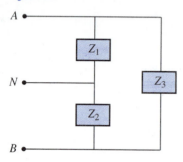

12.2 In Fig. P12.1, $\hat{V}_{AN} = \hat{V}_{NB} = 120\underline{/0°}$ V. If $Z_1 = Z_2 = Z_3 = 30\underline{/16.26°}\,\Omega$, find the total power delivered to the system of loads.

12.3 In Fig. P12.1, $\hat{V}_{AN} = \hat{V}_{NB} = 120\underline{/0°}$ V. If $Z_1 = 8 + j6\,\Omega$, $Z_2 = 8 - j6\,\Omega$, and $Z_3 = 4 - j20\,\Omega$, find the total power delivered to the system of loads.

12.4 In Fig. P12.1, $\hat{V}_{AN} = \hat{V}_{NB} = 120\underline{/0°}$ V. If $Z_1 = 20\underline{/30°}\,\Omega$, $Z_2 = 30\underline{/-36.87°}\,\Omega$, and $Z_3 = 30\underline{/45°}\,\Omega$, find the total power delivered to the system of loads.

Section 12.3

12.5 In the system of Fig. P12.2, the line voltages are 240 V and are in the positive sequence. Find $\hat{I}_b$.

Figure P12.2

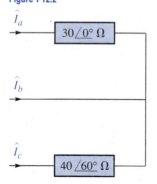

12.6 What are three line currents, $\hat{I}_1$, $\hat{I}_2$, and $\hat{I}_3$, and the neutral current $\hat{I}_n$ in the system shown in Fig. P12.3 if the line-to-line voltages all have a magnitude of 300 V and are connected in the positive sequence?

Figure P12.3

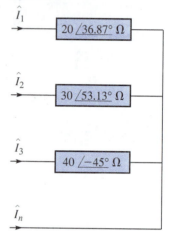

12.7 What are the three line currents, $\hat{I}_1$, $\hat{I}_2$, and $\hat{I}_3$, and the neutral current $\hat{I}_n$ in the system shown in Fig. P12.3 if the line-to-line voltages all have a magnitude of 300 V and are connected in the negative sequence?

12.8 For the three-phase generator connected as shown in Fig. P12.4, the phase voltages have an rms amplitude of 400 V and are connected in positive sequence. What are the currents $\hat{I}_a$, $\hat{I}_b$, and $\hat{I}_c$ if the load is balanced with all $Z = 44\underline{/0°}$?

Figure P12.4

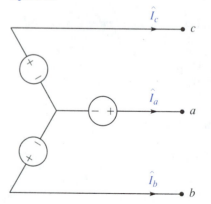

Section 12.4

12.9 If the line-to-line voltages in the wye connected generator that supplies the loads in Fig. P12.5 have a magnitude of 208 V and are in the positive sequence at 60 Hz, show all line and phase voltages in instantaneous form.

Figure P12.5

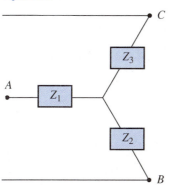

12.10 If the line-to-line voltages in the wye connected generator that supplies the loads in Fig. P12.5 have a magnitude of 208 V and are in the negative sequence at 60 Hz, show all line and phase voltages in instantaneous form.

12.11 If the line-to-line voltages in the delta connected generator that supplies the loads in Fig. P12.6 have a magnitude of 208 V and are in the positive sequence at 60 Hz, show all line and phase voltages in instantaneous form.

Figure P12.6

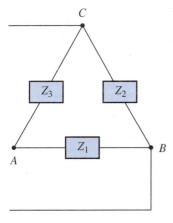

12.12 If the line-to-line voltages in the delta connected generator that supplies the loads in Fig. P12.6 have a magnitude of 208 V and are in the negative sequence at 60 Hz, show all line and phase voltages in instantaneous form.

Section 12.5

12.13 The line-to-line voltages in Fig. P12.5 have a magnitude of 400 V and are in the positive sequence at 60 Hz. The loads are balanced, with $Z_1 = Z_2 = Z_3 = 25\underline{/36.87°}\ \Omega$. Show all line and phase voltages and load currents in instantaneous form.

12.14 The line-to-line voltages in Fig. P12.5 have a magnitude of 400 V and are in the positive sequence at 60 Hz. The loads are balanced with $Z_1 = Z_2 = Z_3 = 44\underline{/-45°}\ \Omega$. Show all line and phase voltages and load currents in instantaneous form.

12.15 The line-to-line voltages in Fig. P12.5 have a magnitude of 440 V and are in the negative sequence at 60 Hz. The loads are balanced with $Z_1 = Z_2 = Z_3 = 88\underline{/0°}\ \Omega$. Show all line and phase voltages and load currents in instantaneous form.

12.16 The line-to-line voltages in Fig. P12.5 have a magnitude of 440 V and are in the positive sequence at 60 Hz. The loads are balanced with $Z_1 = Z_2 = Z_3 = 44\underline{/30°}\ \Omega$. Show all line and phase voltages and load currents in instantaneous form.

Section 12.6

12.17 Use the conditions of Problem 12.13 to determine the power drawn by the load.

12.18 Use the conditions of Problem 12.14 to determine the power drawn by the load.

12.19 Use the conditions of Problem 12.15 to determine the power drawn by the load.

12.20 Use the conditions of Problem 12.16 to determine the power drawn by the load.

Section 12.7

12.21 In Fig. P12.7, the rms amplitude of the line-to-line voltages in the three-phase source is 240 V. The source is connected in the positive sequence and operates at 60 Hz. If the balanced load consists of $Z_1 = Z_2 = Z_3 = 30\underline{/36.87°}\ \Omega$, find the power delivered to the load.

12.22 In Fig. P12.7, the rms amplitude of the line-to-line voltages in the three-phase source is 240 V. The source is connected in the positive sequence and operates at 60 Hz. If the balanced load consists of $Z_1 = Z_2 = Z_3 = 40\underline{/53.13°}\ \Omega$, find the power delivered to the load.

Figure P12.7

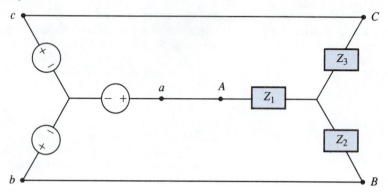

12.23   In Fig. P12.7, the rms amplitude of the line-to-line voltages in the three-phase source is 240 V. The source is connected in the positive sequence and operates at 60 Hz. If the balanced load consists of $Z_1 = Z_2 = Z_3 = 50\underline{/16.26°}\ \Omega$, find the power delivered to the load.

12.24   In Fig. P12.8, the rms amplitude of the line-to-line voltages in the three-phase source is 240 V. The source is connected in the positive sequence and operates at 60 Hz. If the balanced load consists of $Z_1 = Z_2 = Z_3 = 30\underline{/36.87°}\ \Omega$, find the power delivered to the load.

Figure P12.8

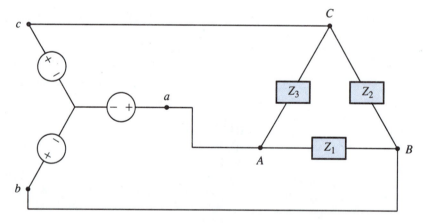

12.25   In Fig. P12.8, the rms amplitude of the line-to-line voltages in the three-phase source is 240 V. The source is connected in the positive sequence and operates at 60 Hz. If the balanced load consists of $Z_1 = Z_2 = Z_3 = 40\underline{/53.13°}\ \Omega$, find the power delivered to the load.

12.26   In Fig. P12.8, the rms amplitude of the line-to-line voltages in the three-phase source is 240 V. The source is connected in the positive sequence and operates at 60 Hz. If the balanced load consists of $Z_1 = Z_2 = Z_3 = 50\underline{/16.26°}\ \Omega$, find the power delivered to the load.

12.27 In Fig. P12.9, the rms amplitude of the line-to-line voltages in the three-phase source is 240 V. The source is connected in the positive sequence and operates at 60 Hz. If the balanced load consists of $Z_1 = Z_2 = Z_3 = 30\underline{/36.87°}\ \Omega$, find the power delivered to the load.

Figure P12.9

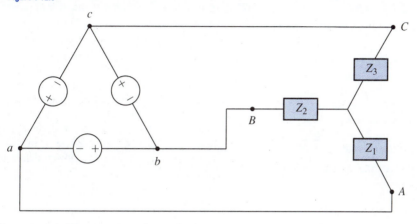

12.28 In Fig. P12.9, the rms amplitude of the line-to-line voltages in the three-phase source is 240 V. The source is connected in the positive sequence and operates at 60 Hz. If the balanced load consists of $Z_1 = Z_2 = Z_3 = 40\underline{/53.13°}\ \Omega$, find the power delivered to the load.

12.29 In Fig. P12.9, the rms amplitude of the line-to-line voltages in the three-phase source is 240 V. The source is connected in the positive sequence and operates at 60 Hz. If the balanced load consists of $Z_1 = Z_2 = Z_3 = 50\underline{/16.26°}\ \Omega$, find the power delivered to the load.

12.30 In Fig. P12.10, the rms amplitude of the line-to-line voltages in the three-phase source is 240 V. The source is connected in the positive sequence

and operates at 60 Hz. If the balanced load consists of $Z_1 = Z_2 = Z_3 = 30\underline{/36.87°}\ \Omega$, find the power delivered to the load.

12.31 In Fig. P12.10, the rms amplitude of the line-to-line voltages in the three-phase source is 240 V. The source is connected in the positive sequence and operates at 60 Hz. If the balanced load consists of $Z_1 = Z_2 = Z_3 = 40\underline{/53.13°}\ \Omega$, find the power delivered to the load.

12.32 In Fig. P12.10, the rms amplitude of the line-to-line voltages in the three-phase source is 240 V. The source is connected in the positive sequence and operates at 60 Hz. If the balanced load consists of $Z_1 = Z_2 = Z_3 = 50\underline{/16.26°}\ \Omega$, find the power delivered to the load.

Figure P12.10

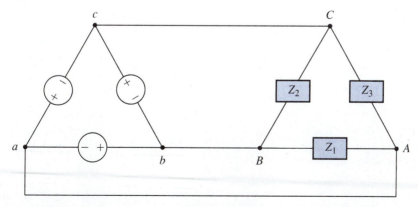

Section 12.8

12.33 In Fig. P12.7, the rms amplitude of the line-to-line voltages in the three-phase source is 440 V. The source is connected in the positive sequence and operates at 60 Hz. If the unbalanced load consists of the three impedances $Z_1 = 88\underline{/0°}\ \Omega$, $Z_2 = 44\sqrt{2}\underline{/45°}\ \Omega$, and $Z_3 = 44\underline{/36.87°}\ \Omega$, find the power delivered to the load.

12.34 In Problem 12.33, what is the overall power factor of the system?

12.35 In Fig. P12.8, the rms amplitude of the line-to-line voltages in the three-phase source is 300 V. The source is connected in the positive sequence and operates at 60 Hz. If the unbalanced load consists of the three impedances $Z_1 = 60\underline{/0°}\ \Omega$, $Z_2 = 75\sqrt{2}\underline{/-45°}\ \Omega$, and $Z_3 = 50\underline{/36.87°}\ \Omega$, find the power delivered to the load.

12.36 In Problem 12.35, what is the overall power factor of the system?

12.37 In Fig. P12.9, the rms amplitude of the line-to-line voltages in the three-phase source is 208 V. The source is connected in the positive sequence and operates at 60 Hz. If the unbalanced load consists of the three impedances $Z_1 = 20\underline{/0°}\ \Omega$, $Z_2 = 30\sqrt{3}\underline{/60°}\ \Omega$, and $Z_3 = 40\underline{/16.26°}\ \Omega$, find the power power delivered to the load.

12.38 In Problem 12.37, what is the overall power factor of the system?

12.39 Figure P12.11 shows a wye-connected generator delivering power to a wye-connected load in a four-wire system. The rms amplitude of the line-to-line voltages in the three-phase source is 240 V. The source is connected in the positive sequence and operates at 60 Hz. If the unbalanced load consists of the three impedances $Z_1 = 48\underline{/0°}\ \Omega$, $Z_2 = 24\sqrt{2}\underline{/-135°}\ \Omega$, and $Z_3 = 60\underline{/36.87°}\ \Omega$, find the power delivered to the load.

Figure P12.11

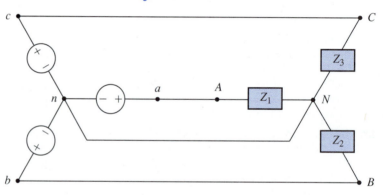

FREQUENCY SELECTIVITY AND RESONANCE

OBJECTIVES

The objectives of this chapter are to:

- Show the variation of network admittance and impedance as a function of frequency.

- Define the resonant or resonance frequency and show how this frequency may be evaluated for any network.

- Discuss and catalog the various symptoms of resonance in *RLC* parallel and *RLC* series networks.

- Define the quality factor for *RLC* parallel and series networks and show why it is considered as a measure of frequency selectivity.

- Consider the bandwidth of a network.

INTRODUCTION

SECTION 13.1

The study of the frequency response of a network is the study of the magnitude and the phase of the steady-state voltages and currents as the frequency varies. It is an extremely useful analysis technique because a knowledge of the frequency response permits the design of frequency-selective networks. Frequency response also lies at the core of analysis by Fourier methods, which is considered in detail in Chapters 21 and 22. Moreover, the knowledge of the frequency response of the system transfer function of a feedback control system plays a significant role in the prediction of the stability of the control system.

The phenomenon of resonance pertains to a particular condition in a network containing *R*, *L*, and *C* elements. A discussion of this phenomenon and its ramifications is the subject of this chapter.

SECTION 13.2   IMPEDANCE AND ADMITTANCE AS A FUNCTION OF FREQUENCY

Figure 13.1a shows an *RLC* parallel network that is driven by a current source. The admittance *Y* is a function of the angular frequency ω

$$Y = G + j\left(\omega C - \frac{1}{\omega L}\right) = |Y|\underline{/\theta} \tag{13.1}$$

where the magnitude is

$$|Y| = \sqrt{G^2 + \left(\omega C - \frac{1}{\omega L}\right)^2}$$

and the *argument* or admittance angle is

$$\theta = \arctan \frac{\omega C - 1/\omega L}{G}$$

Because the voltage is related to the current via $\hat{V} = \hat{I}/Y$, the output voltage, which exists across all of the elements in parallel, is $\hat{V} = V\underline{/\theta}$, where θ may take on positive or negative values depending on the values of *C*, *L*, and ω.

A series *RLC* network driven by a voltage source is shown in Fig. 13.1b. In this case, the impedance is given by

$$Z = R + \left(\omega L - \frac{1}{\omega C}\right) = |Z|\underline{/\theta} \tag{13.2}$$

FIGURE 13.1   (a) An *RLC* parallel network driven by a constant-current source with the voltage across all elements considered as the output and (b) an *RLC* series network driven by a voltage source with a common current throught all elements in series

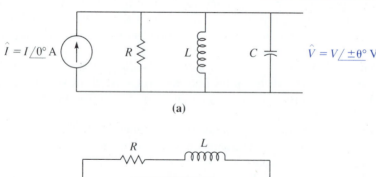

(a)

(b)

The magnitude of the impedance is

$$|Z| = \sqrt{R^2 + \left(\omega L - \frac{1}{\omega C}\right)^2}$$

and the impedance angle is

$$\theta = \arctan \frac{\omega L - 1/\omega C}{R}$$

In this case, $\hat{I} = \hat{V}/Z$, and the current that flows through all of the elements in series is $\hat{I} = I\underline{/-\theta}$, where here, too, θ may take on positive or negative values depending on the values of L, C, and ω.

Consider the *RLC* series network and its impedance given by eq. (13.2). Notice the dependency of the impedance Z upon the angular frequency ω. The two components in the parenthetical term,

$$\omega L - \frac{1}{\omega C}$$

are both frequency-dependent. If these two components and their sum are plotted as in Fig. 13.2a, it is seen that at a certain frequency, called the *resonant* or *resonance frequency*, the two components just cancel; so that the impedance given by eq. (13.2) is entirely real.

The *locus* of the reactive terms of the impedance for a wide range of frequencies is shown in Fig. 13.2b, and this further demonstrates the variation of impedance with frequency. Observe that the impedance is at a minimum at the frequency ω_o. And to be sure, a similar picture can be painted for the admittance of the *RLC* parallel network by using the form of eq. (13.1).

(a) The reactive components of the impedance of an *RLC* series network and (b) the locus of reactive terms of the impedance

FIGURE 13.2

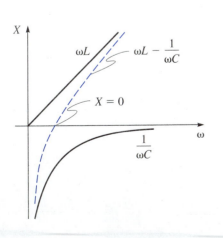

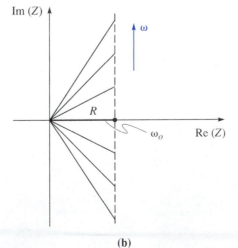

(a)   (b)

FIGURE 13.3

(a) The magnitude of the normalized impedance and (b) the phase of the impedance of an *RLC* series network as a function of angular frequency

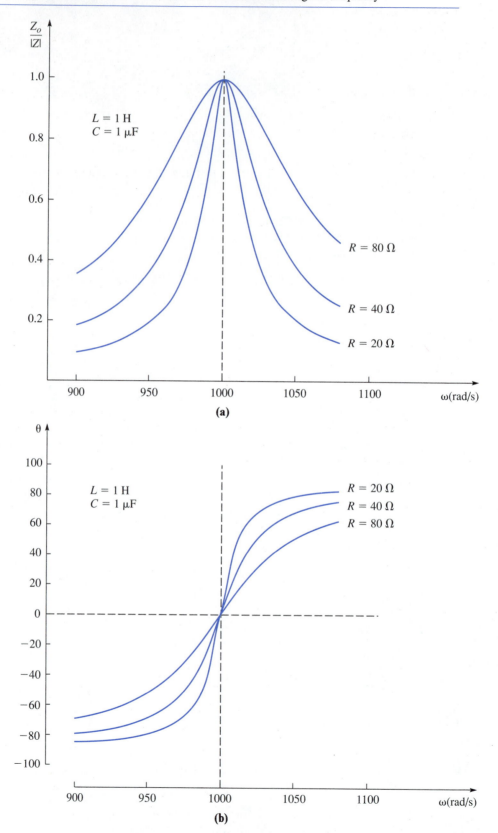

If the impedance at the resonance frequency is defined as Z_o, then the ratio of Z_o to the magnitude of the impedance,

$$\frac{Z_o}{|Z|} = \frac{Z_o}{\sqrt{R^2 + (\omega L - 1/\omega C)^2}}$$

will achieve a maximum value of unity at ω_o. It is instructive to examine plots of this ratio as a function of frequency for particular values of R, L, and C. Three such plots are displayed in Fig. 13.3a for $L = 1$ H, $C = 1$ μF, and for $R = 20, 40,$ and 80 Ω. Notice how clearly these plots show a minimization of impedance at a particular frequency, which, in this case, is 1000 rad/s.

The phase (or phase angle or argument) of the impedance is plotted in Fig. 13.3b for the same component values used in Fig. 13.3a. Here, it is observed that the impedance angle at resonance is zero, which confirms that the impedance is real (no imaginary or reactive component). A similar plot can be generated for the admittance of the RLC parallel network.

The foregoing qualitative approach shows the notion of frequency selectivity. A more meaningful quantitative approach now follows.

THE RESONANCE FREQUENCY

Figures 13.2 and 13.3 show that at a particular frequency, the admittance of the RLC parallel network and the impedance of the RLC series network both achieve minimum values. This frequency is called the *resonant* or *resonance frequency* and is designated by ω_o. The next section will describe several symptoms that indicate the location of the resonance frequency. However, at this point, it is possible to derive the resonance frequency by considering only the development of the previous section.

As indicated by the zero phase angle, ω_o is the angular frequency that makes Y or Z a real quantity. So with regard to eqs. (13.1) and (13.2), if Y and Z are to be real, the imaginary part of each of these equations must be zero. Thus, for eq.(13.1),

$$\omega_o C - \frac{1}{\omega_o L} = 0 \qquad \text{(parallel)}$$

and for eq. (13.2),

$$\omega_o L - \frac{1}{\omega_o C} = 0 \qquad \text{(series)}$$

and for either case,

$$\omega_o = \frac{1}{\sqrt{LC}} \qquad\qquad\qquad\qquad\qquad \textbf{(13.3)}$$

or

$$f_o = \frac{1}{2\pi} \cdot \frac{1}{\sqrt{LC}} \qquad\qquad\qquad\qquad \textbf{(13.4)}$$

Equations (13.3) and (13.4) are applicable to the networks shown in Fig. 13.1. For any other network, ω_o is found by writing the impedance or admittance function, equating the imaginary part of this function to zero, letting $\omega = \omega_o$, and then solving for ω_o.

■ **EXAMPLE 13.1**

Find ω_o for the network in Fig. 13.4a.

Solution   Here,

$$
\begin{aligned}
Z &= \frac{(R + jX_L)(-jX_C)}{R + j(X_L - X_C)} = \frac{X_L X_C - jRX_C}{R + j(X_L - X_C)} \cdot \frac{R - j(X_L - X_C)}{R - j(X_L - X_C)} \\
&= \frac{RX_L X_C - RX_C(X_L - X_C)}{R^2 + (X_L - X_C)^2} - j\frac{X_L X_C(X_L - X_C) + R^2 X_C}{R^2 + (X_L - X_C)^2}
\end{aligned}
$$

When the imaginary part of Z is set equal to zero, the denominator is immaterial. Thus,

$$X_L X_C(X_L - X_C) + R^2 X_C = 0$$

A common X_C may be canceled, and with $X_L = \omega_o L$ and $X_C = 1/\omega_o C$,

$$\omega_o L\left(\omega_o L - \frac{1}{\omega_o C}\right) + R^2 = 0$$

$$\omega_o^2 L^2 - \frac{L}{C} + R^2 = 0$$

$$\omega_o^2 = \frac{1}{LC} - \left(\frac{R}{L}\right)^2$$

$$\omega_o = \sqrt{\frac{1}{LC} - \left(\frac{R}{L}\right)^2} \qquad R < \sqrt{\frac{L}{C}}$$

Observe that when $R = 0$, $\omega_o = 1/\sqrt{LC}$. Networks of this type are known as *tank* circuits and have a certain utility in the tuning of electronic circuits. ■

■ **EXAMPLE 13.2**

Find ω_o for the network in Fig. 13.4b, and evaluate ω_o if $R_L = R_C = \sqrt{L/C}$.

FIGURE 13.4   Networks for Examples 13.1 and 13.2

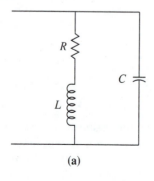

(a)

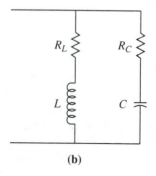

(b)

Solution This time, use the admittance. For the inductive branch,

$$Y_L = \frac{1}{R_L + j\omega L} = \frac{R_L - j\omega L}{R_L^2 + (\omega L)^2}$$

For the capacitive branch,

$$Y_C = \frac{1}{R_C - j(1/\omega C)} = \frac{R_C + j(1/\omega C)}{R_C^2 + (1/\omega C)^2}$$

This makes Y equal to

$$Y = Y_L + Y_C = \frac{R_L}{R_L^2 + (\omega L)^2} + \frac{R_C}{R_C^2 + (\omega C)^2} + j\left[\frac{1/\omega C}{R_C^2 + (1/\omega C)^2} - \frac{\omega L}{R_L^2 + (\omega L)^2}\right]$$

At ω_o, the imaginary part is equal to zero. The denominator is immaterial and

$$\frac{1}{\omega_o C}[R_L^2 + (\omega_o L)^2] - \omega_o L\left[R_C^2 - \frac{1}{(\omega_o C)^2}\right] = 0$$

$$R_L^2 + \omega_o^2 L^2 - \omega_o^2 L C R_C^2 - \frac{L}{C} = 0$$

$$\omega_o^2(L^2 - LCR_C^2) = \frac{L}{C} - R_L^2 = 0$$

$$\omega_o^2 = \frac{L/C - R_L^2}{L^2 - LCR_C^2} = \frac{L/C - R_L^2}{LC(L/C - R_C^2)}$$

$$\omega_o = \sqrt{\frac{1}{LC} \cdot \frac{L - R_L^2 C}{L - R_C^2 C}}$$

If $R_L = R_C = 0$, this reduces to the value of ω_o for the simple RLC series or RLC parallel network. If $R_L = R_C = \sqrt{L/C}$, the network is uniquely resonant at all frequencies. ■

EXERCISE 13.1

What is the resonance angular frequency for the network shown in Fig. 13.5?

Answer $\omega_o = 1/\sqrt{LC - (RC)^2}$, $LC > (RC)^2$

Network (Exercise 13.1)

FIGURE 13.5

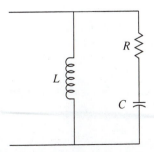

SECTION 13.4   LOCATING THE RESONANCE FREQUENCY

The RLC parallel and series networks shown in Fig. 13.1 exhibit several symptoms at the resonance frequency. These symptoms enable the analyst to quickly diagnose that the networks are operating at ω_o.

It has already been shown that at $\omega = \omega_o$, the admittance and the impedance of the networks are purely real. Thus, symptoms 1 and 2 for both networks are as follows:

  1. The admittance and the impedance are real.

  2a. For the parallel network of Fig. 13.1a

$$\text{Im}(Y) = 0$$

  2b. For the series network of Fig. 13.1b

$$\text{Im}(Z) = 0$$

If Y and Z are real quantities, then Y and Z are both minimum for the parallel and series network, respectively. This is symptom 3.

  3a. For the parallel RLC network, Y is minimum and Z is maximum.
  3b. For the series RLC network, Z is minimum and Y is maximum.

Symptom 3 indicates that for a fixed $\hat{I} = I\underline{/0°}$ in the parallel RLC network (Fig. 13.1a), $\hat{V} = V\underline{/0°}$ at resonance. This indicates that $\hat{V}$ is in phase with $\hat{I}$; and because $\hat{V} = \hat{I}/Y$ and Y is minimum, $\hat{V}$ will attain its maximum value. This is symptom 4.

  4a. In the parallel RLC network driven by a current source, the voltage is in phase with the current and attains its maximum value at the resonance frequency.

Similarly, for the RLC series network (Fig. 13.1b), suppose the source voltage is $\hat{V} = V\underline{/0°}$. Because, at resonance, Z is entirely real, $\hat{I} = I\underline{/0°}$; the current is in phase with the voltage. And because $\hat{I} = \hat{V}/Z$ and Z is minimum, the current must attain its maximum value.

  4b. In the series RLC network driven by a voltage source, the current is in phase with the voltage and attains its maximum value at the resonance frequency.

Symptoms 4a and 4b provide an additional condition that pertains to power dissipation. Because the voltage and the current are in phase in both the parallel and the series configurations, symptom 5 results.

  5. At resonance, the power factor is unity; and for both the parallel and the series cases, the power is real. For the parallel case,

$$P = V^2 G$$

and for the series case,

$$P = I^2 R$$

Now, as a consequence of symptom 2, for the parallel network,

$$\text{Im}(Y) = \omega_o C - \frac{1}{\omega_o L}$$

And for the series network,

$$\text{Im}(Z) = \omega_o L - \frac{1}{\omega_o C}$$

In either case, $\omega_o L = \omega_o C$, or $X_L = X_C$. Thus, for the parallel network,

$$\hat{I}_L = \frac{V\underline{/0°}}{X_L\underline{/90°}} = I_L\underline{/-90°} \quad \text{and} \quad \hat{I}_C = \frac{V\underline{/0°}}{X_C\underline{/-90°}} = I_C\underline{/90°}$$

But $I_L = I_C$ because $X_L = X_C$, and hence, symptom 6 results.

6a. For a parallel *RLC* network operating at resonance, the currents through the inductor and the capacitor are equal in magnitude but are 180° out of phase.

Similar reasoning applies to the *RLC* series network with regard to the voltages across *L* and *C*.

$$\hat{V}_L = I\underline{/0°} \, X_L\underline{/90°} = V_L\underline{/90°} \quad \text{and} \quad \hat{V}_C = I\underline{/0°} \, X_C\underline{/-90°} = V_C\underline{/-90°}$$

6b. For an *RLC* series network operating at resonance, the voltages across the inductor and capacitor are equal in magnitude and 180° out of phase.

THE QUALITY FACTOR

It will be demonstrated in subsequent sections that the quality factor Q is an extremely useful parameter because it provides a measure of the frequency selectivity of the network. It also relates the maximum energy stored to the energy dissipated in a network, and it is this principle that affords a method for its derivation. With this in mind, one defines the quality factor Q as

$$Q = 2\pi \cdot \frac{\text{maximum energy stored}}{\text{total energy dissipated per cycle}} \qquad (13.5)$$

The value of Q depends on whether the *RLC* elements are connected in parallel or series; and although Q has a certain meaning for single *L*'s and *C*'s at all frequencies, its major contribution occurs at the network resonance frequency. For these reasons, the quality factor (or figure of merit) is designated, in general, as Q_o and specifically, when necessary, as Q_{po} and Q_{so} for parallel and series networks, respectively, when the networks are operating at the resonance frequency.

13.5.1 The Quality Factor for the *RLC* Parallel Network

For the parallel network shown in Fig. 13.1a, the current driving the network is $\hat{I} = I\underline{/0°}$. At resonance, the instantaneous form of the current is

$$i = I_m \cos \omega_o t$$

where I_m is the maximum amplitude and is related to the rms value of the current by $I_m = \sqrt{2}I$. The voltage across each element is

$$v = v_R = v_L = v_C = V_m \cos \omega_o t = I_m R \cos \omega_o t$$

The current through the inductor is

$$i_L = \frac{I_m R}{X_L} \cos(\omega_o t - 90°) = \frac{I_m R}{\omega_o L} \sin \omega_o t$$

and the energy stored in the network is

$$w = w_L + w_C = \frac{1}{2} L i_L^2 + \frac{1}{2} C v_C^2$$

or

$$w = \frac{1}{2}\left[\left(\frac{I_m R}{\omega_o L}\right)^2 L \sin^2 \omega_o t + C(I_m R)^2 \cos^2 \omega_o t\right]$$

Equation (13.3) says that at resonance,

$$L = \frac{1}{\omega_o^2 C}$$

so that

$$w = \frac{1}{2}\left[C(I_m R)^2 \sin^2 \omega_o t + C(I_m R)^2 \cos^2 \omega_o t\right]$$

And because $\sin^2 x + \cos^2 x = 1$,

$$w = \frac{1}{2} C(I_m R)^2 \tag{13.6}$$

The total power dissipated in the resistor over one cycle is the average power, computed by using rms values for current and voltage:

$$p_R = \frac{V^2}{R} = \frac{(IR)^2}{R} = \frac{(I_m R)^2}{2R}$$

where again it is to be noted that $I_m = \sqrt{2}I$. The total energy dissipated is p_R multiplied by the period of the impressed sinusoid, $T = 1/f_o$:

$$w_R = \frac{(I_m R)^2}{2R f_o} \tag{13.7}$$

When eqs. (13.6) and (13.7) are substituted into eq. (13.5), the value of the quality factor for the parallel RLC network neatly emerges:

$$Q_{po} = 2\pi \frac{C(I_m R)^2/2}{(I_m R)^2/2Rf_o} = 2\pi f_o RC$$

or

$$Q_{po} = \omega_o RC \qquad (13.8a)$$

Because at resonance, $C = 1/\omega_o^2 L$, it is seen that

$$Q_{po} = \frac{R}{\omega_o L} \qquad (13.8b)$$

Also notice that because $\omega_o = 1/\sqrt{LC}$,

$$Q_{po} = R\sqrt{\frac{C}{L}} \qquad (13.8c)$$

If both the numerator and the denominator of eqs. (13.8a) and (13.8b) are multiplied by the magnitude of the output voltage, then

$$Q_{po} = \frac{V\omega_o RC}{V} = \frac{R}{V} \cdot \omega_o CV = \frac{I_C}{I}$$

and

$$Q_{po} = \frac{RV}{\omega_o LV} = \frac{R}{V} \cdot \frac{V}{\omega_o L} = \frac{I_L}{I}$$

which show that the current throught the inductor and capacitor at resonance will be

$$I_L = Q_{po} I \qquad (13.9a)$$

and

$$I_C = Q_{po} I \qquad (13.9b)$$

where I is both the rms magnitude of the current source and the current through the resistor. This is consistent with resonance criterion 6a in Section 13.4.

There is no inference in eqs. (13.9) regarding maximum or minimum current in the inductor or capacitor. The frequency for maximum or minimum current is derived from the usual procedure of finding maxima and minima. For example, the maximum ratio $\hat{I}_L/\hat{I}$ derives from the current division relationship:

$$\frac{\hat{I}_L}{\hat{I}} = \frac{1/j\omega L}{1/R + j(\omega C - 1/\omega L)} = \frac{1}{j[\omega L/R + j(LC\omega^2 - 1)]} = \frac{1}{1 - LC\omega^2 + j\omega L/R}$$

The magnitude of $\hat{I}_L/\hat{I}$ is

$$\frac{I_L}{I} = \left[(1 - LC\omega^2)^2 + \left(\frac{\omega L}{R}\right)^2 \right]^{-1/2} \qquad (13.10)$$

Now, find where $\partial(I_L/I)/\partial\omega$ vanishes:

$$\frac{\partial(I_L/I)}{\partial\omega} = -\frac{1}{2} \cdot \frac{2(1 - LC\omega^2)(-2LC\omega) + 2(L/R)^2\omega}{[(1 - LC\omega^2)^2 + (\omega L/R)^2]^{3/2}} = 0$$

The denominator is immaterial, and an ω may be canceled from the numerator. The result, after some algebraic manipulation, is

$$\left(\frac{L}{R}\right)^2 = 2(1 - LC\omega^2)LC$$

And if ω_{mL} is designated as the angular frequency that causes I_L/I to be a maximum, then

$$2R^2L^2C^2\omega_{mL}^2 = 2LCR^2 - L^2$$

and

$$\omega_{mL} = \sqrt{\frac{1}{LC} - \frac{1}{2}\left(\frac{1}{RC}\right)^2} \tag{13.11}$$

which is not ω_o.

If eq. (13.11) is put into eq. (13.10),

$$\frac{I_L}{I} = \left\{\left[1 - LC\left(\frac{1}{LC} - \frac{1}{2R^2C^2}\right)\right]^2 + \left(\frac{L}{R}\right)^2\left(\frac{1}{LC} - \frac{1}{2R^2C^2}\right)\right\}^{-1/2}$$

$$= \left(\frac{1}{4} \cdot \frac{L^2}{R^4C^2} + \frac{L}{R^2C} - \frac{1}{2} \cdot \frac{L^2}{R^4C^2}\right)^{-1/2} = \left(\frac{L}{R^2C} - \frac{1}{4} \cdot \frac{L^2}{R^4C^2}\right)^{-1/2}$$

But by eq. (13.8c), $Q_{po} = R\sqrt{C/L}$, so that

$$\frac{I_L}{I} = \frac{1}{\sqrt{(1/Q_{po})^2 - \frac{1}{4}(1/Q_{po})^4}}$$

or

$$\frac{I_L}{I} = \frac{Q_{po}}{\sqrt{1 - \frac{1}{4}(1/Q_{po})^2}} \tag{13.12}$$

Observe that values of Q_{po} greater than 10 will lead to negligible differences between I_L at ω_{mL} and I_L at ω_o.

■ EXAMPLE 13.3

An RLC parallel network is driven by a sinusoidal current source having a peak amplitude of $I_m = 10\sqrt{2}$ mA at all frequencies. If $R = 10\ \Omega$, $C = 100\ \mu F$, and $L = 2.5$ mH, find ω_o, Q_{po}, ω_{mL}, the magnitude of the voltage across the network, the current through the inductor when the angular frequency is ω_o, and the current through both the inductor and the capacitor when the angular frequency is ω_{mL}.

Solution   This example could be worked by using either the rms or the peak amplitudes of the current and voltage sinusoids. The effective (or rms) values will be used. Thus, $I = 10$ mA; and by eq. (13.3),

$$\omega_o = \frac{1}{\sqrt{LC}} = \frac{1}{\sqrt{0.0025(100 \times 10^{-6})}} = 2000 \text{ rad/s}$$

By eq. (13.8a),

$$Q_{po} = \omega_o RC = 2000(10)(100 \times 10^{-6}) = 2$$

The magnitude of the voltage across the network at $\omega_o = 2000$ rad/s will be (to four decimal places)

$$V = RI = 10(0.010) = 0.1000 \text{ V}$$

The magnitude of the currents through L and C will be

$$I_L = I_C = Q_{po}I = 2(0.0100) = 0.0200 \text{ A}$$

The angular frequency at which the current through the inductor is at a maximum is given by eq. (13.11):

$$\omega_{mL} = \sqrt{\frac{1}{LC} - \frac{1}{2}\left(\frac{1}{RC}\right)^2}$$

$$= \sqrt{\frac{1}{0.0025(100 \times 10^{-6})} - \frac{1}{2}\left[\frac{1}{10(100 \times 10^{-6})}\right]^2}$$

$$= \sqrt{4 \times 10^6 - 0.5 \times 10^6} = \sqrt{3.5 \times 10^6} = 1870.83 \text{ rad/s}$$

which can be taken as 1871 rad/s but which is quite different from $\omega_o = 2000$ rad/s.

At $\omega_{mL} = 1871$ rad/s, eq. (13.12) indicates that

$$\frac{I_L}{I} = \frac{Q_{po}}{\sqrt{1 - \frac{1}{4}(1/Q_{po})^2}} = \frac{2}{\sqrt{1 - \frac{1}{4}(\frac{1}{2})^2}} = \frac{2}{\sqrt{\frac{15}{16}}} = 2.0656$$

or

$$I_L = 2.0656(0.010) = 0.0207 \text{ A}$$

This may be compared with $I_L = 0.0200$ A at the resonance frequency of $\omega_o = 2000$ rad/s.

It is useful to look at an alternative method. At $\omega_{mL} = 1870.83$ rad/s,

$$|Y_R| = G = \frac{1}{10} \text{ ℧}$$

$$|Y_L| = \frac{1}{X_L} = \frac{1}{1870.83(0.0025)} = 0.2138 \text{ ℧}$$

$$|Y_C| = \frac{1}{X_C} = 1870.83(100 \times 10^{-6}) = 0.1871 \text{ ℧}$$

The total admittance of the network will be

$$Y = 0.1000 + j(0.1871 - 0.2138) = 0.1000 - j0.0267$$

or

$$Y = 0.1035 \underline{/-14.96°} \ \mho$$

If the signal has a reference of 0°, then

$$\hat{V} = \frac{\hat{I}_s}{Y} = \frac{0.0100 \underline{/0°}}{0.1035 \underline{/-14.96°}} = 0.0966 \underline{/14.96°} \ V$$

so that

$$\hat{I}_L = Y_L \hat{V} = 0.2138 \underline{/-90°}(0.0966 \underline{/14.96°}) = 0.0207 \underline{/-75.04°} \ A$$
$$\hat{I}_C = Y_C \hat{V} = 0.1871 \underline{/90°}(0.0966 \underline{/14.96°}) = 0.0181 \underline{/104.96°} \ A$$
$$\hat{I}_R = Y_R \hat{V} = 0.1000 \underline{/0°}(0.0966 \underline{/14.96°}) = 0.0097 \underline{/14.96°} \ A$$

The total current at the resonance frequency must be I_R, and this can be confirmed by KCL:

$$\hat{I} = \hat{I}_L + \hat{I}_C + \hat{I}_R = 0.0207 \underline{/-75.04°} + 0.0181 \underline{/104.96°} + 0.0097 \underline{/14.96°}$$
$$= (0.0053 - j0.0200) + (-0.0047 + j0.0175) + (0.0093 + j0.0025)$$
$$= 0.0100 + j0.0000 = 0.0100 \ A$$

which is the confirmation.   ■

EXERCISE 13.2

An *RLC* parallel network is driven by a current source and shows a maximum voltage of 12.5 V at an angular frequency of 50,000 rad/s. At this angular frequency, the current measured through the inductor is 62.5 mA. If $R = 1000 \ \Omega$, assume rms values for all currents and voltages and find the resonance frequency, the quality factor Q_{po}, L, C, the angular frequency where the inductor current is a maximum, and the magnitude of this current.

Answer $\omega_o = 50,000$ rad/s, $Q_{po} = 5$, $L = 4$ mH, $C = 0.1 \ \mu F$, $\omega_{mL} = 49,497.5$ rad/s, and I_L at $\omega_{mL} = 62.81$ mA.

13.5.2 The Quality Factor for the *RLC* Series Network

For the *RLC* series network shown in Fig. 13.1b, driven by the voltage source at the resonance frequency

$$v = V_m \cos \omega_o t$$

the current will be

$$i = i_R = i_L = i_C = I_m \cos \omega_o t = \frac{V_m}{R} \cos \omega_o t$$

The voltage across the capacitor is

$$v_C = \frac{V_m}{\omega_o RC} \cos(\omega_o t + 90°) = -\frac{V_m}{\omega_o RC} \sin \omega_o t$$

so that the energy stored in the network will be

$$w = w_L + w_C = \frac{1}{2} Li_L^2 + \frac{1}{2} Cv_C^2$$

This can be written as

$$w = \frac{1}{2} \left[L \left(\frac{V_m}{R} \right)^2 \cos^2 \omega_o t + C \left(\frac{V_m}{\omega_o RC} \right)^2 \sin^2 \omega_o t \right]$$

But at resonance, $C = 1/\omega_o^2 L$, so that

$$v = \frac{1}{2} L \left(\frac{V_m}{R} \right)^2 (\cos^2 \omega_o t + \sin^2 \omega_o t)$$

or because $\cos^2 x + \sin^2 x = 1$,

$$w = \frac{1}{2} L \left(\frac{V_m}{R} \right)^2 \tag{13.13}$$

The total energy dissipated over one cycle is the average power multiplied by the period $T = 1/f_o$,

$$w_R = I^2 RT = \frac{I^2 R}{f_o}$$

or with $I = V_m/\sqrt{2} R$,

$$w_R = \frac{1}{2} \left(\frac{V_m}{R} \right)^2 \left(\frac{R}{f_o} \right) = \frac{V_m^2}{2Rf_o} \tag{13.14}$$

where the rms current has been used.

When eqs. (13.13) and (13.14) are substituted into eq. (13.5), the result is

$$Q_{so} = 2\pi \frac{\frac{1}{2}L(V_m/R)^2}{(V_m^2/2Rf_o)} = 2\pi f_o \frac{L}{R}$$

or

$$Q_{so} = \frac{\omega_o L}{R} \tag{13.15a}$$

which is the reciprocal of Q_{po}. Because $\omega_o L = 1/\omega_o C$, Q_{so} can also be represented by

$$Q_{so} = \frac{1}{\omega_o RC} \tag{13.15b}$$

which is also the reciprocal of Q_{po}. Moreover, it is seen from $\omega_o = 1/\sqrt{LC}$ that

$$Q_{so} = \frac{1}{R}\sqrt{\frac{L}{C}} \tag{13.15c}$$

An individual inductor is sometimes said to have a figure of merit, and the parlance seems to be "the Q of the coil." Such a Q resembles Q_{po} and Q_{so} only in form and does not necessarily refer to Q at the resonance frequency or to Q_{so} or Q_{po} for the entire network at the resonance frequency. If the resistance associated with the inductor is r, then Q_L (not to be referred to merely as Q), the figure of merit for the coil, will be

$$Q_L = \frac{\omega L}{r} \tag{13.16}$$

If the numerator and denominator of eqs. (13.15a) and (13.15b) are multiplied by the magnitude of the current,

$$Q_{so} = \frac{\omega_o LI}{RI} = \frac{V_L}{V} \quad \text{and} \quad Q_{so} = \frac{I}{\omega_o RCI} = \frac{V_C}{V}$$

then two important relationships for V_L and V_C at resonance emerge:

$$V_L = Q_{so}V \tag{13.17a}$$

$$V_C = Q_{so}V \tag{13.17b}$$

Here, V is the rms magnitude of the voltage source, and eqs. (13.17) indicate that the magnitudes of the voltages across the inductor and the capacitor are equal at the resonance frequency. This is consistent with symptom 6b in Section 13.4.

The maximum voltage across the inductor or capacitor will occur at ω_{mC}, which will approximate ω_o only if Q_{so} is high. Use of voltage division in the series RLC network gives

$$\frac{\hat{V}_C}{\hat{V}} = \frac{1/j\omega C}{R + j(\omega L - 1/\omega C)} = \frac{1}{j[(RC\omega + j(LC\omega^2 - 1)]}$$

$$= \frac{1}{1 - LC\omega^2 + jRC\omega}$$

The magnitude of $\hat{V}_C/\hat{V}$ will be

$$\frac{V_C}{V} = [(1 - LC\omega^2)^2 + (RC\omega)^2]^{-1/2} \tag{13.18}$$

The maximum value of this magnitude ratio can be found in the standard manner:

$$\frac{\partial(V_C/V)}{\partial\omega} = -\frac{1}{2} \cdot \frac{2(1 - LC\omega^2)(-2LC\omega) + 2R^2C^2\omega}{[(1 - LC\omega^2)^2 + (RC\omega)^2]^{3/2}} = 0$$

After some algebraic simplification of the numerator

$$2LC - 2L^2C^2\omega^2 = -R^2C^2$$
$$2L^2C^2\omega^2 = 2LC - R^2C^2$$

so that if ω is designated as ω_{mC},

$$\omega_{mC} = \sqrt{\frac{1}{LC} - \frac{1}{2}\left(\frac{R}{L}\right)^2} \qquad\qquad (13.19)$$

If this is put into eq. (13.18), the result, after some algebra, is

$$\frac{V_C}{V} = \frac{1}{\sqrt{(R^2C/L) - (R^2C/2L)^2}}$$

Equation (13.15c) shows that $Q_{so} = (1/R)\sqrt{L/C}$, so that $R^2C/L = (1/Q_{so})^2$. Hence,

$$\frac{V_C}{V} = \frac{1}{\sqrt{(1/Q_{so})^2 - \frac{1}{4}(1/Q_{so})^4}}$$

or

$$\frac{V_C}{V} = \frac{Q_{so}}{\sqrt{(1 - \frac{1}{4}(1/Q_{so})^2}} \qquad\qquad (13.20)$$

Equations (13.19) and (13.20) give the frequency at which the maximum value of the voltage across the capacitor occurs as well as a means for establishing the magnitude of this maximum voltage. Equation (13.20) also indicates that in high-Q_{so} applications, $V_C/V \to Q_{so}$, which means that ω_{mC} is very close to ω_o.

■ **EXAMPLE 13.4**

Figure 13.6 shows a network containing a coil with inductance $L = 1.25$ mH and $Q_L = 20$ at 10,000 rad/s. Measurements taken on this network at 8000 rad/s indicate that $V_C = V_L = 160$ V and that the power drawn is the maximum value, 320 W. Find the magnitude of the current and the source voltage, the values of C, R, and r and the value of the maximum voltage across the capacitor.

Solution   The problem statement clearly identifies the resonance angular frequency. The clue is in the maximum power reference:

$$\omega_o = 8000 \text{ rad/s}$$

and by eq. (13.3),

$$C = \frac{1}{\omega_o^2 L} = \frac{1}{(8000)^2(0.00125)} = 12.5 \times 10^{-6} \text{ F}$$

FIGURE 13.6   *RLC* series network for Example 13.4

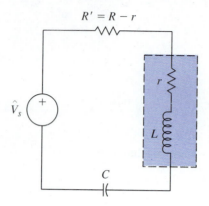

$R' = R - r$

$\hat{V}_s$

r

L

C

The current magnitude can be found from either X_L or X_C. In this case, use $X_L = \omega_o L$:

$$I = \frac{V_L}{\omega_o L} = \frac{160}{8000(0.00125)} = 16 \text{ A}$$

which means that

$$R = \frac{P_R}{I^2} = \frac{320}{(16)^2} = 1.25 \ \Omega$$

By eq. (13.15*a*),

$$Q_{so} = \frac{\omega_o L}{R} = \frac{8000(0.00125)}{1.25} = 8$$

Then by eq. (13.17*a*),

$$V = \frac{V_L}{Q_{so}} = \frac{160}{8} = 20 \text{ V}$$

Observe that $R' = R - r$, and with r determined from eq. (13.16),

$$r = \frac{\omega L}{Q_L} = \frac{10,000(0.0125)}{20} = 0.625 \ \Omega$$

The value of R' is

$$R' = R - r = 1.25 - 0.625 = 0.625 \ \Omega$$

The maximum value of V_C occurs at ω_{mC} given by eq. (13.19):

$$\omega_{mC} = \sqrt{\frac{1}{LC} - \frac{1}{2}\left(\frac{R}{L}\right)^2} = \sqrt{\frac{1}{0.00125(12.5 \times 10^{-6})} - \frac{1}{2}\left(\frac{1.25}{0.00125}\right)^2}$$
$$= \sqrt{64 \times 10^6 - 5 \times 10^5} = \sqrt{63.5 \times 10^6} = 7968.69 \text{ rad/s}$$

This value may be compared with $\omega_o = 8000$ rad/s.

The maximum value of V_C comes from eq. (13.19):

$$V_C = \frac{Q_{so}V}{\sqrt{1 - \frac{1}{4}(1/Q_{so})^2}} = \frac{8(20)}{\sqrt{1 - \frac{1}{4}(1/8)^2}} = \frac{160}{\sqrt{1 - \frac{1}{256}}} = \frac{160}{0.9980} = 160.31 \text{ V}$$

This may be compared with the value at resonance, $V_C = 160.00$ V   ■

EXERCISE 13.3

An *RLC* series network driven by a 100-V-rms voltage source shows a maximum current, a power dissipation of 40 W, and an rms voltage across the capacitor of 800 V at an angular frequency of 100,000 rad/s. Find the resonance angular frequency, the quality factor, Q_{so}, R, L, C, the angular frequency where the capacitor voltage is a maximum, and the magnitude of this voltage.

Answer $\omega_o = 100{,}000$ rad/s, $Q_{so} = 8$, $R = 250\ \Omega$, $L = 20$ mH, $C = 5$ nF, $\omega_{mC} = 99{,}608.6$ rad/s, and V_C at $\omega_{mC} = 801.57$ V.

FREQUENCY SELECTIVITY AND BANDWIDTH

The frequency selectivity demonstrated in Figs. 13.3a and 13.3b is characteristic of all resonant networks, and the exploitation of this selectivity leads to several important applications. The steepness and the width of the impedance and admittance curves near the resonance frequency are important in these applications, and this leads to an investigation of the *bandwidth* of the network, which is the subject of this section.

The parallel network of Fig. 13.1a has an admittance

$$Y = \frac{1}{R} + j\left(\omega C - \frac{1}{\omega L}\right)$$

and at resonance, this admittance is entirely real:

$$Y_o = \frac{1}{R}$$

The ratio Y_o/Y is the reciprocal of Z/Z_o so that

$$\frac{Y_o}{Y} = \frac{1/R}{1/R + j(\omega C - 1/\omega L)} = \frac{1}{1 + j(\omega RC - R/\omega L)}$$

If the imaginary part of the denominator is adjusted by ω_o/ω_o, then

$$\frac{Y_o}{Y} = \frac{1}{1 + j[\omega RC(\omega_o/\omega_o) - (R/\omega L)(\omega_o/\omega_o)]}$$

Use of eqs. (13.8a) and (13.8b), which define Q_{po}, then permit the representation

$$\frac{Y_o}{Y} = \frac{1}{1 + j[Q_{po}(\omega/\omega_o) - Q_{po}(\omega_o/\omega)]}$$

or

$$\frac{Y_o}{Y} = \frac{1}{1 + jQ_{po}(\omega/\omega_o - \omega_o/\omega)}$$

(13.21)

For the *RLC* series network in Fig. 13.1b,

$$Z = R + j\left(\omega L - \frac{1}{\omega C}\right)$$

and at resonance,

$$Z_o = R$$

The ratio $Z_o/Z = Y/Y_o$ is

$$\frac{Z_o}{Z} = \frac{Y}{Y_o} = \frac{1}{R + j(\omega L - 1/\omega C)}$$

and a procedure similar to that just employed for the case of the parallel *RLC* network will yield, for the series *RLC* network,

$$\left.\frac{Z_o}{Z}\right|_s = \left.\frac{Y}{Y_o}\right|_p = \frac{1}{1 \pm jQ_{so}(\omega/\omega_o - \omega_o/\omega)}$$

(13.22)

Equations (13.21) and (13.22) for the parallel and series *RLC* networks are identical in form and differ only in the presences of Q_{po} and Q_{so} in their denominators. If the more general Q_o is used to identify either Q_{po} or Q_{so}, then a general discussion can proceed based upon

$$\left.\frac{Z_o}{Z}\right|_s = \left.\frac{Y}{Y_o}\right|_p = \frac{1}{1 \pm jQ_o(\omega/\omega_o - \omega_o/\omega)}$$

(13.23)

where the subscripts s and p stand for "series" and "parallel," respectively, and where the plus or minus is used in the denominator because of the variation of ω from values below ω_o to values above ω_o.

Now, let

$$\frac{\omega_o}{\omega} - \frac{\omega}{\omega_o} = \pm\frac{1}{Q_o}$$

(13.24)

so that

$$\left.\frac{Z_o}{Z}\right|_s = \left.\frac{Y}{Y_o}\right|_p = \frac{1}{1 \pm jQ_o(1/Q_o)} = \frac{1}{1 \pm j}$$

(13.25)

There will be two angular frequencies ω_1 and ω_2 that make eq (13.24) take on values

$$\frac{\omega_1}{\omega_o} - \frac{\omega_o}{\omega_1} = -\frac{1}{Q_o}$$

(13.26a)

and

$$\frac{\omega_2}{\omega_o} - \frac{\omega_o}{\omega_2} = +\frac{1}{Q_o}$$

(13.26b)

These frequencies define the *bandwidth* of the network,

$$\beta = \omega_2 - \omega_1$$

(13.27)

and are called the *upper* and *lower half-power frequencies* because they yield, by eq. (13.25), magnitude ratios

$$\frac{Z_o}{Z} = \frac{Y}{Y_o} = \frac{\sqrt{2}}{2}$$

which indicates a half-power condition in both the *RLC* series and parallel networks.

The values of ω_1 and ω_2 depend only on Q_o and ω_o. Take eq. (13.26b) and form a quadratic in ω_2:

$$\frac{\omega_2}{\omega_o} - \frac{\omega_o}{\omega_2} = \frac{1}{Q_o}$$

$$\omega_2^2 - \omega_o^2 = \frac{\omega_o}{Q_o}\omega_2$$

$$\omega_2^2 - \frac{\omega_o}{Q_o}\omega_2 - \omega_o^2 = 0$$

This quadratic has two roots,

$$\omega_2 = \frac{\omega_o}{2Q_o} \pm \sqrt{\left(\frac{\omega_o}{2Q_o}\right)^2 + \omega_o^2}$$

and if the positive root is taken (the negative root indicates a negative frequency and is of no interest),

$$\omega_2 = \omega_o\left[\sqrt{1 + \left(\frac{1}{2Q_o}\right)^2} + \frac{1}{2Q_o}\right]$$

(13.28a)

The same procedure applied to eq. (13.26a) will yield

$$\omega_1 = \omega_o\left[\sqrt{1 + \left(\frac{1}{2Q_o}\right)^2} - \frac{1}{2Q_o}\right]$$

(13.28b)

and a simple subtraction provides the very important relationship

$$\beta = \omega_2 - \omega_1 = \frac{\omega_o}{Q_o}$$

(13.29)

This also indicates that Q_o is, indeed, a measure of the frequency selectivity:

$$Q_o = \frac{\omega_o}{\beta}$$

(13.30)

If eqs. (13.28) are multiplied,

$$\omega_2\omega_1 = \omega_o^2\left[\sqrt{1+\left(\frac{1}{2Q_o}\right)^2}+\frac{1}{2Q_o}\right]\left[\sqrt{1+\left(\frac{1}{2Q_o}\right)^2}-\frac{1}{2Q_o}\right]$$

$$= \omega_o^2\left[1+\left(\frac{1}{2Q_o}\right)^2-\left(\frac{1}{2Q_o}\right)^2\right] = \omega_o^2$$

and this shows that ω_o is the geometric mean of ω_1 and ω_2,

$$\omega_o = \sqrt{\omega_1\omega_2} \tag{13.31}$$

All of this discussion is illustrated in Example 13.5, which follows. This example uses the data that was employed in constructing Fig. 13.3a.

■ **EXAMPLE 13.5**

In an RLC series network with $L = 1$ H and $C = 1\ \mu$F, find Q_o, ω_1, and ω_2 for the cases of $R = 20$, 40, and 80 Ω.

Solution   The resonance frequency, as indicated by eq. (13.3), depends only on the values of L and C:

$$\omega_o = \frac{1}{\sqrt{LC}} = \frac{1}{\sqrt{1(1\times 10^{-6})}} = \frac{1}{10^{-3}} = 1000 \text{ rad/s}$$

Then, by eq. (13.15a) for $R = 20\ \Omega$,

$$Q_o = \frac{\omega_o L}{R} = \frac{1000(1)}{20} = 50$$

For $R = 40\ \Omega$,

$$Q_o = \frac{\omega_o L}{R} = \frac{1000(1)}{40} = 25$$

And for $R = 80\ \Omega$,

$$Q_o = \frac{\omega_o L}{R} = \frac{1000(1)}{80} = 12.5$$

Three bandwidths can be determined from eq. (13.30). For $R = 20\ \Omega$,

$$\beta = \frac{\omega_o}{Q_o} = \frac{1000}{50} = 20 \text{ rad/s}$$

For $R = 40\ \Omega$,

$$\beta = \frac{\omega_o}{Q_o} = \frac{1000}{25} = 40 \text{ rad/s}$$

And for $R = 80\ \Omega$,

$$\beta = \frac{\omega_o}{Q_o} = \frac{1000}{12.5} = 80 \text{ rad/s}$$

Finally, from eq. (13.28a), for $R = 20\ \Omega\ (Q_o = 50)$,

$$\omega_1 = \omega_o\left[\sqrt{1 + \left(\frac{1}{2Q_o}\right)^2} - \frac{1}{2Q_o}\right] = 1000\left[\sqrt{1 + \left(\frac{1}{100}\right)^2} - \frac{1}{100}\right]$$
$$= 1000(\sqrt{1.0001} - 0.01) = 1000(0.99005) = 990.05 \approx 990 \text{ rad/s}$$

and

$$\omega_2 = \omega_1 + \beta = 990 + 20 = 1010 \text{ rad/s}$$

The reader may verify that for $R = 40\ \Omega\ (Q_o = 25)$,

$$\omega_1 = 980.2 \text{ rad/s} \qquad \text{and} \qquad \omega_2 = 1020.2 \text{ rad/s}$$

And for $R = 80\ \Omega\ (Q_o = 12.5)$,

$$\omega_1 = 960.8 \text{ rad/s} \qquad \text{and} \qquad \omega_2 = 1040.8 \text{ rad/s}$$

Figure 13.3a is repeated here as Fig. 13.7, with the addition of a horizontal line at $Z_o/|Z| = \sqrt{2}/2 = 0.707$. Note how the numbers generated in this example fit into this picture.

The magnitude of $Z_o/|Z|$ for an RLC series network, where $L = 1$ H and $C = 1\mu$F and where R can take on values of 20, 40, and 80 Ω

FIGURE 13.7

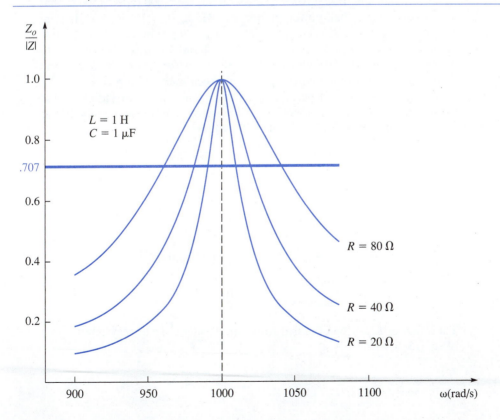

FIGURE 13.8   An *RLC* parallel network driven by a constant current source

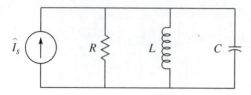

■ **EXAMPLE 13.6**

In the parallel *RLC* network of Fig. 13.8, the output voltage is to have a maximum rms magnitude of 1 V at $\omega = 100{,}000$ rad/s and 0.707 V at 99,500 rad/s. Select the components R, L, and C if $\hat{I}_s$ has the rms magnitude $I = 40$ μA at all frequencies.

Solution   The network is to be resonant at $\omega_o = 100{,}000$ rad/s. At this frequency, the admittance of the network must be real. Hence,

$$G = \frac{I}{V} = \frac{40 \times 10^{-6}}{1} = 40 \ \mu\mho$$

or

$$R = \frac{1}{G} = \frac{1}{40 \times 10^{-6}} = 25{,}000 \ \Omega$$

The condition $V = 0.707 = \sqrt{2}/2$ V at $\omega_1 = 95{,}000$ rad/s implies that ω_1 is the lower half-power frequency. Here, $P = V^2/R$; so that at 100,000 rad/s, $P = 1^2/25{,}000 = 40 \ \mu$W; and at 99,500 rad/s, $P = (0.707)^2/25{,}000 = 20 \ \mu$W. This shows that $\omega_1 = 99{,}500$ rad/s is truly the lower half-power frequency. It is also assumed that the measurement of the power and the measurement of the frequency ω_1 have been made with precision.

Although eq. (13.31) pertaining to the geometric mean indicates that the difference $\omega_o - \omega_1$ does not exactly represent half of the bandwidth, as a first approximation, this example will assume that it does. In this event,

$$\beta = 2(\omega_o - \omega_1) = 2(100{,}000 - 99{,}500) = 2(500) = 1000 \ \text{rad/s}$$

Then, by eq. (13.30)

$$Q_o = \frac{\omega_o}{\beta} = \frac{100{,}000}{1000} = 100$$

The value of C can be obtained from eq. (13.8*a*):

$$C = \frac{Q_o}{\omega_o R} = \frac{100}{100{,}000(25{,}000)} = 40 \times 10^{-9} \ \text{F}$$

The value of L can be obtained from the resonance frequency relationship of eq. (13.3),

$$L = \frac{1}{\omega_o^2 C} = \frac{1}{(100{,}000)^2(40 \times 10^{-9})} = 0.0025 \ \text{H}$$

or from eq. (13.8b)

$$L = \frac{R}{\omega_o Q_o} = \frac{25,000}{100,000(100)} = 0.0025 \text{ H}$$

This concludes the determination of the quantities sought by the problem statement but there is a bit of unfinished business concerning the actual bandwidth and resonance frequency. With $\omega_1 = 99,500$ rad/s and $\beta = 1000$ rad/s, then

$$\omega_2 = \omega_1 + \beta = 99,500 + 1000 = 100,500 \text{ rad/s}$$

Under these circumstances, the value of ω_o cannot be 100,000 rad/s. Indeed, eq. (13.31) gives

$$\omega_o = \sqrt{\omega_1 \omega_2} = \sqrt{99,500(100,500)} = 99,998.75 \text{ rad/s}$$

which is very, very close to, but not exactly, 100,000 rad/s.

This very tolerable and almost immeasurable (and arguably insignificant) discrepancy might have stemmed from a lack of precision in reading $\omega_1 = 99,500$ rad/s. If the network by specification must have a bandwidth of $\beta = 1000$ rad/s with a resonance frequency of $\omega_o = 100,000$ rad/s, then $Q_o = 100$; and by eq. (13.28b),

$$\omega_1 = \omega_o \left[\sqrt{1 + \left(\frac{1}{2Q_o}\right)^2} - \frac{1}{2Q_o} \right] = 100,000(\sqrt{1.000025} - 0.005)$$
$$= 100,000(1.0000125 - 0.005) = 100,000(0.9950125)$$
$$= 99,501.25 \text{ rad/s}$$

which is not $\omega_1 = 99,500$ rad/s and which would make, if the bandwidth is to be $\beta = 1000$ rad/s, $\omega_2 = 100,501.25$ rad/s. Under these circumstances, the resonance frequency could be confirmed by using eq. (13.30):

$$\omega_o = \sqrt{(99,501.25)(105,125.25)} = 100,000 \text{ rad/s}$$

All of this, of course, is nit-picking, but it serves to show that measurement precision is necessary to support a design. If the network is to possess a quality factor $Q_o = 100$ at a resonance frequency of $\omega_o = 100,000$ rad/s, there is no way that $\sqrt{2}/2$ V can be measured across the network at exactly 95,000 rad/s.

EXERCISE 13.4

In a parallel RLC network with $R = 25$ kΩ, $L = 20$ mH, and $C = 5$ nF, what are the upper and lower half-power frequencies, and what is the bandwidth?

Answer $\omega_1 = 96,080$ rad/s, $\omega_2 = 104,080$ rad/s, and $\beta = 8000$ rad/s.

FIGURE 13.9 RLC series network for Example 13.7

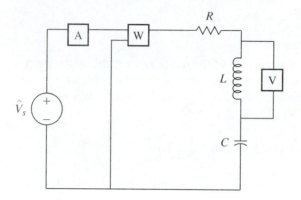

■ **EXAMPLE 13.7**

In the series RLC network of Fig. 13.9, the ammeter (designated by A) reads a maximum value of the magnitude of the rms current, $I = 40$ mA, at $\omega = 100{,}000$ rad/s. At this frequency, the voltmeter (designated by V) reads 400 V, and the wattmeter (W) reads 400 mW. Find ω_o, L, C, Q_o, the rms magnitude of v_s, R, ω_2, ω_1, and the bandwidth β.

Solution   The required quantities need not be determined in the order given in the problem statement. Observe that by criterion (or symptom) 4b in Section 13.4, the condition of maximum current identifies the resonance frequency:

$$\omega_o = 100{,}000 \text{ rad/s}$$

The value of L is obtained from the voltmeter reading:

$$X_L = \omega_o L = \frac{V_L}{I} = \frac{400}{0.040} = 10{,}000 \ \Omega$$

so that

$$L = \frac{X_L}{\omega_o} = \frac{10{,}000}{100{,}000} = 0.1 \text{ H}$$

The value of C is established from eq. (13.3):

$$C = \frac{1}{\omega_o^2 L} = \frac{1}{(100{,}000)^2(0.10)} = 1 \times 10^{-9} \text{ F}$$

and the rms magnitude of the voltage source v_s can be obtained from the power drawn and the current:

$$V_s = \frac{P}{I} = \frac{0.400}{0.040} = 10 \text{ V}$$

Then,

$$R = \frac{P}{I^2} = \frac{0.400}{(0.04)^2} = 250 \ \Omega$$

This allows the determination of Q_{so} from eq. (13.15a),

$$Q_o = Q_{so} = \frac{\omega_o L}{R} = \frac{100{,}000(0.10)}{250} = 40$$

or from eq. (13.15b),

$$Q_o = Q_{so} = \frac{1}{\omega_o RC} = \frac{1}{100{,}000(250)(1 \times 10^{-9})} = 40$$

or from eq. (13.15c),

$$Q_o = Q_{so} = \frac{1}{R} \sqrt{\frac{L}{C}} = \frac{1}{250} \sqrt{\frac{0.100}{10^{-9}}} = 0.004(10^4) = 40$$

or from eq. (13.17a),

$$Q_o = Q_{so} = \frac{V_L}{V} = \frac{400}{10} = 40$$

The bandwidth is determined from eq. (13.29):

$$\beta = \frac{\omega_o}{Q_{so}} = \frac{100{,}000}{40} = 2500 \ \text{rad/s}$$

It would be a mistake to say that $\omega_1 = \omega_o - \beta/2$. Instead, use eq. (13.28b):

$$\omega_1 = \omega_o \left[\sqrt{1 + \left(\frac{1}{2Q_o}\right)^2} - \frac{1}{2Q_o} \right] = 100{,}000 \left[\sqrt{1 + \left(\frac{1}{80}\right)^2} - \frac{1}{80} \right]$$
$$= 100{,}000(\sqrt{1.00015625} - 0.0125) = 100{,}000(1.00007812 - 0.0125)$$
$$= 100{,}000(0.98757812) = 98{,}757.81 \ \text{rad/s}$$

It is not $\omega_1 = \omega_o - \beta/2 = 100{,}000 - 1250 = 98{,}750$ rad/s.
Then,

$$\omega_2 = \omega_1 + \beta = 98{,}757.81 + 2500 = 101{,}257{,}81 \ \text{rad/s}$$

and eq. (13.31) can be used to provide a check:

$$\omega_o = \sqrt{\omega_1 \omega_2} = \sqrt{(98{,}757.81)(101{,}257.81)} = 100{,}000 \ \text{rad/s}$$

SECTION 13.7   **SPICE EXAMPLES**

Reading: No additional reading is required for the understanding of the SPICE examples that follow.

EXAMPLE S13.1

In Example 13.2, it was shown that the networks in Fig. 13.4b (repeated here as Fig. S13.1) is uniquely resonant at all frequencies if $R_L = R_C = \sqrt{L/C}$. If $L = 1$ H and $C = 1$ μF, then

$$R_L = R_C = \sqrt{\frac{L}{C}} = \sqrt{\frac{1}{(1 \times 10^{-6})}} = 1000 \ \Omega$$

and PSPICE can be used to verify that with a sinusoidal current input with a constant magnitude at all frequencies, the network will indeed provide a voltage with constant magnitude across the parallel combination of network elements. The network made ready for SPICE is shown in Fig. S13.2, the input file is reproduced in Fig. S13.3, and pertinent extracts from the output file are presented in Fig. S13.4.

Figure S13.1   Network for SPICE Example S13.1

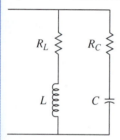

Figure S13.2   The network of Fig. S13.1 ready for PSPICE analysis

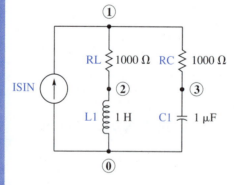

Figure S13.3   Input PSPICE file for analysis of the network of Fig. S13.1

```
SPICE PROBLEM - CHAPTER 13 - NUMBER 1 - RESONANCE
****************************************************
*HERE ARE THE SIGNAL GENERATOR AND THE FOUR COMPONENTS
ISIN      1      0      AC       .04
RL        1      2      1K
L1        2      0      1
RC        1      3      1K
C1        3      0      1U
****************************************************
*HERE ARE THE CONTROL STATEMENTS
.AC       LIN    30     500      100K
.PLOT     AC     VM(1,0)
.END
```

Figure S13.4   Pertinent output for the PSPICE analysis of the network in Fig. S13.1

```
SPICE PROBLEM - CHAPTER 13 - NUMBER 1 - RESONANCE


****     AC ANALYSIS                    TEMPERATURE =    27.000 DEG C


**********************************************************************************


  FREQ         VM(1,0)


 (*)---------   1.0000E+01   1.0000E+02   1.0000E+03   1.0000E+04   1.0000E+05
                 - - - - - - - - - - - - - - - - - - - - - - - - -
  5.000E+02  4.000E+01 .        *     .            .            .            .
  3.931E+03  4.000E+01 .        *     .            .            .            .
  7.362E+03  4.000E+01 .        *     .            .            .            .
  1.079E+04  4.000E+01 .        *     .            .            .            .
  1.422E+04  4.000E+01 .        *     .            .            .            .
  1.766E+04  4.000E+01 .        *     .            .            .            .
  2.109E+04  4.000E+01 .        *     .            .            .            .
  2.452E+04  4.000E+01 .        *     .            .            .            .
  2.795E+04  4.000E+01 .        *     .            .            .            .
  3.138E+04  4.000E+01 .        *     .            .            .            .
  3.481E+04  4.000E+01 .        *     .            .            .            .
  3.824E+04  4.000E+01 .        *     .            .            .            .
  4.167E+04  4.000E+01 .        *     .            .            .            .
  4.510E+04  4.000E+01 .        *     .            .            .            .
  4.853E+04  4.000E+01 .        *     .            .            .            .
  5.197E+04  4.000E+01 .        *     .            .            .            .
  5.540E+04  4.000E+01 .        *     .            .            .            .
  5.883E+04  4.000E+01 .        *     .            .            .            .
  6.226E+04  4.000E+01 .        *     .            .            .            .
  6.569E+04  4.000E+01 .        *     .            .            .            .
  6.912E+04  4.000E+01 .        *     .            .            .            .
  7.255E+04  4.000E+01 .        *     .            .            .            .
  7.598E+04  4.000E+01 .        *     .            .            .            .
  7.941E+04  4.000E+01 .        *     .            .            .            .
  8.284E+04  4.000E+01 .        *     .            .            .            .
  8.628E+04  4.000E+01 .        *     .            .            .            .
  8.971E+04  4.000E+01 .        *     .            .            .            .
  9.314E+04  4.000E+01 .        *     .            .            .            .
  9.657E+04  4.000E+01 .        *     .            .            .            .
  1.000E+05  4.000E+01 .        *     .            .            .            .
                 - - - - - - - - - - - - - - - - - - - - - - - - -



           JOB  CONCLUDED
```

EXAMPLE S13.2

As shown in Example 13.1, the network shown in Fig. 13.4a (repeated here as Fig. S13.5) has a resonance frequency given by

$$\omega_o = \sqrt{\frac{1}{LC} - \left(\frac{R}{L}\right)^2}$$

With $L = 1$ H, $C = 10$ nF, and $R = 5000$ Ω,

$$\omega_o = \sqrt{\frac{1}{1(1 \times 10^{-8})} - \left(\frac{5000}{1}\right)^2} = \sqrt{10^8 - 25 \times 10^6} = 8660 \text{ rad/s}$$

or $f_o = 1378.3$ Hz. This can be verified by using PSPICE.

The network made ready for analysis with PSPICE is shown in Fig. S13.6, the input file is reproduced in Fig. S13.7, and pertinent extracts from the output file are presented in Fig. S13.8. The confirmation of the resonance frequency may be noted.

Figure S13.5   Network for SPICE Example S13.2

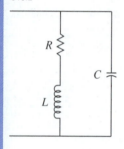

Figure S13.6   The network of Fig. S13.5 ready for PSPICE analysis

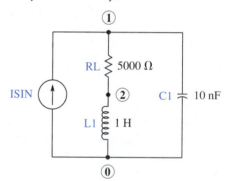

Figure S13.7   Input PSPICE file for analysis of the network of Fig. S13.5.

```
SPICE PROBLEM - CHAPTER 13 - NUMBER 2 - RESONANCE
*********************************************************
*HERE IS THE SIGNAL GENERATOR AND THREE COMPONENTS
ISIN      1       0       AC      .015
RL        1       2       5K
L1        2       0       1
C1        1       0       10N
*********************************************************
*HERE ARE THE CONTROL STATEMENTS
.AC       LIN     30      500     15K
.PLOT     AC      VM(1,0)
.END
```

Figure S13.8 Pertinent output for the PSPICE analysis of the network in Fig. S13.5

```
SPICE PROBLEM - CHAPTER 13 - NUMBER 2 - RESONANCE


****        AC ANALYSIS                    TEMPERATURE =    27.000 DEG C


*******************************************************************************



  FREQ        VM(1,0)


(*)---------    1.0000E+01    1.0000E+02    1.0000E+03    1.0000E+04    1.0000E+05
            - - - - - - - - - - - - - - - - - - - - - - - - - - - -
5.000E+02  9.682E+01 .              *        .              .              .
1.000E+03  1.766E+02 .              .    *    .              .              .
1.500E+03  3.304E+02 .              .        *.              .              .
2.000E+03  2.374E+02 .              .      * .              .              .
2.500E+03  1.486E+02 .              .   *    .              .              .
3.000E+03  1.075E+02 .              . *      .              .              .
3.500E+03  8.477E+01 .             *          .              .              .
4.000E+03  7.036E+01 .           *.           .              .              .
4.500E+03  6.036E+01 .          * .           .              .              .
5.000E+03  5.297E+01 .         *  .           .              .              .
5.500E+03  4.728E+01 .        *   .           .              .              .
6.000E+03  4.274E+01 .        *   .           .              .              .
6.500E+03  3.903E+01 .      *     .           .              .              .
7.000E+03  3.594E+01 .      *     .           .              .              .
7.500E+03  3.331E+01 .     *      .           .              .              .
8.000E+03  3.106E+01 .      *     .           .              .              .
8.500E+03  2.910E+01 .      *     .           .              .              .
9.000E+03  2.738E+01 .    *       .           .              .              .
9.500E+03  2.585E+01 .    *       .           .              .              .
1.000E+04  2.449E+01 .    *       .           .              .              .
1.050E+04  2.327E+01 .   *        .           .              .              .
1.100E+04  2.216E+01 .   *        .           .              .              .
1.150E+04  2.116E+01 .  *         .           .              .              .
1.200E+04  2.025E+01 .  *         .           .              .              .
1.250E+04  1.941E+01 .  *         .           .              .              .
1.300E+04  1.864E+01 .  *         .           .              .              .
1.350E+04  1.793E+01 .  *         .           .              .              .
1.400E+04  1.727E+01 .  *         .           .              .              .
1.450E+04  1.666E+01 . *          .           .              .              .
1.500E+04  1.610E+01 .  *         .           .              .              .
            - - - - - - - - - - - - - - - - - - - - - - - - - - - -



                JOB  CONCLUDED
```

CHAPTER 13

SUMMARY

- Network admittance and impedance both vary with frequency. For simple *RLC* parallel and *RLC* series networks, the resonance angular frequency is given by

$$\omega_0 = \frac{1}{\sqrt{LC}}$$

- Resonance angular frequencies for all but *RLC* parallel and series networks are obtained by developing an expression for the network admittance or impedance, setting the imaginary part of the admittance or impedance equal to zero, and solving for ω.

- The following symptoms indicate a resonance:

  —The admittance and impedance are real.

  —$\text{Im}(Y) = 0$ and $\text{Im}(Z) = 0$.

  —For the RLC parallel network, $Y(j\omega)$ is minimum and $Z(j\omega)$ is maximum. For the RLC series network, $Z(j\omega)$ is minimum and $Y(j\omega)$ is maximum.

  —In the parallel RLC network, V is in phase with the current source and is a maximum at the resonance frequency. In the parallel RLC network, I is in phase with the voltage source and is a maximum at the resonance frequency.

  —At resonance, the power factor is equal to unity:

  $$PF = \cos\theta = 1.00 \qquad \theta = 0°$$

  —In the parallel RLC network, the inductor and the capacitor currents are 180° out of phase:

  $$\hat{I}_L = I_L\underline{/-90°} \qquad \text{and} \qquad \hat{I}_C = I_C\underline{/90°}$$

  In the series RLC network, the inductor and the capacitor voltages are 180° out of phase:

  $$\hat{V}_L = V_L\underline{/90°} \qquad \text{and} \qquad \hat{V}_C = V_C\underline{/-90°}$$

- A quality factor

  $$Q = 2\pi \, \frac{\text{maximum energy stored}}{\text{total energy dissipated per cycle}}$$

  is defined for RLC parallel and series networks. For the RLC parallel network,

  $$Q_{po} = \omega_o RC \qquad Q_{po} = \frac{R}{\omega_o L} \qquad Q_{po} = R\sqrt{\frac{C}{L}}$$

  For the RLC series network,

  $$Q_{so} = \frac{\omega_o L}{R} \qquad Q_{so} = \frac{1}{\omega_o RC} \qquad Q_{so} = \frac{1}{R}\sqrt{\frac{L}{C}}$$

- In an RLC parallel network, the angular frequency that yields a maximum inductor current is given by

  $$\omega_{mL} = \sqrt{\frac{1}{LC} - \frac{1}{2}\left(\frac{1}{RC}\right)^2}$$

  and the maximum inductor current will be

  $$\frac{I_L}{I} = \frac{Q_{po}}{\sqrt{1 - (\tfrac{1}{4})(1/Q_{po})^2}}$$